St. Louis Community College

Forest Park
Florissant Valley
Meramec

Instructional Resources
St. Louis, Missouri

TURN UP THE CONTRAST

TURN UP THE CONTRAST

CBC
TELEVISION
DRAMA
SINCE 1952

Mary Jane Miller

A UNIVERSITY OF BRITISH COLUMBIA PRESS
CBC ENTERPRISES CO-PUBLICATION

Vancouver 1987

TURN UP THE CONTRAST: CBC TELEVISION DRAMA SINCE 1952

© University of British Columbia Press 1987

This book has been published with the help of a grant from the Canada Council.

Canadian Cataloguing in Publication Data

Miller, Mary Jane, 1941-
    Turn up the contrast: CBC television drama
since 1952

    Co-published by CBC Enterprises.
    Bibliography: p.
    Includes index.
    ISBN 0-7748-0278-2

    1. Television plays—Canada—History and
criticism. 2. Canadian Broadcasting Corporation—
History.   I. CBC Enterprises.   II. Title.
PN1992.65.M54 1987   791.45′5   C87-091494-4

International Standard Book Number 0-7748-0278-2

Printed in Canada

CONTENTS

ILLUSTRATIONS

ACKNOWLEDGEMENTS

When I began the research for this book in 1979 I did not know that I would finish it when the issue of whether or not we will continue to have a distinctive Public Broadcasting Service would be in the headlines. Unwittingly my parameters, CBC television drama in English 1952-84, enclose a completed era. In the late 80's the CBC faces a crisis unique in its 50-year history. Whatever comes out of it, there will be changes in how and perhaps whether Canadian television drama is made.

Without the active help of the members of the CBC drama department in Toronto and Vancouver, and the cooperation of other regions there would be no book. To begin with, there would be no programmes. Moreover, if largely unknown people had not kept kinescopes in Vancouver, Winnipeg and above all in Flin Flon, I would not have been able to see many of those that have survived, for accurate description and some kinds of evaluation. No one including myself could anticipate that out of all this would come three books, each with a different focus. Of the next two books, one will present the off-camera voices—interviews with CBC creative personnel past and present. The other is a more detailed account of the internal and external constraints which have shaped the programmes, how decisions were and are made, how dramas were and are made and some reflection on the differences tape, film, High Definition digital television can make to television drama in an aesthetic sense. Despite the current crisis, it is my conviction that by the time those books appear, there will also be more high quality television drama to think about.

At the top of my CBC "without whom" list is John Kennedy, Head of TV Drama. He opened the doors and shared his time and his thoughts with me very generously. Producers Maryke McEwan, Sam Levene, and David Barlow, director George McGowan, co-producers Martin Weiner and Duncan Lamb, directors of cinema Nick Evdemon, Ed Long and Brian Hebb, designers Dan Yarhi and Rudi Dorn, composer Phil Schreibman, and editor Vincent Kent helped me understand what I was observing. Every single member of the technical crews answered questions on the shoots of *Out of Sight, Out of Mind, Ready for Slaughter*, two episodes of *Seeing Things* and a taping of *Backstretch.*

Over the last three years I have formally interviewed on tape: executive Producers Robert Allen, Philip Keatley, Maryke McEwen, Don S.Williams; producer Martin Kinch; directors Mario Prizek, Eric Till, David Gardner; story editor Alice Sinclair; designer Rudi Dorn; Head of Arts, Music and Sciences (and middle level manager for many years) Hugh Gauntlett; actor/writer Hugh Webster; and by phone and letter, Paul St. Pierre; Head of Audience Research Brian Stewart; Dennis O'Neill of CBC Enterprises; Sydney Newman of the NFB, CBC, ITV, BBC and most recently Channel 4. The tapes are on deposit with the Public Archives.

A.W. Ross, Corporate Supervisor, Programme services, one of the few people who has been active in trying to preserve the heritage of the CBC, helped me with the intricacies of the Corporation, both in Toronto and Ottawa, Jim Beckman and Richard Wright made the files of Programme Resources available (when that indispensable department still survived at the CBC) where the catalogues, lists, cards, computers and viewing facilities were gathered before the budget cuts of 1984 and 1986. In Vancouver, Philip Keatley spent hours with me. In Montreal, Jean Jacques Berard, Directeur Assistant Communication and Jean Archambault, Supervisor of TV Broadcast archives, were kindness itself. In Winnipeg, Marvin Terhoch, Director of Television, made his facilities available.

Margaret Lovees who is in charge of the CBC Talent Bank located background files for me, Diane Murdoch solved problems, Bernie Dobbins let me look at tapes and his very accurate records, Oly Iwanyshan of Audience Research explained how his arcane art works. Judy Squires opened up new fields in Newfoundland.

The foundation of my formal research were laid in 1979 when Sam Kula, Director of the National Film, Television, and Sound division of the Public Archives of Canada, listened to my lament that I could not view most CBC programmes—the primary source for any reliable analysis of the CBC drama output. He responded by giving me carte blanche for three months to look at his (then-uncatalogued) CBC collection of drama kinescopes. Since then Ernie Dick, Head of Sound and Moving Image Collections Development has also provided invaluable help in sending on materials. Professor Ross Eamon shared access with me to his Carleton Oral History Project, specifically to tapes of interviews with Robert Allen and Ron Weyman. Professor Howard Fink, director of the Concordia Radio Drama Project and an old friend encouraged and Professor Paul Rutherford discussed methodologies. In its last stages, Alice Sinclair, story editor for CBC TV drama for much of its history, brought her special expertise to the *Ms.* with thoroughness, common sense, and wit. Thanks to Professor Malcolm Page for his invaluable help and to David McCaughna and especially Roy Martin at the CBC and Sylvie Robitaille at the PAC helped locate photographs.

Closer to home and over the long haul, my special thanks to Rebecca Harris, research assistant and typist, and thanks to Kate Hull, research assistant. Katya Davison and Blema Miller were meticulous proofreaders. Throughout many drafts, Elizabeth Miller (whose encouragement spans 45 years) proofread and made helpful comments.

Professor J. M. Miller, information manager, co-researcher on *The Beachcombers*, calm, concentrated and a very present help in times of crisis, loving husband—thank you.

Finally, I would like to thank the Social Science and Humanities Research Council for their support over a three-year period, the J. P. Bickell Foundation for an equipment grant, President A. E. Earp, Maurice Yacowar, Dean of Humanities and the Fine Arts Department of Brock University.

I
INTRODUCTION

1

THE FLOWERING OF CBC DRAMA

In 1952, television drama started on the road to becoming Canada's national theatre. From its beginnings, it has been one of the few elements in Canadian life that marked our southern border, distinguishing "here" from "there." But in fact, very few of its directors, actors, writers, technicians or even planners in television spend much time thinking about this past. Like the bard who, centuries ago, sang for his supper and perhaps fame in his lifetime, the creators of television have to keep singing, night after night, from their corner by the hearth. As with the bardic singer, when television is through with a song, the song is finished in the minds of its creators. Their "whole habit of thinking is forward, never back and then forth! It takes a vast cultural change to develop a new kind of poetic. The opportunity offered in dictating [tape/film/VCR] is not sufficient."[1]

The song may be sung many times, but it is never quite the same twice. Successful formula television series such as sitcoms, copshows, soaps, Westerns depend on variations on the basic set of conventional characters, situations, and narratives, as well as on a common experience shared by the tale's teller and his audience. The ambience for bards, as for Elizabethan playwrights in the golden age of English drama, was one in which the tale and how it was told were expendable.

Television drama in particular is rarely taken seriously in Canada, whether in newspapers and periodicals, in journals, classrooms, among critics, by the general public or the governments of the day, unless they hit a nerve of controversy as happened with *The Open Grave*, *The Paper People*, *Neighbours*, or *The Tar Sands*. Yet no one disputes the fact that most people watch television drama for at least an hour or two most days of the week.

An editor of a learned journal once suggested to me that "drama" is a rather pretentious word for sitcoms, soaps, Westerns, and copshows, but this is not so. Drama denotes a fiction, a story in performance. It encompasses both the predictable "black hat-white hat" formula action-adventure show and the most complex, intellectually challenging and aesthetically beautiful "single" television film. This is all "drama."

Yet critical analysis of the forms and subject matter of television drama has for many years been left to the sociologists, political scientists, economists, education specialists, and to special-interest groups. Until now, as far as I have been able to find out, no one has been taking a systematic look at whether the television drama we make in Canada is distinctive or imitative or innovative; whether it is informative or simply reconfirms our social norms; whether it is censored on a systematic basis or full of grave omissions; whether it is developing in insight, breadth, and sophistication. No one has really looked at how changing technology, the mandate contained in the Broadcasting Act, the cultural and political context, trends in broadcasting and changes in taste and, most important of all, the individuals who actually make the programmes have influenced the tales we tell each other about ourselves.

Of course, there have been many book-length studies of Canadian broadcasting. The sociological, economic, political and historical contexts have been well mapped by scholars as diverse as Paul Audley, Frank Peers, Marshall McLuhan, Herschel Hardin, E. A. Weir, and Paul Rutherford in Canada, and R. Williams, Fiske and Hartley, Self, Murdock, Briggs, Barnouw, and Speradakos in Britain and America (see Bibliography, p. 403).

English Canada, however, has not followed the path of scholars in Quebec, who, since the Quiet Revolution, have rigorously protected, catalogued, and analyzed their culture. Unlike the British and, more recently, the Americans, who have turned their attention to questions of aesthetics, genre theory, and qualitative evaluation of their accomplishments in broadcast drama, English Canadians have simply neglected their remarkable heritage. In this book, I am attempting to begin a scholarly exploration using a set of satellite overviews with closeups of selected areas that might encourage others to join the mapping. It is my hope that the serious study of Canadian and other television drama will form part of curricula offered by departments of drama, communications, and Canadian studies, as well as by comparative culture courses. I am also trying to present a history and analysis of CBC television drama for all the people within the corporation who want to recover a sense of their heritage. Finally, this book is intended for any interested viewer.

Criteria for Analysis

I have selected for detailed analysis those copshows, sitcoms, and family-adventure series that have been characterized by longevity and innovative inflections of the formula or conventions, series which evolved over the years, and particularly those displaying originality in individual episodes. To be specific, this means *The Beachcombers*, not *Ritter's Cove* or *The Forest Rangers*; *King of Kensington*, not *Snow Job* or *Flappers*; *Seeing Things* and *Wojeck*, not *Police Surgeon* or *The Phoenix Team*; *A Gift to Last*, not *Backstretch*; and, in miniseries, *Empire Inc.*, not *Vanderberg*.

Anthology presents a more difficult problem. There is no doubt that I will have missed real treasures, for several reasons. The first is that far too much material from every era, not simply the early days of kinescopes, has disappeared. Next, even though the number of hours of television drama made in Canada has dwindled over time (Telefilm and tall tales told to the CRTC by the CTV, Global, and individual stations and pay networks notwithstanding), there is still far more material to analyze in a single season than this book could contain—and I am trying to give an overview, with selected closeups, of well over thirty years of CBC English television drama.

Finally, you the reader-viewer and I the viewer-critic are on our own. You have not had a chance to look at some of these programmes in years. Some you have never seen, and unless CBC policy and union practices regarding "further use" of broadcast materials change, you never will see most of them in rerun. Still, the very thorough, sensible and sometimes surprising 1986 Caplan-Sauvageau Report[2] (1986) tackles the issues of a second channel for quality repeats, of copyright protection and the potential for a developing market, as well as the astounding profits of most individual stations in the private sector. The economics of television are being changed by the VCR, the direct broadcast satellite, and pay channels. Perhaps we may yet join the other television systems which show reruns as a matter of course—or in Europe, as a matter of pride.

When I selected material to look at in some detail (in a warehouse in Etobicoke, peering at a movieola with an 8x8-inch screen) I chose to begin with the prestige anthologies *Scope*, *Folio*, and *Festival*, concentrating on original scripts by Canadians. From those I saw everything that survives on kinescope—by no means all of the programmes. Many treasures are lost, like Mavor Moore's *The Ottawa Man* or Marie Claire Blais's *The Puppet Caravan*, the first colour drama shot in English and French, directed by Paul Almond and starring Geneviève Bujold. The can was on the shelf, empty. During that summer's work with the PAC collection, I also saw all of *First Person*, a 'fifties anthology of fine Canadian drama used to sell Canada Savings Bonds. However, when it came to *The Unforeseen*, *CBC Television*

Theatre or *On Camera*, I picked plays that sounded interesting from the plot synopsis or because I knew the authors' work. Because of limited time, I could not do justice to *G.M. Presents*, so again I chose a few Canadian scripts which sounded interesting, sparked controversy, or had an author or producer whose work I respected. Since innovation and experiment are directly related I saw more of *Q for Quest*, but none of the *Eyeopener* series.

As a child of the radio generation, I was twelve when a television set arrived in our living room and, through it, the CBC. For a decade, 1959-69, I saw very little television and so I had later to rely first on the kinescopes, then on tapes of films and VTR to fill in that period. In 1969, I began in an increasingly organized fashion to analyze "off air" CBC TV drama. From those notes I chose other dramas to review for this book. The criterion used here was primarily a sense that "there is more to this drama than one viewing can exhaust," together with an interesting script, performance, editing, camera work, theme, or director's concept. When writing this book, I went to Winnipeg, Vancouver, and Montreal where I looked over their rudimentary records or the profile of programmes made by two students travelling on a grant; I talked to people on the spot and then selected as above. In Vancouver, I saw every episode of *Cariboo Country* that I could find time for because of the extraordinary quality of the whole series. Circumstances also presented me with forty-plus episodes of *The Beachcombers* to view within a ten-day period. Finally, a VCR has made my task very much easier in the last five years, allowing for both detailed and comfortable rewind and search—and for fast forward when necessary. The VCR has the power to make every viewer a collector, a critic, and a connoisseur.

It is axiomatic to my approach that I do not analyze in detail any drama I have not seen, any series or anthology I have not sampled. Overall, the criteria I have used to select the material for this book are whether the drama is innovative in form or content; whether its production values enhance the plot, characterizations, and subtext; whether it was a popular success, and on occasion what that might reveal; whether it influenced subsequent programmes; what the programme or series or trend in programming may tell us about perceptions of the audience over a given period; on a few occasions, what the relationship is between the original (novel, play, radio play) and the television adaptation; whether there was any controversy threatening censorship or resulting in self-censorship; and whether the programme deserves attention simply on the basis of its excellence.

Too much of the primary material before 1978-79, that is, the programmes themselves, is gone. "The Ephemeral Art," as critic T. C. Worsely calls it. We are unable to turn to a body of scholarship, or even much in the way of general criticism, for marking out a logical path through this terrain. Yet I hope to explore such a path with you in the full expectation that others will soon branch out in new directions, dispute the choice of route, the stops made, the byways ignored.

The scholarly reader and particularly the student may wish to know which critical methods I have found useful in looking at this mass of material. The short answer is that many approaches have value, depending on what we are looking at and what we are looking for. The more general reader may want to fast-forward through the next few paragraphs.

I have used some of the premises and methods of content analysis, contextualism, archetypal and modal criticism, and structuralism. Each methodology has yielded some interesting observations. I found that the most useful approach to series drama was through audience expectations as defined by genre. Genre analysis depends on such questions as: is it a copshow? a mystery? what kind? if not, why not? Genre is the reference point for the first half of this book. In the second half, I have shifted focus to a chronological look at plays found in anthologies and as drama specials. The chapter on regional drama concentrates on programming made specifically in Montreal, Vancouver, Winnipeg, and St. John's for regional audiences.

I have used content analysis simply to identify certain basic patterns of character or situation, setting or recurring motif, particularly in series television. Content analysis is, of course, one way of identifying systems, or "codes," as the structuralists and semiologists call them. Another way of pinpointing codes is to look for metaphorical and metonymical patterns. Contextualism focuses on what the viewer brings to the programme on initial and subsequent viewings, which includes, first, the age, sex, education, nationality, region, urban or rural way of life, marital status, work, and personal taste of the viewer. Audience research figures on enjoyment indices, who watched, how many, etc. are related to this approach. Then, the viewer's particular training and experience in the medium, formal or informal, have to be taken into account. Finally, the programme has to be seen in relation to the social context and sense of values of its particular time.

Claude Lévi-Strauss, one of the major influences on structural analysis, assumes that the structure of anything studied, particularly the visual as opposed to the verbal, can be both unconscious and rational.[3] In his paradigms, he finds patterns of relation between characters or situations in the variants of a common narrative, which are both like and unlike, by breaking down the elements of characters and situations into binary opposition. He then identifies other terms in the paradigm which mediate, transform, transcend or simply dissolve the oppositions. I have also found this technique valuable for analysis of long-running series.

The most obvious codes or structures of television drama, whether in single programme or series, include the visual information of lighting, costume, gestural (body) language, composition (where every actor is located in space within the frame), and blocking (how they move). Other structures of meaning are created by music and sound effects as recorded and mixed, and by the dialogue itself, including diction and idiom. Equally important is the actor's inflection of the lines. All of these create his or her interpretation of the subtext, which is a larger pattern or structure within the play conveying what is not said but is understood by the audience. Other codes include filmic conventions like the editing of action and reaction shots, the master shots which establish time and place and people, the use of closeups for intimacy and extreme closeups for emphasis, as well as whether the cameras take a subjective or objective point of view. Traditional narrative structures, that is, plots or story lines, are set out in the dialogue and situations, given pace and focus by the director, and then, when the film or, more recently, new kinds of tape are shot, given a flow and a finish by the editor.

Semioticians who work chiefly in film criticism agree that in a performing or collaborative art like television drama, analytical emphasis must fall on the *gestalt* or interaction of these systems. To understand how they interact to shape a given performance, a critic should have some first-hand knowledge about the process of making television drama on film or tape. In another book, I will introduce readers to the world behind the cameras and to the way dramas were and are made in Canada. In a third book, the voices behind the camera—actors, producers, designers, cinematographers—talk about what they do and how and why they do it. Many of the observations in this book are based on what I learned from those conversations. This book is a series of long-shots of CBC television drama, with more detailed medium shots of genres, periods and trends and selected closeups of clusters of individual programmes from every period and programme type.

The Bard of Avon and the Bard of 790 Bay Street

The analogy between the privately owned Globe Theatre (in which Shakespeare was a shareholder) and the CBC as a Crown corporation is as close as that between the "mixed" system of public and private ownership characterizing both programme production and dissemination of television drama in Canada and the special status of the authorized acting companies of the sixteenth century. Plays like *Richard II*, which did not meet official approval, could spell serious trouble for the players. So too the CBC has both extended the limits of taste and taken risks, artistic and political, and has self-censored some of the more innovative dramatic programming—for example, when Mario Prizek's 1965 anthology, *Eyeopener*, was going to air in the middle of a growing uproar about the possibility of the cancellation of the popular and very controversial current affairs programme *This Hour Has Seven Days*. Prizek's first programme, a satire on successful Quebec separatism called "A Borderline Case," was unceremoniously shelved. The issues raised by censorship and self-censorship never go away. In 1985, the CBC chose to broadcast a version of Fennario's stage hit, *Balconville*, with its unprintables intact. However, CBLT Toronto, the network's flagship station scheduled the drama at 11:30 p.m. to minimize the number of viewers who might be offended. Ironically, CKNX Wingham (population 10,000, in rural southwestern Ontario) delayed a few weeks and then broadcast the programme in prime time.

Whatever changes of government, crises, or parliamentary budget cuts intervene, the CBC must also provide education, information and entertainment for a broad spectrum of society at the same place on the dial, seven days a week, fifty-two weeks a year. To survive, the CBC, like Shakespeare's company, has had to be adaptable to changing fashions; serve up both caviar and small beer for its public; evade or risk censorship; try to satisfy audience expectations by returning to the successful formulae which are the staple of any success in the performing arts; provide novelty to keep viewers coming back; avoid perpetuating overly familiar images, and yet develop a visual or aural shorthand—a set of conventions—that would let the makers of television drama get on with the more interesting tasks of storytelling.

The analogy between the Elizabethans and the CBC breaks down when we come to the fact that the Elizabethans were not competing with playwrights from other English-speaking countries with ten times their population or a 500-year head start on creating an indigenous, mature dramatic heritage.

To Imitate or to Be Ourselves?

Two central facts, taken together, distinguish the Canadian Broadcasting Corporation's situation from those of all other broadcasting networks in the world. One is that, like Australia, Canada did not have a distinctive, mature tradition of theatre and film on which to draw when radio and then television drama developed. The other is that, from the beginning, the majority of Canadians have had access to American radio and then television drama, either directly from the American border stations, selectively from the CBC, or indiscriminately from the CTV network, Ontario's Global network, and independent stations. In a golden age of national pride, Elizabethan playwrights could weave the rich tapestries of their lives on an almost empty loom. The CBC more closely resembles Homer's Penelope, weaving and secretly unweaving her tapestry so that she will not have to marry one of her unwelcome suitors. How else can one explain the consistent self-deprecation, the less than confident scheduling, the lack of publicity and the refusal to build stars that characterized the corporation for many years?

At its best, however, CBC television drama has continued in the superb tradition of radio drama developed under Andrew Allan, Esse Ljungh and others. To paraphrase both Allan and playwright Len Peterson, the idea was to "choose a play or (create a series) for adults . . . then get it on the air and answer for it afterwards." Of course, it has never been that simple. Anyone who creates a drama, that collective and "impure" art, must consider many factors: who the audience is; how to tell the story within the physical constraints of the medium; black and white or colour; live or film or tape; studio or location; thirty or ninety minutes; using a well-tried formula or working in a more distinctive mode.

Then there are such variables in quality as the casting of parts; the design agreed to and executed by many technicians; the skill and artistry of the director of photography and the lighting and sound technicians; the creativity of the director who "calls the shots" on the actual production as it is being made; and above all, as costs have risen, the initiative of producers who often think of projects, control casting, editing, and budget. How a TV drama reaches telecast, how it is scheduled and promoted, and what the broadcasting mix as a whole may be, also directly affect what we see and how we see it. These processes have changed radically through the years, and so has television drama. I allude to these factors throughout this book.

Other variables which have shaped our television drama include the CBC monopoly in the early years of television. Equally important, then and now, has been the need of the francophone population for their own services and, in particular, their own drama. In recent years, pressures brought to bear by government on the CBC have become acute. There is still acrimonious debate about the Canadian-content regulations, largely because they have never worked for television drama. There is the inevitably uneasy relationship with the Minister responsible for broadcasting (it varies from regime to regime) and the dozen other government agencies to whom the CBC reports.

People on the creative level at the CBC have also taken seriously the various broadcasting acts with their changing and increasingly demanding definitions of the corporation's functions, which in turn have directly affected the content of TV drama—"mandate work," as John Kennedy, Head of Television Drama, calls it with only a hint of irony. This is a fair enough précis of what the 1968 Broadcasting Act demands: that the CBC be a national service in two languages which "must safeguard, enrich and strengthen the cultural, political, social and economic fabric of Canada," strengthen a sense of regional identity, be the agent of communication between the disparate parts of this land, foster Canadian talent, and keep viewers tuned to the network often enough to justify their very small share of our taxes.

Last but by no means least among these constraints is the decision by successive governments to make the CBC partially dependent on commercial sponsorship, which means that drama series, if not drama specials, have to be sold to sponsors before they go to air. From the beginning, this has created deeply rooted hostility and suspicion between the CBC and independent broadcasters and production houses. With production costs rising beyond the means of even American networks for ninety-minute or two-hour specials or miniseries, it now means that "sustaining" programmes— dramas that cannot find sponsors or else, for some reason, should not be sponsored—are finally extinct. Risk-taking is virtually impossible. Yet somehow, occasionally, the CBC continues to take risks.

The Electronic Border

Perhaps the most significant variable of all, however, is the pervasive influence of the widely advertised, easily available American programmes which can be bought for telecast at a tiny fraction of what it would cost Canadian producers, public or private, to make their own, a factor compounded by the well-deserved popularity of American forms like the television Western, copshow, sitcom, and soap. For three decades, these offerings have directly influenced the expectations of the Canadian audiences. Back in the early 'thirties, when many Canadian radio stations were either joining American networks or were owned by Americans, Graham Spry, one of the founders of the Radio League which lobbied for publicly owned radio, put the case very succinctly: "the state or the United States." CBC executive producer of *The Journal*, Mark Starowicz has pointed out how odd it is that, alone of all industrial nations, we have a quota, not for foreign content, but rather for "Canadian content."[4] Indeed, as Paul Rutherford demonstrates at length in his book on what Canadians watch in prime time, that any distinctive Canadian television exists at all is something of a miracle.

Along the electronic border, another system has developed. To this day there is no government-sustained public network in the United States. U.S. commercial companies export enough material to other countries to fill 22 networks, 18 hours a day, every day of the year. ABC, CBS, and NBC import enough material to fill two days a year.[5] This narrow vision, virtually unchanged since the mid-'fifties, has had a profound effect on our television.

But now, even American independent producers are having trouble raising money for new programming. Co-productions are becoming a reality in the

United States. At the 1986 Banff Television Festival, a whole session was set aside for independent American producers to tell the international delegates just what kinds of projects they would buy. What each described was a project stamped "U.S.A." Other distinctively regional, national, or even personal voices, ideas, dramatic conventions, and visual styles would not sell to American networks, they said. The message seemed to be, "Export to us look-alikes of ourselves and we may buy them." To quote ex-CRTC commissioner and renowned critic Northrop Frye, "In an immature society, culture is an import; for a mature one, it is a native manufacture that eventually becomes an export."[6] In Canada, television drama accounts for 68 per cent of total viewing, yet in a good year only 4 per cent of it originates here. As the figure for TV drama has steadily climbed, the made-in-Canada component has declined.

In the 1970s, CTV tried to break into that lucrative American market by cloning U.S. programmes; *Police Surgeon*, for example, with its American uniforms, flags, laws, and street names was a series commissioned and paid for by Colgate-Palmolive. It didn't work, largely because the series was not very good. In 1981 CTV lost a bitterly fought appeal to the Supreme Court against a CRTC ruling which stated that its next licence renewal would depend in part on an increase in Canadian drama and children's programming. The fact that the very profitable individual stations making up the loose confederation called CTV lost the appeal does not seem to have dimmed their resistance to producing Canadian television drama. However, in March 1987, they proceeded to cut back on their news operations despite handsome profits made by affiliates. At the 1986 hearing, after much handwringing, they promised an extra half hour of drama a week.

The CBC has had some sales to independent American stations of *Beachcombers* and *King of Kensington*; to PBS of some of the *For the Record* anthology, as well as *Anne of Green Gables* and *Chautauqua Girl*. Pay channels like *Arts and Entertainment*, have bought several specials plus series like *The Great Detective*. Independent Canadian producers co-produced *Night Heat* (1984-present) and *Adderley* (1986-present), two action shows for CBS midnight viewing. Until August 1987, when these two copshows were given a prime-time slot (not quite the same as a winter prime-time slot), no Canadian network or independent producer had broken into the prime-time slots of CBS, NBC, or ABC. Sales abroad have always been good, and they are getting better. It is well known in Britain that the export of CBC scripts, kinescopes, personnel, and techniques in the fifties revitalized the stage-bound TV drama telecast on both the private network, ITV, and the BBC.

The CBC cannot develop the sheer quantity of speculative pilots and drama specials that American networks and private production houses have been able to mount over the years. On the brighter side, from the fact that our television drama cannot be slipped into American network prime time without a ripple, I infer, unlike others, not that Canadian TV drama is inferior (despite the mythology, our production values are usually competitive), but that it differs from American television drama. That a significant proportion of our television drama has been and still is distinctive is, in fact, the fundamental premise of this book.

I began to write this book three and a half years ago. In the last few months, it became obvious that 1986-87 marked a major turning-point in the development of television broadcasting. The Caplan-Sauvageau task force recommended an overhaul of both production and delivery systems. Moreover, the 50th anniversary of the CBC in 1986, together with the new technologies of the 'eighties, mark the beginning of a new phase in the story of broadcasting in Canada.

What these changes will mean to Canadian television drama until 2002, which is the year of the 50th anniversary of CBC TV, is unclear. It could well be that new technology will destroy all commercial networks, and with them all television that is not "pay" or state-owned. In the 1980s, production by Crown Corporations or state-owned television has been under attack in every Western industrial nation. On the other hand, if the political will exists to ensure the survival of the CBC, VCRs and pay TV may do for the CBC what the repeated recommendations of parliamentary bodies have not—free the corporation from the need to find sponsors. If the privately owned networks do survive as "free" television, CBC drama may more easily forgo the ratings-share game that has become less relevant in this age of "fragmented audiences." The CBC may instead design television drama for various specific audiences. Experiment may be encouraged again. Glimpses of world theatre may return. Canadian theatre may contribute as much to television drama in the 'nineties as it did in the mid-'seventies. Hits from Radio-Canada may be more often subtitled or dubbed, or even remounted in English. The narrow focus on film and studio naturalism may be extended to the imaginative limits which were explored by CBC TV drama in the 'fifties and 'sixties. The excitement of "live" television might even reappear on occasion, not for the sake of nostalgia, but to make the creative juices flow in a different way.

Despite manifest difficulties, the Canadian television drama we have enjoyed in the last thirty-five years has often been of high quality and deservedly popular. The two are not mutually exclusive. In his 1974 book, *A Nation Unaware: The Canadian Economic Culture*, Herschel Hardin, far from lamenting the "mixed" system of commercial and national interests and the plethora of choices now available to all of us, argues cogently that this system results in "continuous creative tension."[7] He claims that this set of circumstances "has a prodding, integrative effect and makes for a tolerant cohesive society." As he points out, in contrast to the U.S. and the U.K., "Our majorities have never been inclusive and dynamic and conformist enough and the individualists never rich and numerous enough to afford an escape from each other." He goes on to contrast the American norm of smooth, competent efficiency with our norm, which is unresolved and often quite public conflict: "Canada's problem is not to evade or suppress conflict or controversy . . . but to find ways of legitimizing the conflict so that it will add to the tradition rather than take away from it." His more recent book, *Closed Circuits* (1985), however, is a persuasive indictment of the Canadian Radio-Television and Telecommunications Commission for weakening the whole system by continuing to license CTV pay television, the Global network of independent stations, and the superstations, while allowing its regulations on Canadian content to be systematically ignored. Hardin's subtitle is "The Sellout of Canadian Television."[8]

This procedure contrasts with the national content regulations governing the Independent Television Authority in Britain and other systems in Europe. However, with the choices of channel and delivery system now opening up in Europe through direct broadcast satellites, cable and privately owned networks, the fears of change, the resistance to the new technology are all summed up in one word: the dreaded "Canadianization" (as they call it) of television as a medium. We have managed to develop and maintain more or less consistently distinctive television under these pressures for decades. To Europeans, such choices are a new experience.

In 1980, Northrop Frye came to the pessimistic conclusion that the likelihood of a sense of "Canadian identity" so far as television is concerned "begins to look as desperate as a Spartan at Thermopylae," and that, during the past two decades, "the majority of Canadians preferred the flood [of American programmes] to any Canadian ark that would seem to float above it."[9] This is simply not so. Nor will it necessarily be so in the future. As everyone points out (but as everyone who decides on broadcasting policy and budget promptly forgets) the millions of viewers who chose CBC drama did so not because these programmes came from the Canadian Broadcasting Corporation, but because they are good. With the delightful co-production of *Anne of Green Gables* (1986), the CBC drew the biggest audience for programming (other than for the Stanley Cup) in its history: 4.9 and then 5.2 million viewers, or nearly half the people watching television in the country.[10]

Since satellites and community dishes have been bringing the continent to the viewer, the CBC now functions (as, to a lesser degree, it always has) as part of a continental grid. It reacts to the internal and external constraints on programming in various ways; sometimes innovatively, and sometimes retreating into imitation, confusion, didacticism or blandness. CBC television drama also lives with royal task forces, commissions, irate Members of Parliament, the kinds of bureaucracy apparently endemic to a Crown corporation, and ever more restricted, annually fought-over budgets. Up to the late 'eighties there has also been the 1968 mandate, with its blithe expectation that the CBC will work wonders: unify the country, defend regional identity, educate, be midwife and patron to the arts (not to mention training-ground for playwrights, directors, actors), inform, and entertain.

In the interest of accurate scholarship, I must caution the reader about the dates given in this book, particularly with regard to anthology dramas broadcast only once or twice. Problems with the kinescopes vis-à-vis the most recent tapes differ, since the CBC has kept better records since 1973. Even with relatively new material, however, the vagaries of affiliate scheduling and preemption can make a liar of the press releases and newspaper critics, who had the advantage of previews in the 'seventies. I have not been able to locate a complete run of *TV Guide* in any library, including the CBC's, and, in any case, the information it supplies would be subject to the same problems. The *CBC Times* ceased publication in 1970. Thus, Eric Koch's carefully researched book *Inside Seven Days* (1986) assumes from print sources that the controversial drama *The Open Grave* (1964) was broadcast on Easter Sunday as planned. It was not because it was preempted by a game in the Stanley Cup Playoffs.

For material from the 'fifties and 'sixties, I have supplied readers with very precise dates, many coming straight from the film cans I handled while assisting the Public Archives in cataloguing the CBC collection. The date is certainly a telecast date of the programme, but it may represent the telecast date in Flin Flon, where so many of the kinescopes ended up. Before the days of microwave and satellite, programmes travelled across the land in cans, not over transmission lines, so their air dates varied from place to place. The affiliates' freedom to reschedule many of the dramas further complicates the dating to this day.

What this means for the reader is that the year is accurate, the month is probably accurate, and the day means that the programme went out somewhere in the country on that date. One further word of caution is called for here. A year appears on the end of the print of every complete kine, tape, or film. This is the year the drama was completed, not when it was broadcast. Programmes can sit on the shelf for months, even for a year or two, either because of the CBC's strange budgeting rules by which a programme's costs are charged, not to the year it was made but to the year it is broadcast, or, more rarely, because the corporation thinks the programme is "not up to standard" and holds it for an empty slot where few people will see it. (For financial reasons, most programmes do make it to air at some point.) Co-productions intended first for release to cinemas may also take longer to reach the air than expected if they are successful there. This is why Sandy Wilson's successful film, *My American Cousin*, originally scheduled for telecast in 1986, was not seen until 1987. In the 'seventies and 'eighties, exact telecast dates represent dates from tape cans, lists, or my off-air notes. When only a year is given, that is all I have been able to find.

"Reading" Television Is Not Enough—A Warning

The part of the spectrum of television broadcasting that is drama, though usually treated as ephemeral, is by no means the notorious "chewing gum for the eyes," that image perpetuated by so many contemptuous and ill-informed journalists—and academics. Nevertheless, everyone from the president of the CBC down to the production secretary has treated most programmes, once finished, with complete indifference. As it has always been in the theatre, that truly ephemeral art, the focus is on this week's production problems, this month's shoot, this year's ratings, or next season's schedule. The Elizabethans were the same. Shakespeare did not bother collecting his plays together into one volume while he was alive. Luckily, Heminge and Condell, two friends of his did or we would never have known about *Macbeth* and many of his other best plays.

To remedy this ignorance of their own heritage, people who make television are going to have to sit down to look at programmes, just as filmmakers look at all kinds of films going back to the earliest days of silent films, as a matter of course. Then, perhaps, the programme makers will not waste time and resources reinventing forms like the docudrama; they will know what went on before, and may thus develop a perspective on the work they are doing at that moment.

Distinctive Canadian television drama doesn't have to be set in Canada, fly flags, or show Mounties. When we do Molière or Shakespeare well, our way of doing them reflects our different perspective. When we work out a variation on an American or British formula, the result, if it's good, is unmistakably made in Canada, from the first season of *Sidestreet* to *SCTV* and the Mackenzie Brothers. When we do produce original drama, we sometimes strike pure gold. A distinctive culture is not just a matter of economics, politics, or even political will. Above all, it needs talented people to make programmes, professionals who know what they have already accomplished, what they and the medium can do, who are not just interested in "out-Heroding Herod," even when fustian gets the ratings. To do the best possible work, our creative people must recover a sense of their tradition. It is not good that our scriptwriters, directors, producers, designers, technicians work (with a few honourable exceptions) in a critical vacuum. Surely more books on television drama in Canada, by other critics, will follow.

Scholars, graduate students, and undergraduates must also have easier access to the materials that have survived, often by chance, so that they can teach, study, and analyze Canadian television drama with the rigour the material deserves. Perhaps (some day soon?) the general public will be able to buy or rent video cassettes from a full selection of their favourite shows, or watch reruns and low budget experiments on a second CBC channel.

In spite of itself (i.e., despite the wanton destruction or disappearance into private hands of so much of the primary material) the CBC still has a collection of television classics to deposit with the national and provincial archives. When I talked to Eric Till about *The Offshore Island*, which won so many awards and which has survived in kinescope, he said that he would not now remount the play and he is right. We need to rethink and recast those issues for the 'eighties. However, the original *Festival* production should be rebroadcast as often and for the same reasons as vintage films are rebroadcast. How much effort has been put into renegotiating the contracts of those people behind and in front of the cameras who made the outstanding shows? I know that the older contracts were never negotiated with replay in mind. Is it cheaper to fill a prime-time series with contracts renegotiated from the 'fifties and 'sixties, or with all new material? Has anyone really tried? Why can't universities rent or buy VCR copies of the television classics for study after signing a "no-copy" agreement? How will the legitimate reclamation of this heritage proceed in the colleges and universities, in film and drama courses, sociology and history courses, without organized access to programmes which should rightfully belong to us as citizens? I have yet to see a broadcasting study, white paper, Royal Commission, Parliamentary committee, Board of Broadcast Governors, or later, the CRTC address this fundamental question. After the various empires negotiated terms, an orderly deposit of CBC programmes was finally assured at least up to the mid-'eighties. The problem is twofold— preservation and access. Preservation must come first, but without access preservation is meaningless.

Collections of CBC television scripts do exist. The most extensive is in the York University archives and is catalogued by series, title, author, etc. Provincial archives have some related materials and some scripts. There are also a few at McMaster and some at Concordia, uncatalogued. However, the researcher must be warned that these scripts do not provide reliable data about what actually went on the air. Most of the York collection seems to be clean copies of drafts sold to directors or producers. They have not been revised during production by the authors, the directors or the producers.

They do not reflect on-set changes or post production editing cuts. The hazards of using such scripts as a basis for specific comment may be demonstrated by the fact that the only existing scripts of Timothy Findley's excellent drama, *Paper People* (1967), are "clean copies" of this kind. Yet the ending of the film is completely different from that of the script— rethought and rewritten on the spot.[11]

There are no warnings on these collections to indicate to the unwary exactly what it is that has been saved, but if the script is all on white paper (each revision is colour coded), not written over, scratched out, or full of notes, then it is not what reached the screen. To recreate a programme without the actual airtape, one would need the scripts of the director, continuity person, lighting and sound technicians, director of photography, actors and producers. Even then the editing, sound mix and music would not be "on the record," that is, on paper. Trying to recreate *Hamlet* as Shakespeare wrote it and Burbage played it has occupied scholars for generations. It is far more difficult to recreate the lost images of *The Puppet Caravan* or *Slow Dance on the Killing Ground* or *God's Sparrows*, to name only three which have been lost or mutilated.

Why does CBC television drama matter? The Canadian identity is not to be found in definitions or generalizations, although I admit to committing a few in this book. The Canadian identity is to be found in creating art, entertainment, popular culture, among many other activities, in Canada and for Canadians. It is also to be found in the responses of viewers to what they see and hear. If we do not tell ourselves funny or satirical or tragic or ironic stories about ourselves, if the teller of tales in the corner does not sing our songs, speak our sorrows, narrate our wars, then we will not exist as a nation. Television, for good or ill, is the mass medium of this age. Like all cultures, we find our sense of self, not in facts, but in fiction, in the songs and plays that express our view of the world.

NOTES

1. Alfred Lord, *The Singer of Tales*. Atheneum, New York: 1965, 128. Various critics in recent years have noted that television is "bardic" including John Fiske and John Hartley, who approach the term through the semiotics of Barthes and Claude Lévi-Strauss. I prefer to stress the analogy from the maker's point of view rather than exclusively with the function of television in society.
2. A sample: "Although CTV representatives have repeatedly recommitted the network to significant amounts of Canadian content, in the 1984 calendar year about

one per cent of programming on CTV between 7 p.m. and 11 p.m. was Canadian drama while some 50 per cent was foreign drama—overwhelmingly American of course." *Report on the Task Force on Broadcasting Policy,* Ministry of Supplies and Services, Canada, 38. Ottawa: 1986

3. Octavio Paz, *Claude Lévi-Strauss: an introduction,* tr. J.S. Bernstein and M. Bernstein. Cornell University Press: 1970, 50 passim.

4. *This Magazine,* March 1985.

5. Joyce Nelson, "Dumping Ground," in *Saturday Night,* May 1981, 21. The statistics have not altered significantly since.

6. "Across the River and out of the Trees," *The Arts in Canada: The Last Fifty Years,* W.J. Keith and B.Z. Shek, eds. University of Toronto Press, Toronto: 1980, 4.

7. J.J. Douglas Ltd., Vancouver: 1974, 267-71.

8. Douglas and McIntyre, Vancouver: 1985.

9. "Across the River," 4.

10. CP wire service in the *Standard,* St. Catharines, 28 December, 1985.

11. M. J. Miller, "An Analysis of *Paper People*" and Timothy Findley, *The Paper People* in *Canadian Drama,* vol. 9, no. 1 (1983). Compare this article with the script which is reprinted in the same volume.

II
GENRES

2

COPSHOWS AND MYSTERIES
"Check your guns at the border"

As viewers of Canadian television, we feel at times as if we are looking into a mirror pretending to be a window, where the images look very familiar. Yet we have to work out whether these images of characters and situations resemble the lives we really lead beyond that window, or whether they look familiar because they are reflections, borrowed versions of familiar images from other cultures. It is not always easy to distinguish between "actuality" and "iconography," particularly when "actuality" consists of staged photo opportunities. Though television's fictional iconography is often derived from the iconography of politics, sports, advertising, and music videos, its dramatic conventions are also independent of these worlds. Dead men aren't dead; we all know that. Bullets fired at the good guys never hit them—at least not fatally. Serious personal problems can be solved in twenty-six minutes on sitcoms, an hour and forty-six minutes on TV movies, or chewed over for many months in the night-time soaps.

Sometimes the rules of series television change on us abruptly. *Hill Street Blues* (1980-87), a copshow of the 'eighties, adopted the multiple open-narrative structure of soaps, explored the domestic problems of its personnel, and showed some of the cops as making mistakes or even being corrupt —and slowly, the audience has been persuaded to accept a new inflection of the copshow format. Popular commanding officer Henry Blake is shot down in the last few seconds of a *M*A*S*H** episode (1974-75 season) and the most fundamental law of sitcom is violated. "Thou shalt not kill a popular character"—just write him out comfortably, letting your audience make the transition without trauma. No wonder that particular scene is often cut in syndicated reruns. Yet audiences do tire of particular formulae after a while. Trends change.

We will discover that good, usually successful Canadian versions of formula television, copshows, sitcoms, family adventure series, or series about doctors and lawyers and parole officers are distinctively inflected. Irony replaces moral certitude. Open narratives or unresolved emotional conflicts replace happy or poetically just endings. Surprising levels of subtext appear on action shows in scenes which are allowed time to develop. Literate dialogue, allusions to the actual society we are living in, music which counterpoints instead of hyping the emotional temperature or telling us when to laugh or cry, even satire on popular culture appear regularly in Canadian series: for example, the last season of *King of Kensington* (1974-80), or *Sidestreet* (1975-79), every once in a while on *The Beachcombers* (1972-present), or consistently on *Wojeck* (1966-68) and *Seeing Things* (1981-87).

The purpose of the whole intricate structure of television drama production—one too often forgotten by academics, politicians and bureaucrats but never by the audience—is the programming itself. No one argues about the value of news and documentaries to society and their occasional direct influence on events, as in the *fifth estate*'s 1985 fall season opener, an exposé of a rancid-tuna scandal that triggered the resignation of a cabinet minister. But it seems less obvious to the average viewer that the superbly orderly, crazily individualized world of the imagination, where most of us choose to live part of our day, every day, is as essential to our health and welfare as safe tuna for our casseroles. To demonstrate that fact, we begin with one of the dominant conventional television forms, the copshow.

Everybody who watches television in North America knows roughly what kind of show is designated as a "copshow." American radio in the 'thirties and 'forties and television in the 'fifties taught most Canadians to recognize the characteristics of the genre. Canadians responded warmly and in large numbers to many of these shows, yet the CBC did not develop an indigenous copshow series until *Wojeck* (1966-68). Why? Even more interesting, why did our culture transmute some of the basic conventions of the formula copshow or mystery in many episodes of the series which followed: *The Collaborators* (1973-75), *Sidestreet* (1975-79), *The Great Detective* (1979-81), *The Phoenix Team* (1980), *Seeing Things* (1981-87), and *Judge* (1982-84)?

One of the factors that condition what programmes are developed by the CBC is the availability of development money for pilots. Usually, in the tight-money situation that has prevailed since the mid-'sixties, the corporation tests and then produces only one show at a time in a given genre. This is part of the reason why the CBC was slow to produce shows in the mystery-copshow formula. *Wojeck* appeared years after detective series like *The Shadow* (1936-56) and *Sam Spade* (1946-mid-'fifties) had been aired on American radio and *Dragnet* (1951-59; 1967-70) and *Perry Mason* (1957-66) had become established television hits.

Since the CBC has not made programmes for twenty-six- or even thirteen-week seasons since the early 'sixties, shows cannot be "stripped" (run in syndication daily) until they have been produced for four or five years. They are not intended to be sold initially in such markets. Yet the best ones are sold worldwide, usually to other state-owned networks with less rigidly structured schedules. The *Wojeck* series produced twenty episodes in two seasons, *Sidestreet* around thirty-five over four seasons, and the rest fourteen or fifteen. It took four seasons for *Seeing Things* to build up a stock of at least twenty-six programmes. The advantage, of course, is that there should be in principle very few episodes which pad out a season. As John Kennedy, CBC's Head of TV Drama, pointed out to me in an April 1985 interview, given the available technical and human resources, the limitation of making only eight to ten episodes a year should and often does help ensure quality in a series. Even though advertising is sold for this kind of series on the CBC, a series should not have to win in the do-or-die ratings game or fear sponsor pressure on its content. On the other hand, good ratings definitely prolong the life of a CBC series.

Buried in the Audience Relations report on *The Collaborators* (Tor 74-12) is the observation that if *Wojeck* is removed from the statistical breakdown, the enjoyment index for CBC drama is consistent from 1965 to 1974. Therefore, the report concludes, from the perspective of measurable "audience enjoyment" there was no "golden age" of television drama. Clearly, this is debatable when the methodology shifts from statistical to qualitative analysis, or to the consideration of formal innovation, or the reflection of a rapidly changing society, or even to engagement with contentious issues—but those criteria are not part of the "audience research" methodology. By then the audiences for traditional forms like copshows were older (42 per cent over fifty), tended to have a high-school education, and were evenly split between men and women. *Seeing Things* was part of a plan to win back the younger, 24- to 40-year-old viewers. In contrast to the audience profile for copshows, in more recent years, older women have formed the largest percentage of the audience for CBC anthology drama.

In some ways, the structure of CBC drama production resembles that of Britain, not the United States. The corporation has concentrated on making studio drama on tape as well as location shooting on film. As in Britain, eight or ten episodes of a serial or a series (and not twenty-six) have been judged to be about right. Originality and craftsmanship were the solution to tight production budgets, not a reversion to formula. Also, as in Britain, anthology drama had been the staple dramatic fare of the CBC long after the form had died out in the United States. In analyzing series programming,

I have concluded that anthology continues to be an influential factor on the forms of series programming developed in Canada right up to the present day. A few of these series move consistently toward the characteristics of anthology—varying tones, focus on individual characters, explorations of different forms and visual and verbal styles, open endings, irony, ambivalence, surprise. Some even subvert the cause and effect conventions of narrative structure. When that happens, series television on the CBC takes on a distinctive flavour.

GENRES

John Cawelti, in his seminal *Adventure Mystery Romance: Formula Stories as Art and Popular Culture,*[1] makes helpful distinctions between the terms "archetype," "genre," and "formula." "Archetypes" are transcultural, instantly recognizable characters and situations occurring all over the world. One example would be the familiar and universally recognized "star-crossed lovers." For our purposes, a "genre" is fairly narrowly defined as having specific subject matter. When murderers or other criminals are caught and punished by cops or detectives or amateur sleuths, we are dealing with three different but closely related genres; respectively, the "copshow," its brother the "detective show," and the "mystery."

We can find the "star-crossed lovers" archetype in several genres. When they are sailing on a *Loveboat* (1977-present), their problems will usually be happily and comically resolved within the conventions or rules governing the "domestic comedy" genre. When they visit *Fantasy Island* (1978-84), however, the happy ending is not always assured. They could find themselves in a "comedy" or an "action-adventure" story. The ironic, sometimes sad, yet funny Louie-Marge subplot of *Seeing Things*, in which Louie and Marge are a married couple who separate, then reunite on a rather fragile basis, is another variant of the archetype. Given a tragic twist, the "star-crossed lovers" archetype can become either the sentimental stereotype of the 'seventies mini-movies featuring "super athlete comes down with fatal disease," or the genuine francophone-anglophone angst of *For the Record's Don't Forget Je Me Souviens* (1979).

"Formula" is a word that too often connotes "stereotype" with its overtones of "stale" or "unimaginative." But "formula" simply denotes structures or narrative and dramatic conventions that are very widely used, archetypes adapted through specific cultural materials to particular shapes. "The concept of a formula . . . is used primarily as a means of making cultural history and cultural inferences about the collective fantasies shared by large groups of people and of identifying differences in these fantasies from one culture or period to another'' (Cawelti, p.7). The word ''genre'' is often used to denote the aesthetic boundaries which organize our expectations, to some degree shape our responses to characters, settings and plots as viewers, and thus help us to define how formulae and formula variants interact. Like many other critics, Cawelti emphasizes both the reassuring and the self-reflexive nature of formula: ''It intensifies the expected experience without fundamentally altering it. . . . It appeals to previous experience of the type itself: it creates its own field of reference'' (p.10). When that field of reference is consciously drawn to the attention of the audience, as Wayne and Shuster have done for thirty years or as the SCTV troupe did with their barbed satires on television forms, the programme created is self-reflexive. We are made aware of the conventions, which until then are likely to have gone unnoticed.

''By the late 'fifties, Hollywood was the place; the drama series was the form; repetition was the key; predictability was the goal,'' Jeff Greenfield tells us in his survey *Television: The First Fifty Years*.[2] Writers were coming to producers saying, not ''Here's an idea,'' but, ''What do you want me to write?'' The characteristic conflict of much of television drifted from the interplay of character or the clash of ideas to physical action-reaction. Greenfield defines the basic formula of series television, as it evolved, as easily identifiable, humanized, central characters with quirky friends in ''life-jeopardizing or life-saving situations, which often involved a gun, a scalpel or a law-book.'' He quotes ABC audience-research chief Marvin Mord as saying that ''people are not willing to accept real problems in television drama.'' *Wojeck*'s astounding audience figures and enjoyment indices can mean only one of two things. Either Mr. Mord was mistaken, or Canadian audiences were different from American audiences, at least in 1966.

According to Eric Till (interview, 12/9/83), the reason why the CBC did not try to create Canadian variations or imitations of the huge successes of *Dragnet* or *Perry Mason* in the 1950s or even the critical success of the less formulaic *Naked City* (1958-63) and *The Defenders* (1961-65) in the 1960s was that the corporation basically distrusted popular genres, and particularly those originating in America, such as the sitcom, copshow and soap.[3] After all, the majority of Canadians could already get American series on American stations, many American series were bought by the CBC, and the drama programming on the CTV network was overwhelmingly American.

Until 1965, with the independently produced series *Seaway* about policing the St. Lawrence system, series centered on one or more wise authority figures or comfortable, inept, average people were not made in Canada. In *Seaway* (1965-67) the settings were new, but plots were often predictable. The two leading characters, one young and brash and the other older and more experienced as the formula dictated, were two dimensional.

For the most part, consciously or unconsciously, the CBC adopted the quite logical premise that since Americans do their own genres best, they should be left to do them. The CBC's response to the extraordinary popularity of the television "Western" throughout the 'sixties was *Cariboo Country* (1960-67). It was not a series featuring one protagonist plus sidekicks, did not take place in the 1870s, and had no gunplay. Instead, it varied enormously in tone from episode to episode, was distinctively regional in idiom and speech patterns, and treated the Indians of the Chilcotin as intelligent, productive people with different but equally valid customs. Most important of all, like the best of our copshows, it was sometimes morally ambivalent, often open-ended, and just as often ironic. Unlike many other CBC programmes, *Cariboo Country* also had a wonderfully wry sense of humour.

Another factor affecting the development of our series television is that the CBC refuses to develop a star system. Comedy and variety shows like *The Tommy Hunter Show, Don Messer and His Islanders, Juliette,* or *Wayne and Shuster* are named after their leading performers. Dramatic series are not. In 1968, the CBC's refusal to spend money on building its stars was one of the factors that destroyed the most popular series the network has ever had; *Wojeck* was a true social phenomenon, watched and hotly discussed all over the country in its short two-year run. Twenty years later, we have few Lorne Greenes or William Shatners (they were not really "stars" when they left Canada), or actors who enjoy "star" status—not only while they are in successful shows, but also through guest spots on other series in dry periods. Yet many American and British "stars" retain audience interest over many years. There are a few Canadian exceptions: Donnelly Rhodes (*Sidestreet*), who starred in a U.S. sitcom that died, then a U.S. soap, is now the only reason for watching a sadly derivative "Canadian" co-production of a family-adventure series called *Danger Bay* (1984-present). Other exceptions include Gordon Pinsent in *Quentin Durgens M.P.* (1960-69) and *A Gift to Last* (1978-79), or Gerard Parkes in *A Gift to Last, Homefires* (1980-84), and *Fraggle Rock* (1983-87). There is no audience identification with the leading actors in most new CBC series: the CBC has to build up that identification from scratch every time. This is quite a handicap in the ratings stakes when the other networks consistently feature old friends of the audiences as stars of new series.

Wojeck was the first true "series" television created by the CBC. It was a smash hit with audiences as well as a critical success. Steve Wojeck is a crusading coroner—a character inspired by a controversial chief coroner in Toronto in the 'sixties. The plots are usually but not always triggered by various forms of violent death, not by "crime" per se. In fact, what exactly constitutes "crime" is one of the major themes of the series. Are careless construction practices and kickbacks "criminal"? Is homosexuality, in the context of the repressive laws of the 1960s? What about neglect in old-people's homes? Who is guilty of the suicide of an Indian in the Don Jail? This sort of series does not conform to Cawelti's harmonizing function of formula television. In *Wojeck*, we are not "exploring in a carefully controlled way the boundaries between the forbidden and the permitted." Instead, we must deal with the disturbing and unresolved questions raised, the loose ends, the odd and unfamiliar characters. However, as Cawelti puts it, this kind of programme may still help us "assimilate changes in the values of traditional imaginary constructs."

Although I do not concentrate in this chapter on the complementary forms of sitcom and action-adventure, no form of series television in Canada, or anywhere else, develops in isolation. There are times when, in our search for Canadian form and content, we forget that television drama, like all fictional forms, belongs to a tradition of storytelling using archetypes, devices of language, and temporal and spatial conventions which grow and change and form part of a continuum reaching back to the shadow play on the walls of caves. Cawelti's distinction between "invention" and "convention" is a useful reminder of this fact. As we shall see, *Wojeck* and, fifteen years later, *Seeing Things* were hybrids drawing strength from genres as various as mysteries, sitcoms, action-adventure, and even, in Horace Newcomb's phrase, the "counselor and confessor" genre to create unique and successful variations on formula series television.

In fact, the CBC pattern reveals a tendency to be innovative, often by the simple expedient of being truthful to our different culture. Yet in some cases, as in *Sidestreet* or the successful sitcom *King of Kensington*, the corporation appears to lose its nerve before the audience has had time to adjust to the new territory being explored. Even so, though the series as a whole begins to revert to formula, individual episodes may still inflect the conventions and surprise the audience.

After *Wojeck* folded at the height of its popularity, the drama department continued to avoid the more traditional detective or police protagonists by focussing on parole supervisors in *The Manipulators* (1970-71), a series examined in Chapter 5. Coming closer to the traditional formula in *The Collaborators*, they offered a "team" consisting of a forensic scientist and a policeman, and added a paper-doll woman who got all the expository lines and stood around. When *The Collaborators* folded, the CBC came up with *Sidestreet*, which set out to be a distinctive variation on the standard copshow, with protagonists who were "community workers" in uniform. It ended four years later as a mixture, with some episodes much closer to standard formulae (including violent solutions to oversimplified crimes) and other episodes still offering ambiguity and nuance. Perhaps to balance the imitative nature of *Sidestreet*'s later seasons, the CBC then developed a charming oddity called *The Great Detective*, featuring a Victorian detective. Some of the plots were adapted from the files of the first investigator in the modern detective mould to work for the Ontario government. Later, this series overlapped with a highly formulaic male-female spy duo in the deservedly shortlived *Phoenix Team*. Luckily, *Phoenix Team* was replaced by the very unusual and interesting innovations on the formula which distinguished the first four seasons of *Seeing Things*. *Seeing Things* in turn overlapped on the mix of criminal and domestic drama that was *Judge*, examined in Chapter 5. I will give a more detailed account of some of these series later, with special emphasis on the two which I consider to be both innovative and of the highest quality, *Wojeck* and *Seeing Things*.

WHAT IS A COPSHOW? A MYSTERY?

The genesis of the American police show, as suggested by Richard Gid Powers in "J. Edgar Hoover and the Detective Hero," is instructive in this connection because it helps to clarify the difference. For one thing, Canada had no Federal Bureau of Investigation or legends from Prohibition days that were still potent in the 'sixties. Instead, Canada had the legend of the Mounties who always get their man, preferably by sheer force of character. Canadians had always enjoyed the crime movies and copshows while living safely across the border from the culture and the mythology that created it.

The series *R.C.M.P.*, also broadcast in prime time in 1960-61, emphasized human decency and detection rather than violent plot resolutions. Its main protagonist was francophone, its settings both urban and rural, and bullets were rarities on the show. Oddly, in the winter of 1983, *R.C.M.P.* was programmed back to back on Saturday mornings with Hollywood's *Sergeant Preston of the Yukon* and his wonder dog King (1955-58). *Sergeant Preston* is pure children's western formula, much like *Hopalong Cassidy* (1949-51) or *Roy Rogers* (1951-57). *R.C.M.P.* has a documentary flavour with carefully observed subsidiary characters and realistic settings. The contrast is instructive.

The "family hour" that arrived in the United States in 1975 represented an agreement of sorts to keep adult subjects to programmes scheduled after 9 or, preferably, 10 p.m. The "copshow" and "detective show" response was to redirect the violence from people towards objects. This trend has reached its clearest expression in the comic distance from their ubiquitous but object-oriented violence by the producers of *The A Team*, an immensely popular 1980s crime show with some of the goofy flavour of the cartoon series *Roadrunner* (1966-67; 1971-72). David Victor has been a successful producer of series and single dramas since the early 'sixties. His dictum for success in a series goes this way: "The secret of a good series is that you must be able to see episode thirty-five or forty-nine before you begin" (Greenfield, 163-64). Greenfield quotes the producer of *Police Story* (1973-77), a more realistic cop series than many: "A dramatic lead is a function, not a human being. He's not going to die, he's not going to quit his job, he's not going to grow in dimension. So the writer starts off with a leading character who is not interesting. You have to find meaningful problems for him to deal with. So every week you give him a surrogate problem" (p. 154). Note the automatic use of the male pronoun for a crime show. Women cops-detectives who rescue themselves or their male partners and take independent action are very rare, as Angie Dickinson found out in *Police Woman* (1974-78). The newer hits *Moonlighting* (1986-present) and *Remington Steele* (1983-present) appear to address the sexist conventions of the genre. Yet night after night, the idea of equal partners or even equal combatants slips away. Only *Cagney and Lacey* (1981-82-present), which is about two policewomen, has broken away from the mould.

CBC COPSHOWS

The Collaborators

In 1973, the CBC decided to enter the copshow sweepstakes with *The Collaborators* (1973-75); it was about this time that the American networks entered the race with street-wise cops working in ghettoes—*Toma* (1973-74), *Kojak* (1973-78), *Baretta* (1975-78), *Starsky and Hutch* (1975-79)—and in the homes of the upper classes with *Columbo* (1971-78). In its first season, the CBC made eight episodes of *The Collaborators*, which eventually reached an audience of 2 million. The show achieved an enjoyment index average of 70, and a 13 per cent share of all the audience watching television on a given night in that time slot, which it built up to 20 per cent. Audience research does not simply count heads or sets turned on. With the producers it devises questionnaires to find what viewers like, dislike, or specifically think about topics. In this instance, CBC viewers liked the seasoned network actor Paul Harding as the forensic scientist and Michael Kane as the cop, enjoyed the naturalistic action and design, and found the plots "true to life" (Audience Research Report, Tor 74-12). The series may also have been given a boost by being scheduled in the traditionally prestigious 9-p.m. slot on Sunday. Publicity releases emphasized suspense in "the strange world of modern crime detection techniques," as well as the intrigue and intricacies of the plot.

In an interview with Jack Miller (*Toronto Daily Star*, 14/12/73), Kane said of his character that "he comes on as dishevelled as Columbo, as mean as Shaft, as wily as Hawkins and as unglamorous as Cannon [a catalogue of 'hard-boiled detectives'], but with an earthy appeal something like [country and western singer] Johnny Cash." The third "lead," the very fine actress Toby Tarnow, who is not mentioned at all in the CBC Audience Relations report, was given all the dry exposition and apparently told to look pretty, perhaps explaining Jeremy Ferguson's sole comment about her role: "She manages to look gorgeous even though she rarely sheds her lab coat" (*Star TV Week*, 18/12/73).

At the end of the first season, however, before the shows were even aired, Paul Harding left the series. Michael Kane found the single-minded pressure of filming series television too much for him and left after three episodes had been made for the second season. The original producer, Ron Gilbert, who had produced the popular children's series, *Adventures in Rainbow Country* (1968-70), and the derivative, made-in-Canada disaster *Police Surgeon* (1972-73) for private production houses, also left, angered about a salary dispute and the fact that he was not allowed to direct. John Hirsch, who had inherited the series when he took over as Head of Drama in 1973, replaced Gilbert with veteran director but novice producer René Bonnière who, according to Blaik Kirby (*Globe and Mail*, 15/5/74), "wanted to lighten the series." Quebec film star Daniel Pilon took over the lead. The writers provided a brief forty-second scene as Pilon's introduction to the audience but made no attempt to explain Kane's disappearance. Tarnow stayed on in a somewhat enlarged role.

Pilon talked about his role as Richard Tremblay, a policeman from Kirkland Lake, supposedly unable to speak French, though the actor had a francophone lilt to his phrasing and definite Gallic charm and warmth. His comments contrast with Michael Kane's, even though both compare their characters to their counterparts in American copshows. "I'm playing a very ordinary Joe who'll probably bend the rules if he has to, not a Toma with disguises or a Kojak with the tough stuff. What I want him to be is not what people imagine a sergeant of detectives to be, beating up people in the interrogation rooms [note the francophone rather than anglophone perception], but a guy who makes mistakes and sometimes doesn't get his man." That sort of comment links the show to *Wojeck*, not American formula television.

Jack Miller in the *Toronto Daily Star* (29/1/74) called *The Collaborators'* first season "the slickest commercial adventure series [the CBC] has ever produced." These are not adjectives normally associated with the work of Allan King, Eric Till, Don Shebib, Don Owen, and Peter Carter, who directed some of the episodes. The second season received mixed reviews.

Dennis Braithwaite in *The Globe and Mail* called it "a trite Americanized successor to *Jalna*" (1972), which he liked (20/12/73). Ron Base in *The Sun* liked it. Blaik Kirby in *The Globe and Mail* (24/8/74) wrote favourably of both seasons: he liked the "good action in a forensic science framework" and "the tough, intense dominant cop portrayed by Kane," elements characteristic of the first season and the interesting plots of the second. These involved, among other things, a beautiful model blackmailed by two acid throwing thugs, an insane homosexual Nazi who is manufacturing speed, and two harmless guys who figure out a scheme for a robbery so they can buy a hotel in the Caribbean.

I have selected three episodes for detailed analysis, two from the first season and one from the second. Here, as always, I quote verbatim from the scene as broadcast. The first (untitled) episode of the series was written by Grahame Woods and directed by Peter Carter (17/12/73). Good, fast editing establishes a man's murder in the teaser, which is followed by standard opening credits and "copshow" music. The sense of place, Toronto's Cabbagetown, is quite strong, the laconic but passionate protagonist Brewer is interesting. Harding as the cold, crisp forensic scientist is less involving and his wife, a psychiatrist, is provided with little in the way of characterization. The relationship between the scientist and the cop is wary (an interesting departure from cliché), often with one riding the other. The multiple-plot structure (*Hill Street Blues* did not invent the copshow with multiple plots) presents a couple having an affair and debating whether they should go to the police; the sub-issue of whether two police should be in every patrol car; a brief look at an overdose case where a doctor knowingly writes twenty-five prescriptions; and a more substantial subplot in which a deserted Portuguese wife has a baby. However, unlike *Hill Street Blues* ten years later, this copshow stuck to the convention of closure, every plot resolved by the end of the episode.

The episode has a very realistic, often ironic texture: the baby cries throughout a scene where one awkward cop tries to cope with it while his partner calls the Children's Aid. The multiple plots are juxtaposed without much explanation, although it becomes apparent to an attentive viewer that they are interconnected by a thematic focus, in this case the issue of how often evidence is suppressed because ordinary citizens who are witnesses have something to hide. The sets are excellent, the locations well chosen. The corpse looks like one.

By contrast to most American and British copshows, the clues found in *The Collaborators* are the result of tedious work and police routine. The time scheme for a given episode is an unusually long six weeks. There is a generalized sense of tension which is stressed as the price policemen pay in their profession. In the last scene of this particular episode, Brewer, a little drunk and apparently in love with a prostitute, recognizes that he can't really prevent the smuggling of illegal immigrants or get at the anonymous men behind the racket. As I noted when I first saw the series, there are no chases, fights, or rapid or permanent solutions to obvious social ills or instant romantic involvements.

"Whatever Happened to Candy" (2/1/74; wr. Kaino Thomas, dir. Peter Carter) is also from the first season. In this episode, the forensic scientist with his slight British accent introduces the clearly working-class Brewer to the monied classes of Rosedale. The "high life" in Rosedale is treated rather wryly: one woman tries to seduce Brewer, who is interested, and the other tries but does not succeed with his stolid, embarrassed partner. Some of the atmospheric detail is right. Jamaican beer is indeed called "Redstripe." However, Jamaicans do not practice "voodoo." Their "obeah" is quite different. Moreover, the dialect spoken is a Haitian version of English. These lapses may be attributable to dramatic licence, sloppy research, or reluctance to offend a particular group. Nevertheless, the violence is unpredictable, and the plot has some neatly ironic twists.

Even better was an episode from the second season called "Dreams of Things" (1974-75), written, as many of them were, by Grahame Woods and directed by Don Owen, which focused on a pregnant teen who is a failed suicide. It sensitively explores the friendship between Tremblay and the pregnant girl and traces, without sensationalism, the roots of child abuse and her repressed memory of a murder. The suspense in this episode is quite acute, yet there are also several reflective scenes. The selection and editing of shots and sound effects economically shows the fragmented perspective of the teenager as she struggles years later to bring her memories to the surface and make sense of them.

Sidestreet

The Collaborators was cancelled after two seasons. I have been unable to find out why, since both seasons were interesting inflections on the copshow as it then was. Its replacement was *Sidestreet*, featuring community service workers who happened to be police. The press release for its debut (21/8/75) announced that Inspector Woodward (Sean McCann) and Sergeant Johnny Dias (Stephen Markle) were "dedicated officers who deal with small 'i' issues and who attempt to defuse potentially explosive situations before events get out of control." The credits read, "created by Geoffrey Gilbert and John Saxton and developed in association with the CBC."

"This eight-part, distinctively Canadian series is filmed in studios and on location throughout Toronto." Toronto again; Ontario again. Except for Vancouver, regional drama facilities did not then exist for film. By the mid-'seventies, there was no money for shooting away from Toronto on a regular basis. Nevertheless, both Britain and the United States do try to get their cops out of Los Angeles and London some of the time. It would be marvelous to see a cop show set in Montreal or Regina, even if we had to settle for outside establishing shots and a few episodes shot on location.

Brian Walker (an associate producer of Colgate's independent production *Police Surgeon*) was producer, and John Ross executive producer. Directors included Don Haldane, Denis Heroux, Don Bailey, and Richard Gilbert. Among the scriptwriters were Don Bailey, Tony Sheer, and Grahame Woods. The eight episodes included plots about a young man turning to crime to pay off a loan shark; two elderly pensioners who are turned in by an informer before they can commit robbery; the effect of rape on a young woman, her fiancé and friends; and the harassment of a young parolee; and a "professional strikebreaker who leads a group of scabs across a picket line," only to find his son has been kidnapped—"McCann and Markle move in to cool matters."

When I first saw it "off-air," the initial episode made a strong impression which held up in a second viewing years later. "The Holdout" (1975) was about a widow (Ruth Springford) terrorized by a developer for refusing to sell her property. The performances of the subsidiary characters were very good, the dialogue crisp, the subplots interesting: an illegal immigrant from Pakistan is exploited by a slum landlord, and a wino ex-journalist refuses to testify for the widow. The treatment of a major social issue, that is whether developers should be allowed to buy blocks of land and break up neighbourhoods by sharp practices or intimidation was not oversimplified at all. The complex, conflicting interests of the developer, the other homeowners and the woman were clearly outlined. Following formula, the widow's house is saved in the end. Yet once again in an ironic inflection of the genre typical of the better CBC series, the script underscored the fact that the men with money behind the small-time thugs will go on to raid other neighbourhoods unimpeded.

"The Rebellion of Bertha MacKenzie" by Mort Forer (directed by Don Haldane, telecast 28/9/75 and rebroadcast 20/9/79) is very good indeed. The episode starred the well-known Québécoise actor Monique Mercure as the Métis Bertha MacKenzie with Neil Vipond as Tony, her common-law husband. The conflict begins when Bertha's welfare cheque is stopped because she has a man living with her. When the social worker questions her, she says simply, and with utter conviction, "Kids need a man in the house." Tony is a layabout and insulting, but love is not relevant in this context. She is looking for a father for the two children. Bertha to Tony: "You know, my grandmother took a squaw man like you . . . they kicked her off the reserve," a fact that explains her dispossession and her displacement as well as underlining the double standard at the heart of her dilemma. "Big deal," he says. "Too stupid to insult," she replies. To the sympathetic social worker, she says, "Ever been so broke you had to count on bottles for bread at the end of the month?" She runs a house that is clean, no bugs, and she will not move, whether she can pay the rent or not.

Like everyone else in the script, Woodward sees her as an earth mother. Then, in an interesting twist, her brother Oliver tries to turn her genuine rebellion into a self-destructive proclamation of "red" power. He boards up the windows, gets out his rifle, and threatens to shoot anyone who tries to evict her. She is as desperate, scared and ruthless as he is, but she is a study in powerlessness, not the "powerful woman" Woodward perceives. Oliver is a deeply disturbed character, but not obviously so, and a very attractive role model to her young son because of his pride in his Indian heritage. When Bertha overhears Oliver's plan to be martyred on camera, she becomes hostage to his vision of red power just as she had been to the welfare rules and the racism in society. Clearly, the episode can be read on more than the simple level of narrative. The effects of commonly held stereotypes of class and race are implied rather than overtly stated, but the unresolved tensions between her values, her eroding cultural heritage and the system, all serve to undercut the formula rescue and the "happy ending" of the arrival of the welfare cheque.

Neither Woodward nor Dias is in the foreground for most of this episode. Once again, the primary convention of series television, that is, concentrate on the stars, is ignored. The episode mixes effective dialogue with some painfully predictable clichés. But the melodrama of Oliver's surrender, long on emotion and short on coherence, is mitigated by director Haldane, who uses a long zoom shot of Bertha's strong face, frozen in grief as he is taken away. Despite a few weaknesses, this episode, like others in the first season, did make good the CBC's claims that *Sidestreet* was not just another cop show.

The newspaper critics again turned in a mixed verdict. Ron Base (*TV Guide* 3/9/77) claimed that both *The Collaborators* and the first season of *Sidestreet* (which he called "crime-action" shows) suffered from a slow pace, weak scripts and vulnerable scheduling. Dennis Braithwaite agreed, railing at "police stories played as kitchen-sink drama" (*Daily Star*, 16/9/75). The actors were unhappy too: Markle thought the "stories were terrific. . . .—But our relationship to the stories is too peripheral" (Margaret Daly, *Daily Star*, 12/9/75). It is worth noting that the better CBC series often foreground characters who are the focus of a particular story, not the continuing series leads. McCann's comment quoted in the same article was direct and to the point: "The key word is anthology." Just so.

Jim Bawden of *The Hamilton Spectator* liked the series because it was "something new from an old form. It's well assembled and it's Canadian"; Joan Irwin: "much closer to ordinary human experience than its rivals" (Montreal *Gazette*, no date); Clarence Metcalfe, the Ottawa *Journal*, recognized its kinship with the CBC's tradition of good docudrama: "The show is real enough to be a graphic portrayal of a recent news story."

Despite the praise for what makes the series different, in the second season the corporation dropped the community-service context. Press releases emphasized more action and more "Canadian" crimes (hijacks of trucks full of maple syrup?), which directed the show towards an emphasis on the standard action-filled, good guy/bad guy formula. Thus, in the Ron Base article already quoted, the new leading actor (Donnelly Rhodes) described his role in the revamped series as "a stock character."

At least two factors were at work here. One was that the chemistry between McCann and Markle, two fine actors, was not right for a series. If the continuing characters of a series are to take focus, there must be on-screen rapport. People who get along perfectly well out of camera range may not strike sparks on the small screen. Moreover, as executive producer Maryke McEwen pointed out when discussing the cast of her first series (*Street Legal*, 1986-present) with me, leading actors have to be personalities. The grind of television production, even at the CBC's more humane and creative pace of two weeks' shooting plus "down" time as against a new episode for U.S. leading actors every seventh day), dictates that stars must ride the wave with characters based on their most comfortable personae, not ones that challenge them as *actors* every on-screen minute.

The other factor in the decision to find new "stars" was that John Hirsch, then Head of Drama, decided to go for a series much closer to standard format and farther from the dangerously varied, unpredictable, sometimes down-beat content of the first season. Perhaps as McEwen, first a production assistant and then a story editor on the show, remembers it no one consciously abandoned the original premises. They were simply lost, "forgotten," as she put it.

Whereas *The Collaborators* had claimed 15 (later 13) per cent of the available audience for that time period, *Sidestreet* managed to average only an 11 per cent share, or 1.5 million viewers. The Audience Research report (Tor 75-25) points out that the decline could have been in the whole Sunday night audience available rather than a result of declining programme quality. Thus, *The Collaborators* had a higher Enjoyment Index, but those figures had been gathered in September-October. The comparison is not straightforward, because the numbers of viewers in Canada increases with bad weather to a peak from December to February, when *Sidestreet* had been scheduled and tested. To further complicate the picture, the popular American copshow *Kojak* had moved into the same time slot. Also, despite this widespread habit of watching the English network, only 3 per cent of the potential Quebec audience looked in on *Sidestreet*.

The show was particularly strong with the people age fifty and over, in cities with over 100,000 population, rural areas where there was less choice, and with those people who had less education. American figures tell us that crime shows are usually preferred by an urban audience, but figures in both countries show a general trend towards action-crime shows appealing to less educated audiences. *Seeing Things* began as an interesting exception with its allusions, polyphonic plot structure, and varied tones. On the other hand, better educated people saw *Sidestreet* as a more realistic show than *Kojak*, *The Streets of San Francisco* (1972-77) or *Harry O* (1974-76). Reactions from episode to episode fluctuated widely suggesting that the episodes were not all cut from the same pattern. Using a wider range of subject matter than conventional copshows, indeed they were not. All these variables demonstrate why the data from Audience Research must be read with much caution. The biggest problem in the first season seemed to be that the audience could not identify with the protagonists as they had with the two leading actors of *The Collaborators*—a major factor affecting enjoyment indices. On the other hand, audience panels, a demographic cross-section of people questioned in detail about an episode, single, or series, did not find that plots in the first seasons' *Sidestreet* episodes confusing, slow, or without suspense.[5]

In *Sidestreet*'s first season, the CBC proudly announced that it had sought writers new to television for the series. Since the CBC severely underpaid its professional scriptwriters (it still does) finding new writers had often been less a virtue than a necessity. A CBC press release of 15 September 1976 claimed that the second season had been "bolstered by tighter scripting, faster pacing and two strong new leads. . . . Despite these changes [the series] will retain those elements that set it apart from other police series and drew favourable response from press and public last season—emphasis on realistic situations and characterizations and no violence for the sake of violence." The CBC now classified the show as belonging to the "police-adventure" genre. They had also written into the series the character (and actor) of Inspector Bowman from the widely acclaimed, realistic and topical BBC series *Z Cars* (1960-78). I find this particularly odd: *Sidestreet* was moving towards the standard cop show, yet tried to incorporate a character from *Z Cars*, a successful but very different kind of copshow from Britain. Bowman added very little to the series because he was never given a fully developed character.

The plot summary for the season opener illustrates the intended contrast with the previous season: "Sgt. Olsen is accused of knocking down a pregnant woman during a demonstration outside a Toronto hospital. Olsen is suspended, pending an investigation, and Raitt risks antagonizing the press, public and his superiors in a desperate effort to clear his harrassed partner." It was a competent but not particularly distinctive show. Other episodes centred on a cop bent on revenge, a criminal seeking the only witness to his crime, and the threat of domestic violence. I chose to look more closely at "Right to Defend" on the grounds that the subject outlined in the press release, a householder's right to defend his property, is an issue that does not date. Unfortunately, the show was not about that subject at all, but rather about the householder's attempts to cover up the murder.

The show begins with conventional suspense: closeup of feet, pull back to long shot of the kid breaking into the house. Although the lighting is rather bright for the situation, it is unclear who exactly sees what. The householder blasts away, and the kid staggers out into the snow to other kids waiting in a station wagon. The episode comes complete with heavy symbolism, for example, a close-up on hunter's trophies as the householder thinks about shooting the kid. Fifteen minutes into the episode, Raitt and Olsen have yet to make an appearance. Then it transpires that Raitt has "worked his butt off to keep the two brothers out of trouble" (the kid is one, of course) for their mother's sake. After that, the plot proceeds on its predictable way.

Jack Miller, a perceptive television critic before he moved on to other subjects, (*Toronto Star,* 22/10/76) said the series worked, not because of occasional guest stars, but because of Welsh and Rhodes: "The job is simply to portray two policemen who care, and who try, and who sometimes lose, and who make it all seem real. They do." If a series does not mythologize its protagonists, inflate its plots, or oversimplify the issues, it is more typical of our prevailing cultural stance, and it is perceived as such by both the audience and some critics. We may or may not like this perception of ourselves, but that is the way it has been, from *Wojeck* to *Seeing Things.*

By 1977, *Sidestreet* was billed as the CBC "hit police adventure series" (CBC press release, 21/9/77) and was given the money to do ten episodes instead of seven, two of them in Vancouver. Stanley Colbert, who had experience in American series television, became executive producer. As the CBC grew more conscious of ratings, *Sidestreet* moved more and more towards the classic copshow formula: that is to say, society's roughly agreed social norms would be reinforced, however they might be called into question during the programme, by those persons clearly deputized by society to do just that, usually policemen. Thus, the sponsor of "the product," as programmes were now called, was satisfied, and ratings were expected to go up. Mind you, a touch of class was added by the Galliard ensemble whose music enhanced an episode about an illegal immigrant in an episode called "With This Ring" (27/11/77). Another, "Once a Hero" (4/12/77) raised a perennial question in Canadian television drama: is pro hockey worth the human cost? Perhaps this episode was what the CBC publicists had in mind when they claimed that "the scripts themselves—with more action sequences—will also deal with current issues and crimes of importance to the average person" (25/1/77).

At least one episode did break formula. "Stakeout" (11/12/77), directed by Peter Pearson, turned a standard hostage taking incident into a superb study of cross-purposes and tragic irony. The episode's value lies, not in the spurious suspense of a SWAT-type fanatic who wants to have a shoot out, but in the ambivalent human drama of the hostage/captor/husband triangle. We see the wife as hostage, terrified and talking for her life, yet also relating to her captor in a human way. Across the street, the husband watches from the police stakeout, able to see perfectly but unable to hear. The director shows us both the actual conversations between captor and captive and the husband's fragmentary glimpses of them. Conditioned by media reports of hostages coming to love their captors, he assumes that his wife is becoming sexually attracted to the kidnapper. As formula dictates, she is freed unharmed. But the formula is then shattered when we see that the husband's jealousy, unassuaged by any explanation, persists even while the credits roll. By the way, the plot summary the CBC supplied to the newspapers indicates, quite erroneously, that she does form that kind of attachment to her kidnapper. Too often, misleading plot summaries in TV guides are the viewers' and, later, the researchers' worst enemy.

"Between Friends" (23/10/77), written by Peter Verner, also creates real tension. Doug McGrath plays a cop whose old loyalties come back to haunt him in the form of Les, a criminal on the run, in an attractive performance by Calvin Butler. Les is not sentimentalized. In the opening sequence, we see a standard shot of a guy with an indistinct head on his shoulder "snuggling" in the front seat of a car. Cut to a closeup of his head coming slowly up, then of the male body's fixed stare, then of the man unwinding his arm. As he gets out, the body falls out of the car. Yet, when surprised by McGrath, now off duty, Les does not take a clear and easy shot. Instead he turns up in his living room lying on the chesterfield wounded by a shot from McGrath. Behind Les's head is a photo of two soldiers in Vietnam, arms about each other, Les as a Red Cross corpsman, McGrath an infantryman: "There's a dead man down there. That concerns me. You concern me." Les: "I don't concern you. I bother you . . . because I saved your bacon in Vietnam seven years ago, because you found out you didn't belong in that war in the first place." "Nobody belonged in that war." Later, Les comments: "So how is it with you hockey pucks? Try to get out of the box. Back on the ice." In the end, the episode superbly explores a clash of loyalties in a good cop, coming when he is exhausted and frustrated with the police force. In this one the good guys and bad guys look much alike, struggling human beings in a no-win situation.

In the fourth and last season, press releases still claimed that the series did not indulge in "undue violence." The promised Vancouver episode finally appeared. Another, directed by Don Shebib, called "Holiday with Homicide" (1/10/78), used the riding skills of Donnelly Rhodes in a quite novel setting, the quarter-horse show at the CNE Coliseum in Toronto. (Apparently, the horse cast as the star, though not entered in the actual meets, beat all the contestants in the filming of the barrel race.)

The episode from this season which I chose to look at in detail, "Just Another Day" (24/9/78), featured Raitt and Olsen much more than previous episodes. There were lots of menacing shots, menacing music, a padded script, and an outrageously stereotyped transvestite entertainer central to a plot that is quite hard to follow. Basically, a psycho called Andrew is haunting headquarters with a wino named Charlie. Andrew takes all the staff hostage, including the inspector. We get some pop psychology about the psycho as spoiled kid, while a critically wounded rookie is tended by the transvestite. Olsen comes to the rescue dressed up to look like Andrew's wife. In this episode, the plot was all there was to the programme.

Many new actors from the flourishing alternate theatres appeared in later episodes of *Sidestreet*. However, the press releases in the CBC library no longer mention the writers, and rarely the directors. Watching other samples of the third and fourth seasons at the time, my observation was that, too often, the series was typified by formula chases and gun battles, hyped-up music, predictable editing, and camera work demonstrating competence rather than flair. Bonnie Felker and Brian Walker shared production duties, and Colbert continued as executive producer. In 1978, thirteen episodes were sold to Britain's ITV, though they are not mentioned in the 1982 edition of *Halliwell's Television Companion*.[6]

While *Sidestreet* was running, a revival called *The New Avengers* (1978-79) had appeared across the street at CTV. The series starred Patrick MacNee, who had done good work for CBC drama in the 1950s but made his reputation in television as suave master spy and lover John Steed to Diana Rigg's leather-clad Mrs. Peel in the very popular British spy series *The Avengers* (1960-64). The original had been the brainchild of ex-CBC supervising producer Sydney Newman and MacNee, who owed his role in part to this Canadian connection. By an odd coincidence, when, by its second season, *Wojeck* reached 2.5 million and a 78 average enjoyment index, CTV retaliated by scheduling the original *Avengers* in the same time slot.

Seven episodes of *The New Avengers* were shot in Ontario, with all-Canadian crews and supporting casts and Ross MacLean as co-producer. The choice of Ontario was owing partly to the fact that it was convenient for the plots that Canadian police are allowed to carry guns, partly because Canadian-made shows were not counted then as "foreign" in the strict quotas imposed by British broadcasting regulations. *The New Avengers* featured complex though largely imaginary technology, some attempt at sexual tension between the woman lead (asexually and improbably named Purdy) and the macho man of the "I can kick your head in too" variety, as well as a now sadly avuncular Steed. Unfortunately, the scripts were not as witty as those of the original, the personal nuances were missing, and there was little energy or inventiveness in the traditionally improbable plots: a berserk computer is finally quelled by a sprinkler system, an enemy nuclear sub lurks under Lake Ontario, etc. It seems to have eluded those involved in this surrogate that the original was primarily a superb parody of the spy formula.

Back at the CBC, the short-lived series *Phoenix Team* was the next attempt to fill the slot in the schedule now reserved for the mystery/ detective/cop show genre. This series was based on the wholly improbable and rather dated premise that somewhere in Canada lurked a master spy (Mavor Moore) who could foil Russian plots only with the help of the heroine, a retired, married but still beautiful British spy (Elizabeth Shepherd) and, as the hero, a cynical, tired ex-lover, now filed and forgotten by the bureaucracy. Unfortunately, he was played by one of our best, most intense actors, Don Francks. The irony of that casting is that Francks carries with him, indelibly and proudly, the persona of a hippie—quite different from the usual stereotyped loner of the cynical 'seventies. Perhaps that is why there was no chemistry between the two actors. When I describe this series to my students they giggle "SPIES? IN TORONTO?!" The subgenre of spy/cop show has no credibility when set in Canada and I suspect this is because it is not part of our mythology. In any case, nothing could cure the frenetic and essentially silly plots or the cliché-ridden dialogue. *Phoenix Team* disappeared in short order.

The Great Detective (1979-81) repeated the pattern of beginning with new and inventive ideas which were then eroded in favour of formula situations and characters. "More action" was already the watchword in its second season, according to writer Peter Wildeblood (interview reported by the *Globe and Mail*, 1/1/80). *The Great Detective* started out as a mystery series in the British mould, usually without the class consciousness, even though Tory Ontario was certainly class-ridden in that era. The series was a designer's paradise with emphasis on rather exotic but faithfully recreated settings: a nineteenth-century hospital, a photographer's darkroom, the rowing club on Toronto Island, and a private school. The period details were often necessary to the action, as well as adding an atmosphere of menace or sunlit Victorian afternoons as required. The intention was "to showcase the rich countryside of southern Ontario"—which was done, admirably. The pace was appropriately leisurely in many scenes.

According to the press release for the second season, that leisurely reflective mode was to give way to "more action, adventure and visuals" with new emphasis on a younger police assistant to the detective, that is, the cliché of one older, one younger lead again. As far as I could see, the young man simply functioned as a pair of legs in a chase or fists in a fight, not as a fully realized character. Once again, a programme was not allowed to find and build its natural audience, but changed toward the formulaic and then dropped. This time the newspaper critics preferred the shows from the first season.

Throughout the series, violence was kept to a minimum and dialogue was competent, though uninspired. Ancillary characters such as the detective's assistant, the coroner who was a fishing crony, an eccentric housekeeper, and a young niece (who was written out after a while), were not really developed beyond predictable stereotypes. The gifted and experienced actor Douglas Campbell playing the gruff, meticulous inspector carried the show on his ample shoulders. Unfortunately, he was not given much of a character to explore. What came through was his own outsized persona, a mixture of learning, energy, abruptness, and sensuality. Campbell's accent is still British, very much in keeping with the rather British flavour of the plots and characters, place, and period. In the late Victorian era, most people in Ontario still felt deeply about being part of the Empire. The two logical series to which this one should be compared are both British: *Cribb* (1979-81), a working-class Victorian London detective, and the modern-day barrister *Rumpole of the Bailey* (1978/79; 1983-84), defender of the dispossessed, the sleazy, and the betrayer. However, both of these series have a subtle but constant undercurrent of class warfare (and interplay) as their motifs. *Rumpole* also has one of the best series television writers around in John Mortimer, who laces his tawdry ambience and seedy characters with mordant wit and sardonic or touching allusions to some of the best and least hackneyed of English poetry. Halliwell comments on *Rumpole* that ITV's Thames produced "sharply written comedy-dramas which proved rather too literary for the public taste," so *The Great Detective* may have been right in staying with the mainstream of the mystery story.

Nancy Gilbert, in her tightly reasoned argument "Getting Away with Murder,"[7] concludes that "detective stories [and by analogy, television series in the same tradition] satisfy a passion for truth not a passion for justice" (p. 599). Several of the episodes of *Wojeck*, *Seeing Things* (and *Sidestreet* on occasion) make the same distinction between the two, with lashings of social comment, parody, wit, and even romance. Such elements are also found in the BBC adaptations of the Lord Peter Wimsey stories of Dorothy L. Sayers as well as *Rumpole*. These series contrast with the wry but usually up-beat endings of their closest American cousins, such as the delightful 1984 hit mystery series *Murder She Wrote*.

The Great Detective did have some entertaining moments: in "The Photographer," for example, careful direction and camera work made the posing of a young factory girl, one shoulder bare and legs exposed, seem as exploitive as any open-thighed centrefold. The 1980s context in which the viewer perceives the scene deepens the irony. In "The Family Business," David Gardner played an amiable family man and skilled forger whose beautifully polite children were learning his trade. But the man who passes the counterfeit comes from a chillingly accurate recreation of crowded, improvised, dirty tangle of sheds and shacks where "girls" were born cynical and middle-aged.

Plots were usually workmanlike, with stress on the "new sciences" as aids to police work. *The Great Detective* was more episodic than most series of this kind, with few running gags, endearing or annoying habits, or recurring situations. Audiences did seem to enjoy the evocation of period as well as the basic whodunnit ratiocination. Its replacement, *Seeing Things*, must have seemed brash in comparison, but then the target audience in this case was supposed to be significantly younger.

WOJECK AND SEEING THINGS

In a culture given to debunking not building its mythic figures, the only mystery series that have really established characters and actors with *personae* strong enough to create new mythology are *Wojeck* and *Seeing Things*. Coroner Steve Wojeck and newspaper reporter Louie Ciccone are such characters partly because the actors are brilliant enough to really shine as stars. John Vernon's Wojeck is a compelling performance, gritty with truth and savage with sincerity. Louis Del Grande's character was devised in part by the actor himself, who was both writer and co-producer. The character's klutzing, workaholic persona and domestic worries are threaded with allusive wit and literate one-liners, manic energy, and considerable acting skills which blend the character of the actor with the protagonist of the series. In both series, for the most part, the writing is very good indeed.

In both cases, however, the leading actors threatened to become stars. John Vernon ran up against the CBC's policy against building leads into stars and its refusal in those days to pay salaries commensurate with star status, and left for Hollywood to play "heavies" in formula cop shows which included *Quincy* (1976-84), the watered-down imitation of *Wojeck*. He has since appeared in a few single dramas for the CBC, but has never been the lead in another series. *Wojeck* was a programme that could have become a national institution but was never given that chance. Fifteen years later, there are indications that this unspoken policy has changed. For one thing, *Seeing Things* has had excellent media coverage. Del Grande has said in many interviews that he prefers Canada to Hollywood because he finds it more real, more down to earth. Neither the money he is paid nor people's attitudes to an actor create a fantasy world. Yet Del Grande continually attacked our refusal to be proud of our own television drama, to build and enjoy our stars. This ambivalence towards fame appeared as the plot line of one of the 1986 season's episodes.

Seeing Things is much closer to the amateur-sleuth subgenre in English mysteries than it is to the copshow. It is in Britain that the "mystery" formula has been most clearly defined. Deriving his definition from Edmund Wilson and William Aydelotte, Powers says that mystery shows invent a "substitute universe with fixed and reasonable laws, in which all events are significant (because they are clues) and in which knowledge is power since it is the act of knowing that solves the case . . . where thought is action, a paradise for the intellectual" (p. 215). That may be so in the land of Wimsey, Holmes and Miss Marple, but in *Wojeck* and *Seeing Things*, knowledge involves a fair amount of unresolved frustration. In *Wojeck*, with its texture of the random, its subplots and unresolvable social problems, or in *Seeing Things* where the amateur sleuth is an unwilling clairvoyant, knowledge compels action. Yet the act of knowing does not always result in the solution of the case. The problem for city coroner Wojeck is that he often identifies a major social evil as the real killer and is then forced to recognize that he cannot solve the root causes of the problem. Louie Ciccone also sees the motive and human circumstances that create a murder. For the viewer, the episode's ending does not necessarily reinforce the norms of society's social values or even satisfy the intellectual curiosity that has been aroused.

Neither series fits the template of action sketched by Powers, in which the hero overcomes an ordeal and, rather than unravelling the knot, simply slices through it. Both Wojeck and Louie are unorthodox and often obsessive in their methods of finding out who is responsible for deaths, but neither is cast in the standard, heroic detective mould; neither is remotely handsome in the conventional, beefcake sense. "Since the action detective is a projection of the culture, his success is a demonstration of the culture's ability to repel all challenges." (p. 216) That the human decency and hard work of a Wojeck or Ciccone can uncover the causes of murder may be reassuring to Canadian audiences. However, many *Wojeck* episodes are open-ended and occasionally ambivalent in their statement of moral values. *Seeing Things* maintains an integral parallel plot that focuses on a complex, emotionally rich and never-resolved relationship between Louie and his initially "separated" wife, Marge. Thus, the overall effect of the two series is definitely not to demonstrate a culture's ability to repel all challenges. Indeed, both series continuously challenge the society in which they are set—a testing that is one of the functions of a publicly owned broadcasting system.

Dick Hobson, interviewed for *TV Guide* (May, 1977; "ABC's Quarter Million Dollar Man Performs Heroics Too")[8] summarized the American formula for television success: make people laugh; hire or create a star; stress the positive rather than the negative; concentrate on the common man and make the star familiar, so that he is acceptable to a wide audience. Programme planners should take scheduling (but not programming) chances and go for "pandemic promotion." The overall strategy is to keep a hard action line, remember that "cartoons are not only for kids," and grab the audience while they are young. The formula for television does not appear to have changed in the 'eighties. To that general formula, Martin Esslin would add a Manichean universe of good and evil in which "TV, with its unending stream of characters conveyed dramatically (whether fictional or 'real') is the most perfect mechanizer of . . . gossip."[9] Regarding the appeal of "copshows," he identifies two contributing factors: our need for reassurance that good will triumph and our suppressed delight in crime.

Critics of the genre emphasize this need for reassurance again and again. Maurice Charland, in "The Private Eye: From Print to Television"[10] (writing before *Hill Street Blues*), says that "on television, things are as they appear to be: the police are decent, working to keep us safe, the villains are villains and the innocent appear so. There is little ambiguity." The RCMP and the police in *Sidestreet* do fulfil this myth in part, but the "hard-boiled" freelancing loners, "powerful anti-heroes in a land of sin" as he calls them, do not exist in the history of Canadian television series. "Mean streets" are not part of our mythology, probably because our cities are and have been comparatively safe, day or night. Tracing changes in the mystery form, Horace Newcomb, one of the pioneers in serious analysis of the aesthetics of American television drama, argues that the prevailing pattern over the years has been that of a "wise father figure [as] the chief implement for change in this world. [Authority] serves to preserve order where the chaos of criminality threatens to touch individuals and the society they constititute."[11] Using the approach of semiotics rather than generic criticism, John Fiske and John Hartley[12] also read most television in Britain, including the relatively innocuous and immensely popular *Dixon of Dock Green* (1955-76) and the innovative *Z Cars*, in the same way: "Myth is validated from two directions, from the specificity and iconic accuracy of the first order sign [in their particular communications model] and second from the extent to which the second order sign [what I have identified as the interplay of context and subtext] meets our cultural needs."

Powers observes that "motive is one of the puzzles in the mystery." To create suspense, "The motive, like the solution, must not be immediately obvious. [However] in the action story, evil character is the only permissible motive" (p. 216). *Wojeck* in the 'sixties and *Seeing Things*, its counterpart in the 'eighties, both present this emphasis on motive. As they became more action-oriented, 'seventies' series like *Sidestreet* too often lost sight of this factor. Morris Wolfe, in his survey of Canadian television, *Jolts: The TV Wasteland and the Canadian Oasis*, disagrees: "Like almost all [series] drama on the CBC, *Sidestreet* was mostly about social issues."[13]

In the only Canadian article addressing the specific question of genres in a Canadian context that my research has uncovered, "TV Formulas: Prime Time Glue,"[14] media critic Joyce Nelson explores, in a different way, observations by Newcomb and others that copshows, mysteries, and spy shows are fantasies of reassurance which acquire a new emphasis on personal relationships and topical issues in the 'seventies. The core of her argument is that "the sitcom and the crime series are closely interconnected. In an important sense they build on one another, deriving added meaning through combination. In many cases, the gaps in one formula are filled in by the conventions of the other. . . . The conventions of sitcoms and crime series function as a set of oppositions which mutually reinforce an ideology that is particularly well suited to the advertiser's message."

She points to the strongly masculine, authoritarian titles of copshows: *Ironside* (1967-75), *Columbo* (1971-78), *Magnum P.I.* (1979-present). Compare with these *Wojeck* (1966-68) and *MacQueen* (1969-70). *The Collaborators* (1973-75) was clearly derived from such American titles as *The Defenders* (1961-65). With its flippant title, *Seeing Things* (1981-87) appears to come from the sitcom land of *Diff'rent Strokes* (1978-present) or *Too Close for Comfort* (1980-present), whose titles suggest informality. It may be the title, or the developing and continuous subplot about Louie's relationship with his wife Marge, or else the increased emphasis on slapstick farce in the last two seasons that made enthusiastic reviewers mistakenly classify this distinctive variation on the mystery genre as a sitcom.

Film and tape create different visual conventions. Nelson reminds us that most crime shows are shot on film, while sitcoms "make do" with tape. In the 'eighties, it is increasingly difficult to tell the two media apart, but in the late 'sixties and throughout the 'seventies tape gave the viewer significantly less visual detail than film.

For Nelson, the important things about film are its crispness of image, depth of field, and hard-edged, dramatically lit quality. These visual conventions are what an audience has come to expect in crime dramas. Part of the success of *Wojeck* and *Seeing Things* is owing to the fact that they were shot on film. *Wojeck*, the CBC's first national film drama series, also had the mobility and the elision of time and space made possible by this medium. In 1966, tape was still very difficult to edit. However, the NFB style of "direct cinema" used by the *Wojeck* crew, whose experience was in documentaries and news, was the antithesis of the slick, hard-edged look that Nelson rightly associates with most filmed Hollywood copshows though the direct-cinema style was rediscovered by the producers of *Hill Street Blues*. This "look" was encouraged by executive producer Ronald Weyman, who had worked at the NFB. Direct cinema typically shows the filming process within the film itself, through hand-held cameras, awkward framing, ragged editing rhythms, harsh lighting and imperfect sound, grainy film stock, and sometimes a sense of improvisation in dialogue. Direct cinema also tends to focus on unknown or forgotten people in a more personal treatment of subject matter. It is a matter of principle to present the subject from a clearly stated point of view. Allan King's television documentaries in the 'sixties used similar conventions and had the same sense of commitment to comment on contemporary Canadian society. *Warrendale* (1967) was made at the same time as *Wojeck*.

The *Collaborators* and *Sidestreet* had a completely different look. Nelson quotes Brian Walker, the producer of *Sidestreet* (still in production when she was writing), as saying that directors for that series were asked to "make sure that the action moves toward the camera. The camera is always being pushed by the action coming into the foreground, rather than moving away from the camera." The former is obviously more "dramatic" creating a sense of urgency, excitement, suspense, or threat, and it is the prevailing style for American action shows. With its roots in the documentary style of the National Film Board, *Wojeck* did not look like standard Hollywood mysteries or copshows.

Nelson describes the typical sitcom or family show videotape image as flat and soft at the edges, moving horizontally in the frame. Viewers will find that if they watch a lot of sitcoms they tire of the horizontal blocking dictated by the standard living room set, with its central couch or pair of chairs centre stage, often a staircase moving from left to right, upstage left and downstage right doors, and so forth. The theatrical nomenclature is appropriate here, because the basic sitcom set and the blocking which results from it are purest nineteenth-century proscenium arch theatre. The television frame serves as the lighted box provided by the standard proscenium arch. All that is added is that our sight lines are perfect, and we can see the actors' faces in the overused closeup. Overall, the effect is a kind of shorthand, standardized "naturalism" of domesticity contained, concentrated and fully available in every detail.

Medium and closeup shots are the standardized vocabulary of the sitcom without the elaborate tracking and zooming and without the subtle tricks with space allowed by the one-camera technique of film, where many angles are shot in sequence and then edited together.

Until recently, it has not been possible to record on videotape in low light. Film's light sensitivity means that one can get superbly moody shots indoors or in dull, rainy conditions, or great beauty in "the magic hour," as cinematographers call the time around twilight. "Dramatic" or subtle lighting is much harder to achieve on tape, primarily because three or four cameras are in action at once from different positions. Thus the lighting on the set must be uniform so that the producer can switch from camera to camera, yet achieve visual imagery. In *Seeing Things*, the writers', producers' and directors' highly developed sense of affectionate allusion and parody also produced a varied visual style that was quite often romantic rather than crisp.

Film is much more often used on location than tape. *Seeing Things* makes the Toronto of the 1980s an unexpected series of wonderful discoveries, a trend begun by *Wojeck* with its grainy and particularized sense of place and continued by *Sidestreet*. As *Sidestreet*'s producer, Brian Walker, told Nelson, "Toronto is another star of the show." Before television, Canada did not really have landscapes that had been fictionalized on film, except for *Cariboo Country*'s Chilcotin plateau and the limited filming of outside shots explored by Weyman in *The Serial* (1963-66). With *The Manipulators* (1971-73) shot in Vancouver, not to mention *The Beachcombers*, filmed for the last fifteen years at Gibson's Landing, or the Ontario landscapes of *A Gift To Last* and *The Great Detective* (tape) and the various miniseries shot in Montreal and Calgary, that has changed.

Wojeck and *Seeing Things* do share one characteristic of the formula for crime drama, what Joyce Nelson calls "a sense of righteousness, personal moral principle," as exemplified in the Louie and Wojeck characters. However, in CBC crime dramas, it does not follow that (as she defines the formula) "in every case the law backs up the hero's action and provides the clear-cut motivation for the decisiveness of the main characters." Often, Wojeck has no hope at all of effecting changes in society. Not only is social order rarely restored, but its values are also often questioned.

Wojeck's closest American cousins are *The Naked City*, filmed entirely on location in New York (1958-63), which retained the gritty texture and ironic mode (though not the unrelieved cynicism) of *film noir*, and *The Defenders* (1961-65), distinguished by early scripts from Reginald Rose, one of the best writers of American anthology. In the latter case, the subjects were controversial: abortion, euthanasia, black listing, civil disobedience. Moral issues were treated with some ambivalence, and the father-and-son team occasionally lost a case. Both are series to which American academics and critics point with justifiable pride. Both overlap the death of American anthology drama and may be said to reflect some of the values of anthology within the series formula. *Wojeck*, however, was conceived and broadcast in the late 'sixties when a new sense was abroad of major change in the social values of the Western world and self-confidence and nationalism in Canada was at a peak.

THE MYSTERY: WOJECK

Wojeck and *Seeing Things*, the latter a product of the less confident and more pragmatic 'eighties, are the two most outstanding and innovative series in the genre. Both reinterpreted the basic mystery format in startling ways. The two are completely dissimilar in many respects, products of different times, tastes, circumstances; one in black and white and one in colour; one charged with the energy and commitment of personal documentary subjects and a direct cinema style, the other maintaining a highwire act which, at its best, balances comedy and serious moral overtones, soluble crimes and insoluble personal and moral dilemmas. All of the series between these two inflected the formulae of the copshow/detective/mystery to a more or less interesting degree or else imitated it outright, but in either case the familiar shape of the formula was apparent. *Wojeck*, however, was qualitatively different from the copshows and mysteries of its era, a vigorous hybrid, if short-lived. *Seeing Things* too is qualitatively different. New conventions—visual, verbal, musical—have been grafted on to the sturdy old stock of the conventions, adapted to our winters, our circumstances, our tastes. These programmes have not been transplanted from south of the border or across the ocean but devised, nursed and finally exhibited here. That is why I decided to concentrate on *Wojeck* and *Seeing Things* in my analysis.

Wojeck: "The Last Man in the World"

"The Last Man in the World" (13/9/66) was directed by Ron Kelly. Like all the episodes in the first season, it was written by Philip Hersch, whose grasp of characterization through idiom, individual choice of words and phrases, and whose ear for contemporary 'sixties speech ranging from dialect and working class to bank clerk, teacher and professional was the strong, straight backbone of the series. In my view, this is the outstanding episode of the series. The audience did not agree, giving it an Enjoyment Index of 59. However, it was the first episode to be shown and, if my premise about Canadian series is accurate, that is that the successful ones do not follow formula—then it follows that they will need time to find, or educate, their audiences. Even though the series was something unfamiliar, the average enjoyment index for *Wojeck*'s the first season rose to 74. Over the season, the audience nearly doubled, outdrawing programmes like *Gunsmoke* (1955-75), *Secret Agent* (1961; 1965-66), *Love on a Rooftop* (1966-67, shown on rival CTV and created by CBC alumnus Bernard Slade), *The Fugitive* (1963-67), *Red Skelton* (1951-70), *Petticoat Junction* (1963-70), *Peyton Place* (1964-69), and Tuesday night movies. *Wojeck* also won the 1967 Wilderness award (internal to the CBC), the Golden Nymph award at the Monte Carlo film festival, and several other distinctions.

"The Last Man in the World" is about an Indian who hanged himself in Toronto's Don Jail. We find out why through extended flashbacks which shift the focus almost entirely away from Wojeck for most of the episode, a not uncommon phenomenon in this series.[15]

The episode opens with an extremely violent fight. The camera picks up one man, lunges, blurs out of focus; over a swiftly moving effective montage, we hear a series of short crunches as blows land. No one in the all-white crowd moves to help either the policeman trying to intervene or the Indian in the fight. Cut to a scene of the same Indian, well dressed, in a sleazy bar. Cut to Wojeck interrogating prisoners in a jail trying to establish that the Indian did hang himself, that it must have taken ten minutes with all the others in the drunk tank watching, and finding that no one will say how he got the belt to do it. All this happens before the titles begin. Wojeck approaches the blanket-covered body, touches it very gently. The head moves, loosely, heavily.

The victim's name is Joe Smith; we learn his Indian name later. Eventually, we learn that his mother was Mary Smith because "all the 'status' in the mission are called Mary. When she died, they didn't know her name so they called her Mary. A lot of women are called Mary up in Moosonee." In a flashback, we see a derelict who envies Joe because he can always go back up north, but the programme shows us over and over why he cannot.

The entire episode is structured around a series of flashbacks of "Joe Smith's" time in the city, intercut with Wojeck's efforts to reconstruct how he came to be there and who gave him the belt to hang himself with. Writer Philip Hersch is also showing us why he committed suicide, why someone helped, and how all of us are implicated in his death.

There are no mitigating romantic or lyric moments in "The Last Man in the World," even though Joe does find a friendly prostitute for a while. We first see her tired face looking for customers in a restaurant, so bored that she squeezes a pimple and dabs it with kleenex. The trivial gesture defines the camera as a privileged eye looking at the ordinarily hidden, the usually overlooked. Sets and locations are also vital to the subtext; recognizable rundown restaurants, a rented room that crowds the camera into a corner. In the frame of this scene, we see her eyes in extreme closeup and a door with a padlock half off the hinge, dirt along the edge, and the look of many other locks busted off it. The camera pulls back to show her cheap blouse and tight pants, then a shot of his bare foot. She is shy, hardly looks at him, says little. The door squeaks. Finally, she asks if he will eat meatballs. "Sure." The awkward pauses in the dialogue end with a rather vulnerable exchange of smiles. He changes his shirt, bouncing on the bed in and out of frame because the camera playfully focuses on the mirror. The mirror reflects glimpses of a friendly, shy, easy-going guy; a superb performance from Johnny Yesno. Then his white shirt cross-fades into the harsher light of the kitchen, and his bouncing becomes the quick up-and-down movement of the step basic to much of Indian dancing. The influence of direct cinema shooting techniques is evident here.

When Joe goes into a park to relax and sing a little, his Indian friends break the mood with, "Hear you got a white girlfriend. You steal from us and starve our kids." His reply, "What are you getting so mad about?" expresses his honest confusion. He goes into a friendly European immigrant's store to get his new friend some costume jewellery. In the upper corner of the frame are Halloween masks of stereotyped "dumb Indians" which Joe doesn't see. "How much are the beads?" he asks, and the viewer hears the ironic echo of the stereotype. In "The Last Man in the World," stereotypes are lethal.

When his girlfriend asks what it is like "up there" on the reservation, he answers without hesitation, "Not as nice as here." He talks about the reserve in spring, but most of the terse detail reinforces the despair, a sense of hopelessness far worse than what he finds in the whore's drab kitchen. However, "here" for her is a trap. Though her mirror is framed with faded icons of American popular culture, she looks at her image in the muddy glass with dead eyes. His discovery of what she does for a living is handled very simply with a lingering shot of her bitter, defiant, weary and rather lovely face. Her attempted reconciliation, an offer of whiskey and friendship, is rebuffed. As with any exceptional piece of drama, the episode's power rests on the unspoken subtextual nuances of performance as well as on the words. Both constitute "dialogue."

The scenes between Joe and the woman are juxtaposed with Sergeant Banas's discovery that, because he is not a treaty Indian, no one will pay to bury him. Wojeck is determined "not to hand [the body] over to the medical school . . . not this one." Eventually, his Indian friends raise the money, yet we have also seen them calling him "just a farmer" over and over. Although beer and alcohol appear, the film implicitly rejects the drunken Indian stereotype, and then ironically comments on it. Late in the play we discover that a perfectly respectable "regular," a harmless drunk, had been sent to share Joe's cell. Since he was a man the police know well, no one bothered to take his belt when he was booked. He promptly passed out. A belt from a "harmless" white drunk was thus available to Joe.

Now we are back at the episode's beginning. After the fight, Joe is taken off to jail and put in a cell struggling all the way. Then the other prisoners bait him. Camera angles emphasize the vertical lines of the cell. Joe's eyes go from the others to the belt on the drunk. A closeup on his hands undoing it as others watch. The hands make a decision. The belt is drawn through a tightening fist, slowly. Cut to realization on the faces of a few. Their eyes look up as his hand loops the belt over the upper bars. The racket quiets. Off screen, we hear the line, very quietly, "Hey Chief. No, Chief," and another voice, quietly, "Call somebody." We hear mumbling, a soft thud, but all we see through the few inches of barred opening is a piece of the belt swaying, no sound at all. Cut to a shot of the men, whose very ordinary faces are now grim, watching, absolutely silent, then to the inquest, where we see the same faces, including the drunk with his belt now hitched (a salesman of church supplies) and his hand half over his face. We also see Wojeck's grim face and an indifferent coroner's jury. As Wojeck reads the charge to the jury, there is a slow zoom in to an Indian watching in the corner of the courtroom. The credits run over the Indian's profile looking out at the trees, seeing them and not listening to the proceedings, finally dozing.

There is no jury finding, no wrapup, no explicit accusation, not even a cathartic explosion of anger from either the Indians or Wojeck. The audience is left to make its own judgments on people, both Indian and white and none really "villains," and to reflect on the various degrees of responsibility viewers share in "Joe Smith's" death. The characters cannot be shelved as stereotypes, and the events have the dignity and the coherence of tragic inevitability. Yet writer, director, and producer do not turn Joe into a helpless victim of a faceless society. This society does have faces and names, and that particularization is the series' most fundamental strength. It is also interesting (and damning) to note that this episode is as relevant today as it was in 1966.

"Listen, an Old Man Is Talking"

Frederick Mueller in "Listen, an Old Man Is Talking" (20/9/66, wr. Philip Hersch, dr. George McGowan) has always been a socialist radical with fire in his belly haranguing people from church steps about the oppression of religion and the brotherhood of man. The camera observes the sympathy of an on-duty policeman, the avid curiosity of a few, and the embarrassment of others. It is jostled by the crowd, switching from his point of view to theirs and back again. When he calls on them to storm the locked church doors and begin the revolution right then and there, they either laugh or drift away; there are no "brothers and sisters" here. When sympathetic police urge him to go home, Mr. Mueller becomes a little vague and seems frail. His daughter shelves her father in an old people's home. The night he is admitted, a man he has barely met chokes to death—graphically on camera—while an inexperienced nurse panics and the annoyed doctor-owner on the other end of the telephone makes a hasty and wrong diagnosis.

Mr. Mueller leaves the place and heads for the coroner's office to protest the death certificate's statement of "heart attack" and is befriended by Wojeck and his wife. He rewards their overnight hospitality by making fresh bread and picking roses for breakfast (is this an echo of the old suffragist-union song "Bread and Roses?"). Up to this point, his only other refuge has been the local library. Out working in Wojeck's backyard vegetable patch, Mr. Mueller turns down his daughter's offer of a home. He has made other arrangements over a store. He also makes friends with a young girl, whom he perceives as innocent by standards long since vanished. Eventually she goes for a walk with him to the botanical gardens. Her disappearance makes Wojeck wonder if he has misjudged the old man, raising the spectre of child molestation which has become the urban reaction to a child's unexplained absence, and we wonder with him. Our innocence is long gone, Frederick Mueller loses the last vestiges of his when he finds that the young girl, like most youngsters, actually enjoys feeding flies to Venus' fly traps. In disgust, he leaves her without even explaining his anger. She is street-smart enough to take a taxi home. The old man, taking the child's pleasure in the death of the flies as the final sign of the unregeneracy of human beings, heads back to his only other refuge—the library—and simply wills himself to death in the arms of a confused but sympathetic librarian.

In this episode, the focus is not only on accidental death through irresponsibility but also on human displacement, a man outliving his time in a world growing colder, more suspicious, that welcomes only young 1960's radicals with hip ideas, not old men with visions forged in the 'thirties. All that remains in the end is the impersonal kindness of a librarian whose job is to look after books, the repositories of the dead past, as well as the currently fashionable and the permanently valuable. Yet even she had betrayed him earlier, when he walked out of the home, by phoning his daughter to tell her that her father had been by the library. In the end, neither Wojeck's kindness nor his daughter's inarticulate love can keep Mr. Mueller alive. Reform of licensing procedures for homes of the aged is only part of the answer here. There are also questions about broken human relationships, misplaced values, impossible ideals. The dilemma of these particular people, as opposed to more stereotyped characters in a formula show, is not easily resolved by the obvious solution of "better retirement homes."

Other Episodes

"After All, Who's Art Morrison" (23/1/68, wr. Grahame Woods, dir. Ron Weyman) dealt with male homosexual prostitution. In deference to the sensitivities of the mid-'sixties, the credits stipulated that "all roles in the film are portrayed by actors." The film also had a voice-over announcement that it was "recommended for adult viewing only." As so often in this series, the victim was randomly murdered, accidentally caught in crime. The emphasis is on various attitudes to homosexuality and not on the killing itself, as when Wojeck explains to Sgt. Banas the law as it then was in all its punitive horrors.

Starting with a long, remarkable sequence that looks and feels like a documentary on street people, "All Aboard for Candyland" (9/27/66, wr. Philip Hersch, dir. Paul Almond) proceeds to show how nice people become heroin pushers. All the unromantic details of the addict's life are presented without sensationalism. The episode is basically an indictment of society's ignorance about and barbaric treatment of addicts, attitudes that create a criminal environment in which people are robbed, maimed, and killed.[16]

Wojeck is short on humour, though long on irony. Rather atypically, however, "Another Dawn, Another Sunrise, Another Day" (15/11/66, wr. Phillip Hersch, dir. Ron Weyman) is a very Hitchcockian tale in which an inoffensive man murders his wife only to be seduced by his landlady, who then kills him. She fails to get his money, though, because the bank teller absconds with it. The teller and his girl do get away; so much for the formula requirements of poetic justice. On the whole, comedy, even in this macabre vein is as rare in *Wojeck* as it is common in *Seeing Things*, reflecting both the different times and the very different personalities responsible for these series.

SEEING THINGS

According to Nelson's definition of the heroes in copshows, Coroner Steve Wojeck conforms to the protagonist as a "forceful figure of authority," quick and decisive in his actions, impatient, obsessive, "forcing the action," according to Nelson's definition of the copshow heroes. Louis Ciccone in *Seeing Things* is propelled by his "visions," which he considers an affliction because they obsess him with the need to make sure that the "right person," not the innocent suspect, is nabbed for the crime. In the first few seasons, however, Louie was also obsessed by the hope of getting back together with Marge, his wife. Often, the two obsessions collided, so that his need to see murderers caught continually belied his claim that Marge and his son, Jason, came before the newspaper, his "visions," and everything else.

His new vocation for crime solving is inflicted on Louie without explanation after his separation from Marge, almost as if the fates sending him the visions wanted to put yet another obstacle in the way of their reunion. The handicap of being visited by psychic visions was one based on both conscience (thus being potentially tragic) and a nose for news (ironic and comic). Marge and Louie were reunited in the fourth season, but new tensions and jealousies about Redfern, Louie's continuing obsession, and Marge's new job as a real estate agent kept the pot boiling.

Louie is not a bionic man or some sort of warlock: he must piece together the real facts with his own brain. He carries no gun, he cannot have visions on command, and has to talk his own way out of each formulaic confrontation with a killer—or, of course, be rescued. This standard mystery-cop show convention is often parodied; the killer has been disarmed by objects as bizarre as a hockey puck or hous-warming cake. Sometimes, Louie has no proof and must bluff the murderer into confessing and surrendering to the police.

Moreover, Louie sometimes has doubts about who is really "guilty" as opposed to "who did it." One such case involved a rabbi (in one early episode called "An Eye for an Eye," 24/10/82, wr. Sheldon Chad) who impulsively murdered a Czechoslavakian Jew turned Nazi and now posing as an Israeli fund-raiser. Louie's visions establish who did the murder and why, despite the confession of the neo-nazi punk, who has confessed to the crime in order to satisfy his urge for public martyrdom. The rabbi, a member of the Hassidic movement understands and accepts Louie's visions as a gift as well as the tension between the gift and the demands of the law so central to the series as a whole. The rapport between them bridges the gap of culture and age. The rabbi in anger has avenged a massacre of forty years ago. Louie wants the neo-nazi to be found guilty instead and sentenced to life in prison. But the rabbi, who is a teacher in a Yeshiva, has a stricter moral code; he chooses to give himself up. The last shot shows this dignified aged figure in black walking down a long hall past the young neo-nazi, who does not meet his eyes but whose hate seemed to me to be undiminished. That episode achieved a low Enjoyment Index, according to CBC audience-relations figures. Episodes that run counter to formula are always risky, particularly near the beginning of a series. Nevertheless, this episode is one of the best in the series. Co-producer David Barlow agrees, but points out that the series could not take that kind of risk very often.

On a lighter note, when Louie as protagonist "moves the action" as the medium and genre demand, it may be a terror-filled chase by motor bike, a motor boat out of control in a hilarious yet deadly sequence or, since Louie does not drive—heresy for any North American protagonist in series drama—giving chase in a VW driven by Marge (Martha Gibson). Louie doesn't drink either, or bed a gorgeous girl in every episode; when he does stumble into Redfern's (Janet Laine Green) bed by mistake, the fact that Marge kicks him out of their own room is quickly subordinated to the episode's whodunnit aspect. Predictably, and perfunctorily, the women make up while he is busy solving the crime.

Nelson contrasts the lovable losers of sitcoms, who reinforce viewer ideas of the average person's incompetence, particularly in the world of work, with the authority figures of the copshows who solve every mystery and right every wrong. Her argument is that this sitcom reinforces the message that the sponsor's product will perform similar miracles for the average incompetent. Characters in sitcoms are often "bumbling," "wacky," "egotistical," or "neurotic," but invariably lovable, always indecisive and, above all, frustrated by their daily problems. She quotes John Leonard, TV critic for the New York *Times*, as saying that in sitcoms, "To be funny, you can't be wise. I mean wise in the sense of wisdom not in the sense of 'wiseacre.' . . . You have to stumble through to the right solution so that you are not presented as being better than anyone else." Here is the answer of the 'sixties and 'seventies to the all-wise fathers and all-patient mothers of *Father Knows Best* (1954-62) and *Mama* (1949-57). Perhaps, as she suggests, this is why so few policemen and detectives are shown with their families.

Fathers or daughters or off-screen wives may appear in minor mystery and copshow roles, but the formula demands that these relationships never take full focus. Even in *Wojeck*, Patricia Collins, as the coroner's wife Marty, seldom had more to do than look beautiful, be loving in bed (a 'sixties breakthrough), provide coffee and sympathy and go to night class. By contrast, Marge, the wife in *Seeing Things*, is continually in focus. Her relationship with Louie evolves as the series progresses. As Jason grows to be a teenager, the writers find more interesting things for him to do and say as well. But above all, unlike the authority figure central to most copshows, Louie appears to be the sitcom klutz who drops and breaks and puts his foot in it. Through parody these characteristics demythologize the traditional stereotype of the strong, silent loner. Louie does not belong to the romantic Baroness Orczy's Scarlet Pimpernel prototype, who uses foolery as a disguise—and whose contemporary descendants are Sayer's Lord Peter Wimsey and Columbo. Our protagonist's physical and social ineptitude is not a deliberate disguise, but rather an intrinsic and funny part of the character. He is also self-conscious, pushy and vain, indecisive, inventive, courageous when really in a corner, and compassionate; altogether, a credible character rather than the formula hero of a copshow, mystery, or sitcom.

"Instead [of the family], the hour-long crime show centres upon work (the solving of the crime) done effectively and without compromise." Nelson tells us this is certainly true of CBC crime shows from *Wojeck* to the first episodes of *Seeing Things*, in which Louie had lost Marge because the paper always came first. An oblivious workaholic, he did not need murders to aggravate his obsessiveness. *Seeing Things* shows, in both a comic and rather sadly ironic light, the cost of that ethic when taken to the lengths usual in formula crime series. Another aspect of the formula in most crime shows is the focus on loners. Despite some delightful but brief scenes with his wife and the constant legwork of Sgt. Banas, Wojeck is clearly such a loner. Louie has few friends who are series regulars, but his family and past provide intricate relationships with which he must cope.

Nelson sees a parallel between the laugh-tracks of sitcoms and the clichéd scores of crime shows, both of which try to control audience reaction. Most CBC series, including *Wojeck* at times, were guilty of this; *Seeing Things* is not. Some music cues are standard: "Louie sees a vision," "Louie klutzes again." Most, however, are tailored to the specific rhythms and dramatic effects of each scene, often with ironic, parodic, romantic or funny allusions tossed in. Of course, the music also performs the standard film, radio, and television tasks of creating suspense, building excitement, establishing mood and pace, supplying atmosphere, and bridging scenes. Nevertheless, Phil Schreiber's music is often contrapuntal and always a distinctive element in the show.

Sitcoms are edited to the needs of dialogue, with specific emphasis on timing of jokes and reactions to them. Crime shows are edited for action, movement, and building suspense. As a crime show, *Wojeck* certainly does not follow formula. Episodes actually use slow verbal and visual rhythms— even silence once in a while. *Wojeck* viewers may often be left with a resonant and highly ambiguous image to resolve for themselves. In both series, of course, "acts" before a commercial usually, though not always, end with someone in peril or, in *Seeing Things*, a gag or a throwaway line. Invariably in both series, the quality of the writing and, in the case of *Seeing Things*, the company's ability to deal with Del Grande's continual improvisation, are in contrast to the formula shows around them.

Nelson concludes that the formulaic conventions of both sitcoms and crime shows reinforce sales of products. Most sitcoms focus on klutzes to reinforce the insecurities of viewers. Copshows reinforce the authority of the problem solvers through the copshow genre. Since the same viewers reject people "better" than themselves when they seek out sitcoms, what are they to do when they are faced with a klutz protagonist in a copshow or episodes featuring old men, Indians, drug addicts, or gays, instead of reassuring authoritative protagonists like doctors and policemen? Somehow, *Wojeck* and *Seeing Things* found sponsors, even though both series routinely subverted the message of authority and the goal of easy audience identification with the stars. More important, no easy answers are given for the human dilemmas which, in the best episodes of every CBC series, displace the chases and fights, meaningless little domestic crises and exaggerated double-takes of the formulae which govern sitcoms and crime shows. In both *Wojeck* and *Seeing Things*, sponsors were eventually repaid for taking a chance with shows which became hits.

A Pair of Contrasting Writer-Producers and How They Mesh

The personalities and backgrounds of the two men who created and co-produced *Seeing Things*, Louis Del Grande and David Barlow, present a paradigm of contrasts that helps explain the unique flavour of this particular series. Del Grande is an American expatriate from New Jersey with credentials from Toronto's alternate theatres in their early, highly experimental and proto-nationalist period. He is also of Italian descent, a Catholic, and proud to be both. His personality is not unlike that of Louie in the series, and he is a brilliant improviser. So is Barlow, who nonetheless appears to be the quintessential, arrow-straight Canadian WASP, understated and friendly, organized and articulate; the behind-the-camera partner who claims to be less analytical and intellectual than Del Grande.

After a stint on *King of Kensington* first as writers and then as part of the production team, Barlow and Del Grande went to Los Angeles to "make it" writing American sitcoms. Although success came their way, it came encumbered by the strictures that are givens on the average Hollywood sitcom, and Del Grande, thoroughly alienated by the working environment, decided to come home to Canada. The two returned with an idea for a series about a workaholic newspaperman, estranged from his wife, who has visions when he is in contact with murder. The Head of CBC Drama, John Kennedy, bought three hour-long pilots, made them co-producers, gave them Robert Allen as executive producer and allowed them to select Dan Yarhi as designer, Nick Evdemon as director of photography, and George McGowan, as director, recalled from working full-time on action series in Hollywood. All turned out to be on their distinctive wave-length. Six years later (not without some argument from the corporation, which is no longer structured to keep together teams of people who understand one another) —the group, except for Yarhi, finished their last season still together. Throughout the history of *Seeing Things*, Barlow and Del Grande exercised an amount of creative control over all its aspects that is unusual even in the relative freedom of the CBC. Together, they set a highly distinctive imprint on the series that is one of the most important reasons why it does not conform to the formulae of the standard mystery show or sitcom.

Elements of Popular Culture

Seeing Things is also a fascinating text for the critic of popular culture. The mix of parapsychology and murder is an old one in popular mystery. This series, however, subverts the associated conventions. Louie does not want to have visions; they complicate his life and never prove "who did it." A modern newspaper office and an ordinary townhouse do not recreate the atmosphere of old houses, old money, storms, fogs, and heaths traditionally used to set "atmosphere" for this kind of story line.

This series also breaks new ground with its continuing development of a domestic subplot, usually but not always treated comically. The complexity, irony, comedy, sadness, exasperation and jealousy inherent in the separation and even in the subsequent reconciliation of Louie and Marge anchored the series in a very specific contemporary experience. They acted out a situation that is now familiar in everyday life. The woman wants to grow and change and be independent, but cannot find her own path, or even a steady job. The man wants to move back home, but repeats the behaviour that helped cause the break. His son and his parents want them to get together again. She dates, and he resents. Louie is mildly attracted to Heather Redfern, the younger, very cool and controlled, blonde, single, prim assistant Crown attorney and Marge's complete opposite. Even though Heather's disinterest is obvious to the viewer, Marge sees her as a threat. Money is a problem; a growing teenaged son presents other difficulties; the list goes on. For the first four seasons, these domestic tensions were not used primarily as sitcom material, although the elements of broadly farcical comedy increased in the last two seasons. Specific domestic problems may or may not be solved, depending on the episode, but the tension in the relationships are essentially unresolved, as they would be in everyday experience.

Seeing Things is very Canadian in its self-conscious, yet off-hand references to elements of American popular culture. These are sometimes deliberately subverted, and sometimes mocked, sometimes used in a straightforward but fully conscious way as intrinsic parts of the texture of our lives. Another distinctive feature of the series which is atypical of most cop shows and mysteries, though characteristic of some American sitcoms, is its steady stream of ad-libbed one-liners. The timing is crisp, and the delivery always throwaway, on-the-run. Often, jokes are topical, political, or social in their thrust. Some are "in-jokes" about Canada, Ontario, occasionally other places in the country, and Toronto. One episode alone

("Someone Is Watching," 23/1/84, wr. Anna Sandor and Bill Gough) is in part a parody on the familiar "haunted house" variant of the mystery genre and a catalogue of popular cultural references. It offers gags about the film classic *House of Blood*, the *Little Rascals* catchphrase "Feet do your stuff," 'fifties Cuban heels, 'eighties concerns about cakes without additives, 'forties-style sensationalist newspaper prose, the platitudes of fem-lib chic, a mad housewife obsessed with her indoor-outdoor carpeting; and a portly man with impeccable diction who passes by in the rain saying, "Good evening," just as he did on 'fifties television before introducing some macabre tale on *Alfred Hitchcock Presents* (1955-65).

Above all, the series consistently juxtaposes topical and nostalgic references, American and Canadian, popular and high culture. The last sequences of the show on psychics and high finance was a typical mixture of all of the above ("Second Sight," 26/2/84, wr. Larry Gaynor). In the chase scene called for by the formula, the gum-chewing thugs after Louie are straight stereotypes. However, they are chasing Louie through a forest of table legs on the floor of a psychic nightclub called *The World of Wonders*, a rather wicked send-up of the title of one of Robertson Davies's highly acclaimed novels that focuses on vaudeville, carnival, and theatre. Finally, Louie runs into the legs of one of thugs made colossal by a low-angle camera shot representing Louie's point of view. Louie tosses off a flip line about the shoes and their metal toes, then unheroically bites the bad guy on the leg and gets away.

Later, when the real killer unburdens herself according to formula, Marge draws attention to the convention by wondering nervously why she is spilling all this information. Atypically for this series, a gun is actually fired, but the real climax comes when the prime suspect, an old drunken ham actor, rushes Louie with a dagger that turns out to be a rubber prop. The actor then bops the gun-toting lady on the head with his empty bottle because she has always upstaged him off-stage. The take-out gag which replaces the usual music running under the credits focuses on a three-way argument between Marge, Louie, and the actor about the last line of *Hamlet*. Louie, in the middle of the cacophony, fires the gun for quiet so that he can have the last word on exactly what the line is, which, as the viewer may by now know or surmise, turns out to be wrong.

Seeing Things catches some fundamental truths about Canadian sensibilities. It revels in irony and ambivalence towards some elements of American popular culture that saturate Canadian culture, but does not spare home-grown mythologies. There have been programmes that deal ironically, or comically, sometimes satirically with subjects like the Canadian military; the survivors of the romanticized 'sixties; the RCMP and its cosy relationship with the CIA; the world of pop music, where "made in

Canada'' sounds exactly like "made in Los Angeles" and is indistinguisha-
ble from the tunes in lifestyle ads; psychic fairs; the chic of skiing; a satire on
television reporting vs. print reporting, in which Louie becomes a TV
personality; and the macho myths of hockey.

Often, what the show does is juxtapose the products of American popular
culture with the realities of Canadian life. One example, "Looking Good"
(wr. Sheldon Chad), focused on that quintessentially North American
phenomenon, the televised beauty pageant. The title for which the girls
compete is "Miss Cosmos," a characteristic double reference to the
inflation of "Miss America" to "Miss World" to "Miss Universe" and the
values upheld in that swinging singles magazine, *Cosmopolitan*. But the
Canadian reality is also satirized by the preliminary titles of the girls; "Miss
Mississaugua," "Miss Kitchener-Waterloo." There are also comic turns
based on the "talent" segment of these pageants. One is a good-natured
take-off on high-school public-speaking contests which offer inane topics
like "Canada—our land of bountiful resources." The blonde contestant
with a huge red sequined maple leaf on her ample bosom begins with
"British Columbia, land of the noble Haida," and ends with the phrase,
"from sea to shining sea," an inadvertently Americanized and thus typically
Canadian paraphrase of the actual "a mari usque ad mare." The parody
finishes with a burst of "O Canada" from the inadequate band while the
contestant flourishes a lobster and salt cod in either hand.

There is a special irony in the fact that most Canadian viewers do not
notice Del Grande's unreconstructed New Jersey accent and idioms; they are
conditioned to hearing American accents on every other copshow or mystery
series made on this continent. Yet they do notice with pleasure, and some
surprise, the jokes about Canadian politics, Toronto—the butt of humour
in every other part of the country—the CBC, Anne Murray, Stratford, lunch
at Winston's, and retirement in Florida. One of the more clever parodies was
in the "psychics" show, where a tycoon called White played by Alan Scarfe
running the "Titan" Corporation, did a very funny, scary take-off on
Conrad Black's mannerisms. Canadian TV in the 'eighties, like its
American counterpart, has been having a bit of a love affair with the
tycoons of industry. Typically, *Seeing Things* picked up on that trend yet
achieved a parodic level that will also work for people who have never heard
of Conrad Black or the Argus Corporation.

The parodic shades into satire in the scene in "Looking Good" where
Sgt. Brown discovers that his daughter, one of the contestants as well as a
prime murder suspect, has actually been sleeping with the dead promoter.
By including his disillusioned, angry, sad acceptance of the fact that his little
girl is not a virgin beauty queen, the episode makes yet another wry
observation on how and to whom the pageants appeal. As with the pop

music episode, this one points repeatedly to the fact that the pageant is an imitation of an imitation, thus satirizing another colonized area of our culture.

Through Louie's son, Jason, the influence of American popular culture on Canadian kids emerges as a minor running motif. He gets a Darth Vader helmet, is given the huge TV-presold LEGO set, plays video games, listens to American rock on a Walkman, and is addicted to American television just like his counterpart across the lake in New York State. These facts of Canadian life are treated wryly, as a familiar problem, related more to what the kid is doing with his time than to whether the game or activity reflects "Canadian culture." It is treated as a barely noticed fact of life for which Marge and Louie have no solution.

In "An Eye for an Eye," Marge makes an unsuccessful attempt to combat Jason's TV consumption by trying to get a piano which a nice elderly widow is giving away to make room for a 31 " colour set with oak console. Struggles with moving the piano are a running gag in that episode. Yet she also greets the latest trends in American popular culture with open-mouthed, comic innocence. Once a hippie, though she seems never to have smoked a joint, Marge is still looking for herself in a series of disastrous jobs: nurse's aide, television continuity girl, and real estate saleswoman. She is not, however, the familiar "dingbat" of sitcom, since she is the one who basically parents, knows how to drive, figures out the finances, and occasionally rescues Louie from the bad guys. In the 1983-84 season, Marge was defined as an expatriate American who has never taken out citizenship papers. She is thus displaced in every sense, yet looking for herself outside the traditional roles.

As I have already pointed out, English Canada does not have a star system in the mass media sense that French-speaking Radio-Canada does. There are few opportunities to recycle well-known stars or enrich other series with their hard-won skills, their glamour, or simply nostalgia. In this series, however, there appears to have been a conscious effort to mine Canadian television's almost forgotten past. Some of our best actors from the 'fifties and 'sixties have been featured, their most successful appearances being when the characters they play run counter to their recognized personae. Lovable but honest "rowdyman," TV Member of Parliament and veteran of the Boer War in *Gift to Last* (1978-79), Gordon Pinsent appeared as a conservative writer of influential stock exchange letters. Bruno Gerussi turned up as a con-man gypsy instead of the heroic father figure Nick Adonidas of *The Beachcombers*. Barry Morse achieved American stardom as the implacable hunter in the very popular *The Fugitive* series, as well as being the voice of the American Express Card. Canadians with longer memories would connect him with dozens of serious anthology dramas in the 'sixties and early 'seventies. In the psychics episode of *Seeing Things*, he sends up his aristocratic persona as a classical actor by playing a ham-actor who has done Shakespeare "directed by a teenage catamite—in the nude—for pay TV." Other fine actors from the earlier days of television appeared in that particular season. In the haunted house episode, Kate Reid played the brains behind the killer instead of her usual neurotic victim role. The brilliant comedienne Barbara Hamilton was wasted in one of the less successful, because too formulaic, episodes about skiing and resort developments as environmental threats. Don Francks is still Canada's best-known survivor and proponent of the more enduring values of the 'sixties, which gave special poignancy to his appearance, in a very good episode (by Larry Gaynor, called "I'm Looking through You," 29/1/84) on the death of the peace and love era, as Sunshine Magee, the aging, possibly stoned, definitely ironic and defeated owner of a 'sixties nostalgia bar. "Bull's Eye" (31/3/87, wr. Bill Lyn and Bill Hartley) featured Rompin' Ronnie Hawkins as con man Country and Western artist and bar-owner. To some viewers, his persona of wild man of Country music and refugee from far south of our border interplayed nicely with his character in this clever parody of the Sounds of Nashville North. The effect of casting these actors against type or commenting on their public personae is not to create another *Loveboat* or *Fantasy Island*, but to use figures both Canadian popular and high culture self-reflexively, in a tribute that is both affectionate and ironic.

Class Conflict—in Canada, You Say?

Very often, the private-eye or police protagonists of mystery shows demonstrate class divisions in the culture from which they come. We can see this clearly in the British series *Cribb*, *Rumpole of the Bailey*, and *The Chinese Detective*, and there are hints of it in such American series of the mid-'seventies as *Baretta* and *The Rockford Files* (1974). In the radically changing population mix of Ontario in the 1980s, class conflict is the source of much of the comedy in *Seeing Things*. It centres on Louie, a second generation Italian immigrant, and his often unwilling source of inside information, WASP assistant Crown attorney Heather Redfern (Janet Laine-Green). Redfern got her job through family connections but hangs on to it because of her ability, despite both gender harassment and explicit, if comic, sexual harassment from the member of the old-boy's network who is her boss. She also wins every cup in her class of yacht-racing, as is her right, both through natural ability and life-long training by the other members of the "family compact." Ironically, she likes Country and Western music because it is the "last refuge of the lyric in popular culture;" Louie hates it. The programme builds sharply satirical one-liners and running gags around differences of attitude, taste, and background. Louie usually comments on whatever the point of difference is: food, parties, friends, wallpaper, stock portfolios. Redfern tends to bite her tongue, and sometimes envies the chaos that follows Louie around.

Whenever Redfern and Louie meet on social rather than professional grounds, comic misunderstandings multiply as they do in the haunted house episode. Redfern's boyfriend can, however, cross class barriers with ease. He is a handsome, singularly clueless, NHL hockey player, and thus part of Canada's most basic myth of superiority—a myth that is taken for a merciless ride in this series. Moreover, the idea that Redfern enjoys having a handsome stud about belies her straightforward, lady-like stereotype. Whatever subtextual sexual tension is implied in Louie's constant involvement with Heather Redfern is entirely on Louie's side, and more in Marge's imagination than in his behaviour. He treats Redfern the way he treats Marge, as a useful appendage when he is on the track of story or a murder, and their reactions are identical. Both get angry, but eventually show a reluctant, somewhat bemused willingness to cooperate.

Louie Ciccone, the protagonist, is a very Canadian Canadian who hates having visions but gets them anyway, then acts on them with a bulldog tenacity that is rarely rewarded, publicly or privately. He is puritanical in morality, a workaholic. When looking at Canadian manners and mores, Del Grande as co-producer starts with the advantages of the double vision of an American-Canadian and the continuous input of producer David Barlow. Like many Americans who become devoted to Canada, he is more articulate, more vocal, and more aggressive about the survival of a separate culture north of the border than are many of the native-born. *Seeing Things* is a distinctive contribution to that end.

In the early seasons, *Seeing Things* did not always reassure its audience by having the good guys always win or safely distance them with a comic, wise-cracking klutz from sitcom. It is worth noting that, in the uncertain climate of the mid-'eighties, after more than twenty years of very profitable existence, private Canadian television networks still would not originate such an unorthodox and expensive hybrid. The competent and marketable *Night Heat* (1985-present), co-produced with CBS for the late-night movie slot, but shown on CTV at 10 p.m. is the private networks' moderately successful entry into the world of mystery and copshow. The irreverence and eccentricity of *Seeing Things* could find an outlet only on the network owned by a Crown corporation with the money and willingness to take risks.

Whenever the CBC encourages personal vision and talent to inflect, subvert, or ignore the formulae that dominate the rest of the channels on the dial, as in the "Western" *Cariboo Country*, some of the docudramas on *For the Record* (1976-85) or, in this case, *Wojeck* and *Seeing Things*, something fresh and original emerges, something that is also distinctively Canadian.

The result is often an episode, or even a series, which is as close to anthology as it is to series television. The American genres of copshow/mystery have produced some very good television drama. They form the context for the CBC's comparatively late entry into the field of series television and the current context for the 95 per cent of Canadian viewers who can get both CBC and CTV. Fortunately, blatant imitations like CTV's *Police Surgeon* subside into deserved oblivion, without reruns. Variations that slide into imitation, like *Sidestreet*, range from good through adequate to poor. Yet the two I discuss at length were truly innovative: *Wojeck* and *Seeing Things*.

I do not think it is an accident that a programme which is consistently polyphonic in structure, weaving strands of murder mystery, satire and sitcom, Canadian and American topical references, popular cultural references and allusions to high culture, Chekhovian subtext and improvised one-liners, and devised by one born-and-bred and one adopted Canadian came into being in the 1980s. *Seeing Things* catches the mood of the times just as *Wojeck*, with its gritty urban landscapes, use of direct cinema conventions, and moral outrage belonged to the mid-'sixties. Northrop Frye's familiar insight that, in Canada, "first we have to find out where 'here' is" finds confirmation in these two shows, and the "here" both changes and stays the same. In the looking glass chess game of television, where ratings and formulae write the game rules, it is often difficult to tell the pawns from the queens, the black squares from the white. It is even harder to change the fundamentals, the details of our mental and emotional geography. Yet in a culture where mosaic, not melting pot, is the ruling metaphor, where the tensions between regionalism and nationalism constantly strain and enrich the fabric of our lives, and television serves a population so small and so diverse that specialized audiences and the mass audience have to fight it out for air time on a single channel, both *Wojeck* and *Seeing Things* manage to address things that unite us, things that divide us, as well as the multiplicity of identities that make us what we are.

NOTES

1. University of Chicago Press, Chicago: 1976.
2. Harry Adams Inc., New York: 1977, 139.
3. *TV: The Most Popular Art.* Anchor Books, Garden City, New York: 1974, see 27-8.
4. In *The Popular Culture Reader*, eds. Jack Nachbar and John Wright. Bowling Green University Popular Press, Bowling Green: 1977.

5. CBC research panels were defined in the report covering 7-13 April 1967, as "2,000 respondents selected on a probability basis to be representative of all persons in Canada with television, over the age of 12 . . . [who are] English speaking, excluding Newfoundland and Quebec, outside Montreal." The questions now asked are more sophisticated.

6. By Leslie Halliwell with [television critic for *The Sunday Telegraph*], Philip Purser, Granada, London: 1982. Entries are alphabetical by title. This useful reference work omits "shows which are not preserved," xix. The tone of the comments is breezy with a populist slant.

7. *Journal of Popular Culture*, 12:4 (Spring, 1979), 581-601.

8. *TV Guide: The First 25 Years*, Ed. Jay S. Harris, New American Library, New York 1978, 257-8.

9. *The Age of Television*. W.H. Freeman and Co., San Francisco: 1982, 30. Esslin is the critic who first named and described the theatre of the absurd. He was a distinguished Head of Radio Drama at the BBC in the 'sixties.

10. *Journal of Popular Culture* 12 (Fall, 1979), (10), 21.

11. *TV: The Most Popular Art*, 109.

12. *Reading Television*. Methuen and Co., London: 1978, 42.

13. *Jolts: The TV Wasteland and the Canadian Oasis*. James Lorimer and Co., Toronto: 1985, 48. I think "serial" is a typographical error here, since this word ordinarily denotes a programme in which the narrative is not closed at the end of each episode.

14. *In Search*, Department of Transport, Ottawa (Fall, 1979). A revised version of this thesis appears in *Canadian Drama*, Vol. 9, No. 1, 1983 as "TV Formulas: Notes for a Zeitgeist of Prime Time," 30-39.

15. The episode is even more powerful than George Ryga's well-known and influential play *The Ecstasy of Rita Joe*, first performed in 1967. Ryga had already written a superb half-hour play about the same dilemmas called "Indian" (*Q for Quest*, prod. Daryl Duke, dir. George McGowan, 25/11/62).

16. For a useful account of the interrelationships of documentary traditions of photography and the dramatic, see *Richard Leiterman* by Alison Reed and P. M. Evanchuk in the *Canadian Film Series*, ed. Piers Handling, Canadian Film Institute, Ottawa: 1978.

3

FAMILY ADVENTURE
"In every box, a prize!"

Why link a distinctive, innovative crossover between series television and television anthology like *Cariboo Country* (1960-67) with an example of the more familiar family-adventure series *The Beachcombers* (1972-present)? Start with the fact that Philip Keatley produced and directed all of *Cariboo Country* and developed and produced the early years of *The Beachcombers* as well. His interests, values, and point of view on specific issues run like threads through both. Then there is the medium of film and a sense of place which make both series distinctly regional (whether West Coast or Chilcotin), as well as actively outdoors series that focus on people living particular lives in specific times and places. Both series have a documentary approach to details of work and play. Both were made away from the close supervision of Toronto and Ottawa and both have been popular with national audiences.

Cariboo Country was absolutely distinctive, evolving from a couple of programmes in a West Coast anthology into a conscious alternative to the pervasive form of the American television Western.[1] All the scripts were written by Paul St. Pierre, whose particular sensibility makes the anthology what it is. He knows the Chilcotin, his eye and ear are keen, his respect for his varied subject matter evident and his wry, elliptical style well suited to the intimate television scale.

The Beachcombers, on the other hand, was intended from the beginning for a more specific market, that of children and their parents, using the well established formulae of the family-adventure genre. Yet the range of tone, the inflections of formula, the occasional experiment in form and the continuity among several skilled scriptwriters under the supervision of story editor Suzanne Finlay also made this series distinctive as well as very successful with audiences. Both have helped keep CBC network television drama from being Toronto television drama with regional interludes.

Finally, I have chosen these two related series to show the reader rather different facets of series television. *Cariboo Country* is the idiosyncratic, wonderful 'sixties series that shows how film created new possibilities for story-telling on Canadian television, how strong regional contributions to the national scene could be, what the right partnership of writer and producer could do over several years, and how topical, anecdotal material can still read as current and interesting twenty years later. Because I had the opportunity to view over forty episodes from the first few seasons of *The Beachcombers* over a ten-day span as well as seeing another thirty off-air over the years, and because I have had access to the "treatment" of the show which outlined the original ideas for management, I am using it to demonstrate how a series evolves. I will also try to identify for the reader the elements in this series that make it one of the longest running and most successful in North American television.

CARIBOO COUNTRY

Many readers are old enough to remember when the Western dominated prime-time television: *Bonanza* (1959-73), *Gunsmoke* (1955-75), *Have Gun, Will Travel* (1957-63), *The Virginian* (1962-70), and literally dozens of others. *Cariboo Country* (1960-67) was the Canadian response, sent up seven years running against the top guns from the south. Even though the series was not always shown on the full network, which includes the large numbers of affiliated stations as well as those owned by the CBC itself, and though it began on the network as a summer replacement before being absorbed into *The Serial*, it managed to draw enthusiastic audiences at the time. Eventually, executive producer Robert Allen recognized its value. A pair of episodes called *The Education of Phyllistine* (1964; wr. Paul St. Pierre, prod./dr. Philip Keatley) were put together and featured on his flagship anthology *Festival* (1960-69). It won several awards. *How to Break a Quarter Horse* (1966; wr. Paul St. Pierre, prod./dr. Philip Keatley) was written specifically for this prestigious anthology. Most important of all, many episodes from the series itself stand up superbly to viewing twenty-five years later.

Cariboo Country began as a live studio anthology series in the summer of 1960. (Summer was not then the throw-away season of reruns in television scheduling that it is now.) By the Fall of 1960 there were eight sixty-minute Westerns and twenty-four half-hour series available to Canadians on U.S. border stations in prime time, including *Cariboo Country*'s rival in the 7:30

Saturday night slot, *Bonanza*. Initially, the CBC showed only two American Westerns in the 1960 schedule, but the network eventually picked up several of the more popular ones during *Cariboo Country*'s seven year run. American TV Westerns ran roughly twenty-six episodes in a season; *Cariboo Country* averaged seven or eight.

Continuing characters included Ken Larsen, a successful rancher (Wally Marsh), Arch MacGregor, the general-store and tavern-keeper (Ted Stidder), Smith, a struggling rancher (David Hughes) and his wife Norah Smith (Lillian Carlson), Morton Dillonbeigh (Buck Kindt), a seventy-year-old failed rancher and successful eccentric and his wife Mrs. "D" (Rae Brown), Ol' Antoine, an elder on the reserve (Chief Dan George), Walter Charlie, a half-breed (Merv Campone, a good actor and a skilled writer), Sarah (Jean Sandy), her husband Johnny (Paul Stanley), and Frenchie (Joseph Golland). The format consisted of stories told by or to Arch McGregor about his friends and neighbours. Narration serving as a bridge was a common television convention in anthology during the 'fifties and early 'sixties. A host would introduce the story, then summarize it after each commercial. Sometimes the narrator was a figure slightly more integrated into the story line or even central to it. Note, however, that all the characters I have listed were protagonists in at least one programme of *Cariboo Country*, in complete contrast to the overriding convention of one or two overall leading characters that prevails in formula series.

Anthology? Series? TV Western?

Richard Carpenter, in an article examining "Ritual Aesthetics and TV," states that the hero of a Western "is assigned his duties by a higher authority, undertakes it skilfully at great risk to himself and his followers, contends with the powers of evil in their most dynamic form and by the end of the programme has succeeded in overcoming them and restoring peace and harmony to the community."[2]

The makers of ritual, like the makers of commercial television, assume that their audiences share a set of recognizable (and marketable) preferences, values, and expectations. Unlike the American networks, the CBC, even in 1960, did not see its audience as homogeneous. Moreover, the anthology form was still alive and well in Canada. Thus, the CBC audience could not be said to be addicted to Carpenter's "large, simple, general patterns." Its expectations of the Western, created by forty years of movies and a few years of American adult Westerns, were tempered by the history, literature and differing mythology surrounding the Canadian West, and by a broadcasting system and viewing habits that were unlike those of the United States.

It was within this context that producers Frank Goodship and Philip Keatley and writer Paul St. Pierre developed *Cariboo Country* from two scripts called "Window at Namko" and "Justice on the Jawbone" presented on a Vancouver anthology in 1959. Goodship had to give up supervising the programme owing to illness in 1960. *Cariboo Country* was not a series, with a star and range of stories limited by a genre, and yet it was not quite an anthology either. Keatley and St. Pierre decided to develop a multi-focus series to range widely in tone and to evolve a highly individual, semi-documentary yet poetic visual style to complement its contemporary, regional flavour.

Arch's stories were initially taped in Vancouver's very limited studio facilities. Two seasons later, despite the CBC doctrine that film drama was the exclusive domain of the National Film Board, Philip Keatley persuaded eastern management to let him film the series, and its sense of place and a particularized way of life grew even stronger. In 1962, Ron Weyman (eventually the executive producer of all film drama in the 'sixties, and *Wojeck*'s progenitor), was doggedly trying to get the CBC to allow him to present filmed drama on *The Serial*. At some sacrifice of precious hours in the schedule, Weyman also found a home for *Cariboo Country* under this umbrella title.

Early press releases for the series take pains to call attention to the differences that set this series apart from the average formula TV Western: "[The series] features authentic Cariboo characters and will show the present day Canadian cowboy in his true light [thereby preparing the audience for the documentary emphasis insisted upon by Philip Keatley]. The series has a marked lack of women and no love interest, preachers or parsons. Most of the plays are set in Namko, a mythical community populated by a few wives and children and by Indians" (17/6/60). Actually, women and love both appeared regularly, though in this respect the series was ahead of its time; the episodes are not particularly sexist in tone. St. Pierre also did not use the predictably romantic plots that became the staple of the later, more domesticated seasons of the front-running American television Westerns, *Bonanza* and *Gunsmoke*. The press release is otherwise quite accurate.

Equally self conscious is this programme note from Paul St. Pierre in the *CBC Times* (9-15/7/60): "It is a measure of how casual a land is Cariboo that no man can say exactly where it is or what its boundaries are except that it is up on the plateau country of British Columbia. It is in truth less a geographic location than it is a condition, a state of mind. Through it all, the Indian, gentleman in rags, has endured. . . . Now it is changing as everything must." The pressure of change is one of the dominant themes in the series. St. Pierre continues, "This may be a curious dramatic series, since almost all of the people are singularly undramatic, given to understatement and casualness, to indirection and private humours. Probably the country makes them that way—strong, self-reliant, hospitable, individualistic, unpredictable.'' The effect of geography and history on men and women, another running motif in the series, is occasionally found in movie Westerns (*Shane*, for example) but rarely in TV forms.

One of the main characters of the series, Smith—no first name—says in ''Sale of One Small Ranch'' (2/3/66): ''We are on the 4500 foot level at the 52nd Parallel of latitude. . . . It means that we are too far up and too far north to ever run a good ranch. The summers are too short and so is my life. There have been two bad summers for hay and two long winters in a row. She is a hard country.''[3] Yet, in the end, he does not sell the ranch. Staying put to see what will happen despite continuous hardship seems to be a very Canadian response—just as much a part of our mythology as ''goin' down the road!''

Unlike its U.S. counterparts, *Cariboo Country* is set in the 1960s, not the 1870s. It does not fall within the largely urban tradition of the pastoral forms of literature and film. It is not centred on one or two stars supported by a small running cast and weekly guests. Above all, it is not about law enforcement in the traditional sense. Missing are the favourite plot devices of sheepherders vs. ranchers, or the law vs. rustlers, gunmen, or Indian raids. Nor is it about the coming of the railroad, the struggles of wagon trains, the vicissitudes of bounty hunters, the aftermath of the Civil War, the exciting ambience of gamblers and saloons, drifters who are settled down by spunky schoolmarms, or the temptations of whores with hearts of gold—and it is certainly not about the loners who haunt the Western in most of its guises.

The perfect summary of the conventions of the American adult Western as widely perceived in the glory years of the mid-'sixties was offered by that well-known expatriate, Lorne Greene, on an NBC special called, without conscious irony, *Lorne Greene's American West* (1964). The programme was an ''episodic exposition'' (*CBC Times*, 1-8/5/64), with songs, dances and dramatic scenes showing ''the fact, fantasy and free spirit of American Westward development. Here are the buffalo men, the scouts, the railway workers, the wide open saloon, the church social and finally the ghost town.''

When viewers remember Westerns, whether the ''horse'' Westerns of Hopalong, Cisco, Gene, and Roy, or the more adult territory of Bat Masterson or Cheyenne, they inevitably think of the ''good guys'' and the ''bad guys.'' Even *Bonanza* and *Gunsmoke* share what Horace Newcomb correctly identifies as the pervasive norm in the thematic structures and the plots of the genre, ''the white hat syndrome.''[4] A look at the selection of

Westerns available in my own area in very late-night reruns confirmed once again these generalizations.

Cariboo Country, however, was characteristic of much of Canadian television drama in that it never promoted a black and white system of values. Canadian television producers and writers have often preferred the enigmatic, the open-ended, troublesome or ironic ending to their tales. In that sense and in the delight of the allusiveness, wit, and inversion of formula, *Cariboo Country* could most usefully be compared to *Have Gun, Will Travel* (1957-63). Often, *Have Gun* seems deliberately to subvert the conventions of the TV Western. For example, in one episode about a mineowner's daughter who tries to buy an election for her feckless husband, it is not Paladin who stops the hired guns from disrupting the funeral of the miners' leader, but his widow, who walks up to the drawn gun of the "Kid" and slaps his face.[5]

Kathryn Esselman, in "When the Cowboy Stopped Kissing His Horse,"[6] divides Westerns into two categories: the lone wolf and the communal. She points out that *Have Gun*" avoided the hero's inner conflict between wilderness values and civilized values by making the hero a commuter between the two value systems, burdened by no values of his own." I would argue that Paladin does have a set of values, as in this line delivered to a killer after he saves him from a lynch mob: "I do this for myself and for the man Jim Toby used to be. What happens to you later is unimportant." Nevertheless, she draws a useful distinction.

In "Savagery, Civilization and the Western Hero,"[7] John Cawelti points to an interesting anomaly about the traditional content of the Western when he writes "the complex clashes of different interest groups over the uses of Western resources and the patterns of settlement surely involved more people in a more fundamental way than the struggle with the Indians and outlaws. Nevertheless, it is the last which has become central to the Western formula." *Cariboo Country* often concentrates on the complex clashes Cawelti refers to as being oversimplified in TV Westerns. It is about a community, Namko, the Indian reserve, and the huge region of the Chilcotin, as well as the loners within and without the community and the rugged individualism that characterizes them all and the clashes of two or three distinctively different Indian cultures, Scots, Irish, English, French, Swede, male and female, young and old.

As is often the case, analysis of television drama from other countries is more useful to Canadians for pointing up differences than for identifing similarities. Quite helpful, often by way of contrast, is Ralph and Donna Brauer's *The Horse, the Gun and the Piece of Property: Changing Images of the TV Western.*[8] Their categories of "Horse Western," "Gun Westerns," "Transition," and "Property" Westerns all coexisted with

Cariboo Country in the early to mid-'sixties.

"Horse," "Gun," "Transition," and "Property" Westerns—Contrasts to Cariboo Country

According to the Brauers, Westerns are anti-technology, against the complex social order needed for technology to develop and survive, ambivalent to and caught between the values of savagery and civilization, emphatically pro law and order, and against intellectuals and politicians. They also claim that heavy-handed, sadistic violence is what defines the adult Western. By contrast to the early "horse" Western of Roy Rogers and the Cisco Kid, the emphasis shifted to the hero's gun, and often on the fusing of man and gun as in *Yancy Derringer* (1958-59), *Colt 45* (1957-60), or *The Rifleman* (1958-63). Since explicit sexuality was still forbidden on American television in the 'sixties, they argue, along with many other critics, that violence was often a substitute. They theorize that this emphasis on violence was also, in part, a response to the arms race. Would Canada's deliberate refusal of that race, despite its position as a middle power with nuclear capability and its choice of a prime minister who had won the Nobel Peace Prize help explain Canadians' distaste for guns in both our copshows and Westerns? Yet we certainly enjoyed American TV Westerns in the 'sixties, just as we enjoy the reruns of *Starsky and Hutch* (1975-79) and continue to watch new series like *Miami Vice* (1984-present). Why was *Cariboo Country*, which is full of action and comic or wry anecdote but largely free of on- or off-screen violence, so popular with Canadian audiences?

Clearly, *Cariboo Country*'s success is an argument for diversity and choice for the viewer. When the Western disappeared, many of its conventions were subsumed into the copshow. The line of Canadian copshows also absorbed a few of the characteristics of *Cariboo Country*, and specifically the refusal to use shootouts as plot devices or denouements. The documentary shooting style, the emphasis on a particularized environment, irony, downbeat endings, ambivalence and social comment in *Cariboo Country* were also characteristic of *Wojeck*, *The Manipulators*, *The Collaborators* and some episodes of *Sidestreet*. Highly urban *Seeing Things* is a long way from the Chilcotin, and yet it too is a self-conscious, often funny, reaction to prevailing television formulae. *Cariboo Country* has pride of place as the first of these conscious variations on familiar television genres.

The Brauers quote John Evans' article, "Modern Man and Cowboy" (1962),[9] which suggests that the gunfight is modern man's "substitute in fantasy for the grand confrontation scene which in real life is impossible." He suggests that Westerns appealed to "the alienated . . . organization man'' of the 'fifties. Through his vicarious positon in the powerful and final act of the gunfight, the factory worker [or the] organization man symbolically shoots down all the individual officials and impersonal forces that restrict, schedule, supervise, direct, frustrate and control his daily existence.'' Perhaps we Canadians rarely use gunfights as a plot device in our anthology drama, Westerns and copshows because we do not believe in and, therefore, seldom accept as credible, ''final acts'' in any of our art forms. Canadian culture has never been a good hunting ground for grand certainties.

In the continuing search for novel settings and plots for new Westerns, ZIU-UA sold NBC a Western ''based on'' Pierre Berton's *Klondike Fever* (1960-61). Interviewed by *Variety* (15/2/61) Berton pointed out that only two out of twenty-two episodes were actually based on his book. Although NBC had announced Klondike as the start of a new, ''non-violent'' era in television, what Berton saw was ''just kicking, punching going on and smashing people over the heads with chairs. After [the first episode] the gunplay was almost continuous. . . . We will not see [my] stories on television now''—and we never have, in the U.S. or Canada. Later, Disney took all the resonance out of St. Pierre's Chilcotin characters in *Smith*, starring the genial Glenn Ford. The fact is that St. Pierre's Smith's gritty, laconic character, so clearly etched by David Hughes, would not have fitted into a 'sixties Disney movie. It seems that the traffic of genres across the border is not always profitable for either importer or exporter.

The Brauers characterize the ''lone hero'' category as associated wide-open space and time, which they call a ''pre-adolescent'' concept. Eventually, however, as the form evolves, the hero's role and the job, the code and the job become one—producing the stereotype of the ''lawman.'' *Cariboo Country* has no lawman but the RCMP constable on patrol, and codes, personal and communal, are constantly evolving. In the copshows that followed, code and job were often in conflict, so too, with comic twists, in *Seeing Things*.

When the Brauers turn their attention to ''the male group and the emerging community'' as a late-'fifties phenomenon that shifted the focus from loner to community, thus creating the potential for a series-anthology perspective, we would seem to be getting closer to *Cariboo Country*. Not so: even in this category, individualism is still suspect. Admittedly, the gunfighter (though not the gun) is abnormal in these new series. For the loners, however, this variant of the TV Western substitutes a free-roaming community, mostly male of course, and not at all democratic, though with strong emphasis on team play. As the Brauers note, this is also the era of the astronaut and what Tom Wolfe calls ''the Right Stuff.'' From the U.S. came the ''giant step for mankind.'' From Canada (on American booster rockets) came the satellites designed to conquer the vast, inner spaces of "the Unknown Country," or deliver more and more American television signals, to Canadian cable antennae and earth dishes depending on the will of the legislators.

The Brauers' example of the communal Western, *Wagon Train* (1957-65), "raises questions about the individual's role in the community, about dissent, about the community's response to outside pressure, about how much authority the community should delegate to elected officials, about how much reliance should be put on experts, and about the relationship between the members of the 'government'. . . . In all the episodes, these conflicts are clearly resolved in favour of the community" (p.96). Many of these themes also appear in some form in *Cariboo Country*, but such conflicts are very rarely "resolved." As the Brauers point out, the protagonists and most of the characters in the average Western are lily-white. Dissent, even from the protagonist, is seen as deviant. Not so in *Cariboo Country*, where eccentricity is admirable, the protagonists are often Indian or half-breed, and consensus is only occasionally achieved. Most important of all, in *Cariboo Country* the price of consensus is always clearly spelled out.

In American adult Westerns of the 'sixties, a "deviant" is defined as anyone who is black, Mexican, long-haired, eccentric, old, or female. Such people had to be either funny or co-opted or helpless victims. They were seen to be irrational people who had to move on, be cast out, or even destroyed so that society's norms would prevail.

The analogue to the Mexican in Canada's western lore is "the Frenchman." The stereotype is that of a conquered, very emotional peasant (habitant) or priest, sometimes married to an Indian or Métis. *Cariboo Country's* Frenchie Bernard was not a stereotype, but a fully realized character like most others in the series. "Frenchie's Wife" (17/2/66) is one of the best episodes of them all with its mixture of laconic love, grinding poverty, jealousy and never fully revealed secrets (who Frenchie's wife really was is never fully explained), its sparse dialogue and evocative images.

In *Cariboo Country*, the old are honoured. The character "Ol' Antoine," played by Chief Dan George—then new to the craft of acting—ensured that Indian respect for age and wisdom was credibly presented. Morton Dillonbeigh, his friend, is nearly seventy. Age is a subsidiary theme in episodes like "A One Man Crowd" (23/5/66) and "Ol' Antoine's Wooden Overcoat" (13/5/66). A new character, the recluse in "One Man Crowd," is particularized as an old man who exploits his deafness to avoid unwanted conversation, cadges drinks, is grossly rude to all women and alone in the world, having lost two sons. This is just the sort of friend Morton would have. He is a crotchety old survivor loyal to a code of his own. He is not an outcast, mad, a victim, or used as comic relief.

WHAT CARIBOO COUNTRY IS

Cariboo Country was very specifically regional in its setting, plots, characters and dialogue. Namko is not Dodge City or Tombstone, or even Moose Jaw or Calgary. Canada's intense regional rivalries emerge in this series in a variety of ways. All other places are referred to as "outside." Vancouver is a seductive lotus land for a number of series' characters, who are tempted to sell out and leave. They never do. In "The Sale of One Small Ranch" (3/2/66), agribusiness is kept at bay. In "Morton and the Slicks" (20/5/65), it is the American sportsmen who lose out. Another episode satirizes the CBC itself, when a film crew from the city comes to capture some background colour for a celebrity who came originally from the Chilcotin. The comic results of their efforts to fit Namko and its people into the limitations of a documentary clip defines among the "outsiders" in a nice bit of self-reflexive irony, the CBC itself.

In "Sarah's Copper" (7/4/66), Sarah's priceless heritage is symbolized by the highest-ranking object in her coastal tribal culture—a very old, pre-contact, shield-shaped object made of native copper, which she and her husband Johnny want to sell to make a new start. The Vancouver antique dealer, however, has a cupboard full of them owing to the poor fishing season and refuses to buy. Besides, the copper will not sell as well as poor copies of old masks carved for the curio trade. He has the name of a private collector who may want the copper for his wall, but when Sarah sees the collector's modern penthouse and senses what sort of meaningless trophy it represents for him, she performs the ultimate act of humiliation her culture can inflict on an enemy; she breaks the copper in front of him, thereby purchasing for all time the highest honour the old culture afforded, and making a different kind of "new start" for herself and her husband.

In another episode, "All Indian" (3/3/66), Sarah's coastal customs clash with those of her husband to the confusion of all. She is "kidnapped" by the Spirit Dancing Society because to be unwilling is an essential qualification of a spirit dancer. It is both a burden and an honour, and above all, a calling. Her inland husband eventually tracks her down, but finds out as he watches her dance that he cannot be part of that phase of her life. Thus, the viewer and Johnny are placed in the same situation. Not all misunderstandings run along racial lines; that would oversimplify the complex lives depicted. In this episode, all the actors were members of various Indian bands. The fact that many of these amateur actors were related to one another may have helped them achieve their very convincing performances. It is part of the relationship between St. Pierre and Keatley and the people whose customs they show that the audience for this episode is specifically told that no sacred dances are actually shown.

When Bonaparte John kills his wife after talking all night with his clan totem, a wolf, the tragedy is partly the result of a clash of cultures resulting in a displacement of values, and partly an archetypal triangle. Human jealousy (she is with another man) also contributes to the murder, and so does the fact that she blames Bonaparte John's mixed blood for her baby's withered arm. Yet elsewhere in the series he is presented as a rescuer and a chief. In this episode, the voice of the old beliefs is present and powerful. It is not treated as a decorative piece of nostalgia or a supernatural device to pump up suspense, but simply as part of the intense, almost surrealistic atmosphere of the piece.

Most critics of the Western acknowledge that, among its various stereotypes, Indians are a special case. For them, the deck is already stacked: they are feared by audiences, they are perceived to have mysterious knowledge about the natural world, and the white middle-class viewer is likely to be guilt-ridden about them. Also, "we" think that "they" hate us. Historically, however, Canada did not have Indian wars on the western Plains. The Six Nations were our Loyalist allies in the Revolutionary War and later the War of 1812, settling alongside the other Loyalists in tracts set aside for them in Upper Canada. Our earlier French-Indian conflicts have not been mythologized. Moreover, we handle our very real guilt somewhat differently from Americans. In fact, *Cariboo Country* looks directly and often at the tensions between native peoples and whites. Ironically the CBC focused more on this issue, in the 'sixties, than in the more "liberated" 'seventies and 'eighties have. As we shall see, only intermittently in *The Beachcombers* does Canadian television consistently explore the tensions, losses, and gains of contemporary native culture.

Regarding the role of women in the American television Western, the Brauers, among other critics, point out that "no married woman on the TV western is shown doing anything besides cooking or 'taking care of the kids' " (p. 188). Unmarried women are suspect, although admittedly, there are a few scattered superwomen and tomboys. Another familiar stereotype is the meddling female from the East, always interested in gun control or some other impractical reform but who does not disarm or change anyone. St. Pierre's depiction of women in *Cariboo Country* is years ahead of its time. Cariboo women are not passive victims awaiting rescue, or whores with hearts of gold, or stolid "kinder, kirche, kuchen" backdrops to the action. They sell the food at the rodeo, put the babies safely in the cloakroom at the dance, where inevitably, the men drink too much and get sick or silly or belligerent. They shoulder their full share of responsibility and the heavy work on a working ranch, leave and come back, like Frenchie's wife, when the isolation becomes too much, or, like Smith's wife Norah, endure and stick it out despite endless winters, unexplained absences, and idiosyncratic neighbours. Norah stays with Smith as Smith stays with his small ranch, to see what will happen next—and out of unspoken love for him as much as he unconsciously loves the land and her.

The traditional masculine values of the Western are often called into question, deromanticized or subtly undercut. Hard drinking, feckless, nearly seventy years old, Morton has been the most inept rancher in Cariboo country. In the episode, "Morton and the Slick," he switches babies at a party and in the process finds himself with a handicapped half-breed baby. For some reason, his wife Mrs. "D" is willing to take on another child (beside Morton), even at that late stage in their lives. Her fondness for Morton is clear in gestures and subtext, but never stated in the dialogue. In a later episode, she is shocked by his death—he has insisted on riding and is trampled in the rodeo—but stoically carries on, saying at his grave simply that she will miss him; she is not the least bit sentimental about it. In the aftermath, with a four-year-old son to care for, she refuses the ranchers' awkward offers of assistance, but does accept as her due the help of the completely misogynist recluse who was her husband's friend. She is clearly a woman with a strong will of her own.

Other aspects of the masculine world are not romanticized either. Arch's beer parlour is typically 'sixties Canadian, with all types of regulations on the walls, ridiculous hours, no music, beer out of bottles, cases stacked all around, and women and unaccompanied men separated as the law then required. Arch does this by drawing a white line on the floor, walls and ceiling. His bar is complete with pickled eggs and comfortable chairs. The drinking is as macho as it is in any Western or for that matter in the rest of Canadian life, but it is not made glamorous—though we are told that Morton is the only man in Chilcotin who drinks Australian rum, and the only man who can.

Fully accepted in the community, Morton has aspects of the licensed fool about him. His little joke of switching several babies at the party amuses many of the men, though not the women. As he treks around swapping the children back, one baby is left over like a "slick"—an unclaimed cattle beast without a brand. The conventions of the usual television Western would dictate that he and his wife adopt the child. This does happen, but not before two specific points are made: first, that the Indian band will always give a homeless child a home, calling into question whether the solution offered by the plot is even necessary, much less ideal from a cultural standpoint; and second, the baby has been deserted by Bonaparte John's wife because it has a withered arm. Bonaparte John himself is now in jail for shooting her: the tragedy here is cross-cultural and again particularized. The comic tone of this episode is coloured by the tragic tone of the earlier one. In a later episode, when Morton has been crushed by a rodeo horse, Ol' Antoine intervenes to protect the child, now four years old, from the sight of his body. Thus people's lives and the passage of time intertwine in this long-running series.

The dialogue in *Cariboo Country* is also crisp, idiomatic, particularized; for example, when Morton and Larsen discuss whether Morton should try to adopt the baby:

> Morton: Bonaparte John's kid. Bonaparte was a real man.
> Larsen: He was a good Indian all right.
> Morton: [without undue emphasis] He was a good man.
> Larsen: Yeah, I kinda liked Bonaparte. If he'd a gone white he'd a been a big man in this country. But they never do. They always drift back to the reserve.
> [The stage directions represent my interpretation of the performances, and are not derived from a script.]

The baby's inheritance is described by Larsen with unselfconscious prose-poetry as a "few square miles of grass with the wind blowin' on it . . . some Jack Pine. . . . Swamp and the coyotes own it. Just grass with the wind blowin' on it." Mrs. "D" replies that "the Bible says men's days is grass. . . . So all he inherits is a shack on the reserve." The point is then made that a ranch cannot be said to exist in the Chilcotin unless at least one generation of men have worked on it: "She's a cruel country. You make a joke at this country and she turns out meaner than a bitch wolf in pup," says Morton.

Inevitably, it's Mrs. "D" who makes the decision to keep the baby. Morton's response is typical of the series: "Hell of a late time in life to get pregnant . . . after twenty-seven years she presents me with bow-and-arrow blood." "That Indian blood," says the realistic Mrs. "D," "means no retirement to Cloverdale [i.e., "outside"]. I'm taking no Cariboo slick out amongst the old milk cows in the Fraser Valley." Morton agrees. This means he will have to phone the Americans who want to turn the ranch into a hunting base and be evasive. If that doesn't work, "I'll tell them to go to hell."

By now, it is apparent that writer Paul St. Pierre and the actors have closely observed the idiom, accent, and dialect of the Chilcotin locals and that the CBC afforded the series an authenticity of language rare in that day on either side of the border by including both individual and idiomatic usage, so that the dialogue sounds like it is derived from the language of adult working men and women, both white and Indian.

Costumes are stereotyped in every TV Western—Paladin with his all black gunslinger's image, Bat Masterson's dandified outfits, Kitty's saloon owner's finery, and the standard gingham of the rancher's wives. In *Cariboo*

Country, each five gallon hat is subtly different, and the Indian western hats are different again. All the cowboys have work shirts and jeans that look as if they had been worn by that particular individual for years. In "The Education of Phyllistine" (1964), the final alienation of Phyllistine, a girl without family who lives with Ol' Antoine turns on what the white rancher's kids wear to a school party. When she arrives reluctantly at the party, she is wearing a cast-off dress. A stupid, though unmalicious, girl announces to all present that the dress once belonged to her. Phyllistine walks out the door leaving her dumbfounded, ignorant city teacher, the school, and formal education of any kind, with all this implies. Perhaps by default, she chooses her people's freer, and in many ways more valuable, way of life. Yet the viewer is forced to acknowledge the cost in self-esteem for this shy and gifted child. When Ol' Antoine reaches a hand out to her, she shakes it off with, "Leave me 'lone. You ain't nothin' but a dirty Siwash, same like me." She has learned, among many other things, that clothes do make the child to some degree, even in the Chilcotin. She has also learned her first lessons in the self-hatred that threatens the lives of so many of her generation.

Cariboo Country was the first CBC series to be derived from the NFB's documentary camera style, thus tapping the country's only indigenous film-making tradition. The CBC had no drama film units at the time. News and current affairs teams learned on the job, as they did in Regina, Calgary, Edmonton, and St.John's later when regional television drama was in the works. The camera shot is sometimes jerky, looks as if it is often hand-held, is not always carefully framed. The image itself is grainy and edited with a more uneven rhythm—an appropriate visual vocabulary which matches the characteristically open-ended plots and idiosyncratic characters.

In 1963, when they started to shoot on film in the Chilcotin, the whole company and crew also had a continuous check on their authenticity, because for weeks they were part of community life. Ice on the puddles in July instead of the balmy Vancouver summer added a distinctive flavour to the series' look and sound.

Cariboo Country does not use music as standard formula television uses it, that is as a bridge, as atmosphere, to heighten suspense, convey sentiment, create the illusion of action to underline the moral stance of characters or, as in older forms of melodrama, to pad the story or tell the viewer how to think and feel. Typically, there is no lone harmonica or guitar at Morton's funeral, no strings, just the ragged, uneasy voices of men unused to such rituals ("He's the first white man who didn't die outside") in a half-remembered version of "Don't Fence Me In" which, a neat touch of dramatic irony, comes from a popular American musical about the West. However, our last vision of Morton's grave, as the credits roll, is of Ol' Antoine singing slowly and with great power an Indian chant for the dead.

The last image is particularly memorable, not least because the song Chief Dan George sings over Morton belonged to him personally; it was the death song sung by his father over his mother, and somehow that authenticity reads through. It is also worth noting that for Chief Dan George to sing it in this context confirmed his trust in the integrity of *Cariboo Country*.[10]

The *CBC Times*, paraphrasing the show's creators, said of Ol' Antoine: "There is a rumour abroad that he will never die, but will, on hearing Gabriel's trumpet, just disappear in a puff of his own dust." Usually, Ol' Antoine discusses the dust or the heat, the lack of deer meat, his lack of horses, his shortage of food or the general shortcomings of the federal government. In his view, the government is like God: "[You] never get to see Him." It is not always clear to the actor (or the director or the writer) when he is living fully in the present and when he is somewhere else. This ambiguity lends the character a considerable resonance.

Describing new episodes of the series shown in 1966, the *CBC Times* (29/1-4/2) declared that, the Chilcotin is "the only place in the world where white settlers have assumed the mannerisms and characteristics of the original natives, rather than changing the Indians to the white man's way of life, probably because the Indians have found a way of coping with the landscape and its climate." The Chilcotin is not, we must note, the open, arid grandeur and epic sweep of the American plains. *Cariboo Country*, with its bitterly cold, long winters, the enclosing mountains with the dry plain between, and the plants and animals of the north is not a land where, as Lohof puts it of American western iconography, we "escape not from the responsibilities of civilization but from its frustrations."[11] Namko is very short of Marlboro men with their "innocence and efficacy." Contrast St. Pierre's description of Smith: "Even his wife doesn't know his first name. . . . He has the smallest herd of cattle of any Cariboo rancher and fully realizes that eventually he will go bankrupt. However, Smith is interested enough in life to stick around and find out what method the fates have chosen to send him down the chute."

The Canadian West, and particularly the Chilcotin Plateau, is not "an environmental memoir," pastoral, pragmatic, vigilant, full of action, purpose, and conflict (as Lohof defines Marlboro Country), perhaps because neither our western past nor our Indian wars have been made to serve as ideal, essentially optimistic prisms of our past.

The West As We Saw It

Even in the 'twenties, when our cultural voice was still a whisper, we did not produce popular novelists like Zane Grey, but rather Presbyterian minister Charles William Gordon writing under the pen name Ralph

Connor. He wrote well-loved books about pioneer life like *Glengarry Schooldays: A Story of Early Days in Glengarry* (1902), heroic tales of the British in the War of 1812, and books like *The Prospector: A Tale of the Crow's Nest Pass* (1904)—not, as the title might suggest, a tale about the Yukon gold rush, but the story of a Victorian missionary who is also a first class University of Toronto Blues football player and a prospector for the souls of miners, ranchers, and cowboys of the west. Connor in the early 1900s and St. Pierre in the 1960s both had a firm sense of identity and place.

Canadians have always envied the legends of the old American West but not at the price of what the *British North America Act* calls "peace, order and good government." The thrust of the *Cariboo Country* series is that government is distant, eastern and a nuisance, but that a Canadian is likely to subvert it rather than defy it.

Posters recruiting immigrants in Europe during the early years of the century called Canada, "the last, best west." This was not the old, wild West of the penny dreadfuls. Our Northwest, sparsely settled by Indians, had been opened up by fur traders, the French and then the Scots who often married their Indian wives. The next wave were Ukrainian, German, Scandinavian, Slavs, and other European peasants. The buffalo were gone before most of them got there.

Canadians did mythologize their West, but rather differently from their American cousins. In the heyday of the early television "horse" Western, a 1954 Canadian grade school text painted the following picture. The prime minister "decided to trust the preservation of law and order in the North West Territories to . . . three hundred men, a tiny body to patrol such an emormous territory. 'There will be', said Sir John A. Macdonald, 'as little gold lace and fuss and feathers, as possible.' Almost the only military feature of the Police was its uniform, which, with its scarlet tunic, was intended to impress the Indians, many of whom had a kindly memory of the red-coated British regulars [and] to remind them that the Police were, in fact, 'soldiers of the Queen.'"[12]

More modern versions of our West included W.O. Mitchell's very popular radio and television series *Jake and the Kid* (1961 on CBC television), which was purely and memorably anecdotal, all about Prairie farming life in the 'forties and 'fifties. There was also lots of country and western music, including the fall replacement for *Cariboo Country*, *Red River Jamboree*. The CBC also did a few other plays that might be called Westerns, about which very little information has survived. A play called, "Man with a Rope" (5/4/59) was advertised as the first CBC "adult Western" and remade for *Playdate* (March 1964). Both versions starred Jack Creley and received excellent reviews. Hugh Kemp's "The Cowboy and Mr. Anthony" (7/3/63), produced by Norman Campbell for *Playdate* (1961-65), starred

Ian Tyson and Sylvia Fricker, then just reaching the status of folk-country stars. In this light comedy musical, a rodeo rider is hurt and the devil makes a deal with his wife—country music stardom for her husband in return for her soul. The cowboy sells out to Hollywood and the various kinds of devils there. However, his wife plays Gospel music and saves them both. In the 'eighties, Nelvana made a cartoon about mice aspiring to be rock and roll stars using the same plot. It would seem that in Canada, we turn aspirations for fame in the U.S. into variations on the prototype of *Dr. Faustus* although, as W. O. Mitchell's popular and enduring fable *The Black Bonspiel of Wullie McCrimmon* demonstrates, Canadians might also sell their soul for a shot at a sports trophy.

As the Western movie faded and the television western reached its zenith, our ambivalence towards American imagery of the West was caught in this brief excerpt from Al Purdy's book of poems, *The Cariboo Horses* (1965):[13]

> At 100 mile house the cowboys ride in rollin'
> Stagey cigarettes with one hand reinin'
> Restive equine rebels on a morning grey as stone
> —so much like riding dangerous women
>   with whiskey coloured eyes—
>   ... But only the horses
>   waiting in stables
> Hitched at taverns
>   standing at dawn
> Pastured outside the town with Jeeps and Fords and
>   Chevys and
> Busy muttering stake trucks rushing
> Importantly over roads of man's devising
> Over the safe known roads of the ranchers
> Families and merchants of the town

The poem juxtaposes the glancing metaphor of wild, wild women, dangerous as the horses that are put on Earth to be tamed by heroic cowboys and the reality of the Canadian West in the 1960s. Audiences watching *Cariboo Country* also saw images of the American West colliding with Keatley-St. Pierre's present-day Chilcotin side by side on the dial.

Peter Pearson's film *Paperback Hero* (1973) took the clash of pastoral image and modern reality in the Western even farther. This classic Canadian film shows how the impoverished language, lack of employment, and the daunting landscape of a Saskatchewan town push the aging town stud and "top gun" on the hockey team farther and farther into the fantasy of being

Gunsmoke's Marshal Dillon (American actor Keir Dullea). When the team is shut down by the town's businessmen and his success with women diminishes, he provokes a confrontation with the RCMP in a desperate move to make everyone "remember me." It is unclear whether he is "only playing," as the town "easy girl" who loves him says, but his sudden bravado move of shooting out the red light on the police car precipitates a real shootout. The local constable fires into the air. He fires at the constable. The other RCMP officer kills him. The last image is well known in film circles: a shot from the ground up, low angle . . . the empty foreground . . . the body in long-shot and, behind it, the grain elevator with the company name PIONEER painted on it.

The ironic lyrics to the song "Paperback Hero," written by Gordon Lightfoot for the movie, make the point that the protagonist has lost the feeling and just can't get it back. In Canada, that John Wayne Western feeling was never ours to begin with. Hockey heroes, yes; gun-toting marshals, no. To me, these examples suggest that Canadians cannot produce conventional Westerns of any hue any more than the Italians or the English, just as we cannot produce "mean street" copshows.

In reminding his readers on 4 February 1966 about the history of *Cariboo Country*, Roy Shields, TV columnist for the *Toronto Daily Star*, said: "Paul St. Pierre . . . is a playwright with the instincts of a poet. Though his plotting is somewhat careless [I would dispute that from the episodes I have seen], he has that rare ability to catch the mood, mind and speech of the simple people with age old problems in a land that makes the old West antiseptic by comparison." (Simple? Ah well, it is a Toronto newspaper.) Jon Ruddy, his rival on *The Telegram*, thought otherwise. In a review of *The Serial*, where many of *Cariboo Country* episodes appeared, he wrote (15/4/64) that it was "a cut-rate, low-aiming excuse for TV drama." The remark was typical of that newspaper's general hostility to CBC drama of all kinds. Fortunately, the audience was more perceptive and continued to watch.

The Strong Ones

I have chosen to analyze one example of the series in detail, a late episode called "The Strong Ones" (10/2/66). Although the central narrative is set in 1910 as a flashback and is thus atypical of the series, I chose it, in part, because it is a link with the more conventional Westerns of movies and television. Even here, however, the conventions shift and the documentary flavour is preserved.

One of the traditional plots in Westerns deals with the key moment of crossover, the meeting of native and white cultures. This is often counterpointed by a clash between western and eastern values. Adding to the thematic richness in this episode is that it is set in 1910—the moment when the first-generation tamer of the land meets his bride, affirming his right to sons and daughters, inheritance and possession.

Philip Keatley feels very strongly that this episode could not, in twenty-five minutes, encompass all the potential cross-currents in this story. He points to it as the crux that caused the St. Pierre-Keatley partnership to go after an hour-long series, an expanded Cariboo Country. They did not get it. They did make *Sister Balonika*, for *Festival* in 1969. It was shot on location in Richmond, B.C., as well as in the studios. The story is an interesting one about a native Cowichan nun in a not quite godforsaken Indian school. Nevertheless, the play did not work, partly because it was miscast.

St. Pierre was afterwards elected as a Member of Parliament, breaking up the collaboration. Keatley and St. Pierre have only worked together a few times since. "The Strong Ones," then, represents a turning-point in their thinking about the series. As far as the episode itself goes, though it could have been expanded to an hour without padding, I think it worked well in the more compressed time period.

The framing narrative, set in the present, is as important as the central story. It draws our attention to other ironies in the complex relationship between the white settler and the half-breed cowboy who is both his son and his employee. As Keatley says, most cowboys in the Chilcotin are actually Indians. We first meet a restless, young, white cowboy who looks much like the "kid" of the traditional Western. This "kid," however, is the half-brother of Peter James, the now fifty-ish half-Indian son of rancher Robert James. He is working for the narrator, Ken Larsen, just as Larsen's father once worked as a foreman for Robert. Ken Larsen is a modern, successful rancher who serves as the social norm of the series. He is sometimes, not always, a leader who can make irretrievable mistakes, as in *The Education of Phyllistine*, which he also narrates.

Talking to Arch McGregor, Larsen remembers living as a child with Robert James and his Indian common-law wife and family, including young Peter. His story centres on the arrival of Robert and his new eastern bride at the ranch. Through flashback we find out why Peter, the older Indian cowboy, has been willing to lend his young white half-brother the money to make a temporary escape to Williams Lake. Here, St. Pierre inverts two of the most common Western stereotypes. The Indian is the responsible member of the community. However, the now aging Peter recognizes his half-brother's need to escape responsibility, even if the kid will once again land in jail. The same kind of inversions also reverberate through the central situation.

In the flashback, Ken Larsen reappears in the privileged position of the innocent boy, a conventional figure in movie and TV Westerns. He watches, listens, and earns his right to observe the adults by taking a licking from his father in front of the lovely new bride, Mrs. James, for being where he shouldn't be. One of the film's major questions is, in fact, whether she is where she should be. Young Ken's stoicism echoes hers when she is confronted with a situation as strange to her as she is to the motherless child.

In the script's narrative structure, the freedom of the old ranching West, the bachelor existence symbolized by the mountains and the unkempt garden meets the constricting female East with its ignorance of the western facts of life. These are signified comically on the natural level by cow dung on the path to the house and seriously on the much more complex social level by Robert's Indian common-law wife and family. Also implicit in Mrs. James English origin is a clash between a scarcely articulated nationalism with the mythology of Mother-country. Again, the stage directions represent my reading of performances and camera shots.

> Jeremiah Larsen: This is Mrs. James. (with a slight emphasis) This is Peter.
> Mrs. James: (warm smile) How do you do, Mr. Peter.
> Peter: It's not Mister Peter. It's Peter James. [closeup of white glove in brown male hand] "How do you do" (in a slightly bemused tone). [Leaves].
> Mrs. James: [to Jeremiah] Do they often take the name of the rancher for whom they work?
> Jeremiah: (deadpan) It ain't uncommon.
> Mrs. James: My husband has no gardener.
> Jeremiah: (still deadpan) No ma'am, no gardener.

Eventually, she learns to use cow pats to fertilize the garden she is determined to plant.

Her adaptability also becomes evident in a much more important area of ranching life. After the initial shock and a confrontation between Peter and Robert which nearly ends in tragedy, she learns Peter's identity as Robert's son. She then insists that the cowboys pay respect to the other "mistress." "Mistress" is the only word she has in her vocabulary for someone like Peter's mother, who is not a wife, ludicrous though it sounds to a viewer remembering the dignity of the figure on horseback glimpsed earlier. As Mrs. James prepares to mother her husband, young Ken Larsen, the Indian family, and her own family-to-be, she clearly symbolizes the new West. The

Canadian response to the American Western was the West of *Cariboo Country*; problems were tackled courageously, but not wished away into a fantasy world or resolved in the last thirty seconds of a half-hour play.

Throughout the episode, the camera quite unobtrusively takes the perspective of various characters; it also frames the second Mrs. James against the unkempt house as she tries to deal with the knowledge of her husband's wife and children, against the fence as she faces her dilemma, and against the trees as she decides to stay.

In the paradigms of opposite terms that give the film its thematic structure—East vs. West, Indian vs. white, 1910 vs. 1960s, owner vs. employee—the eastern English bride mediates and resolves conflicts. She is not the protagonist of this narrative, however. The central figure is sixteen-year-old Peter, eldest son of Robert James. Peter forces the issue. Unarmed, he confronts his father in, appropriately enough, the corral among the now-tamed horses. He tells his father that his mother will live in a house on the ranch and not somewhere out of sight, whether Robert likes it or not.

> Peter: No, it's you that's got to understand. You married that white woman. That's the way it's got to be in this country now. Ain't nothing you can do that thing. I understand that.
> Robert: (relieved, about to dismiss the issue) Thank God for that, boy.
> Peter: But you've got to understand something. My mother, she comes back to this place. She lives right in that yard over there. You build house right beside that big house you got that new woman in.
> Robert: Believe me, that just isn't possible. It can't be. Look, will you trust me?
> Peter: [extreme closeup] All you got to do is understand. That's all. You've just got to understand. My mother, she comes back to this place. She comes back today. I bring her back and you put her in that place.
> Robert: (reasoning, almost pleading) Peter, Peter, Peter, my boy. . . .
> Peter: (completely serious, intense yet matter-of-fact) Suppose my mother don't come back this place. Then you stay this place, right in the ground of this place.

The adult in this scene is not the father, who did not honestly face the situation he had created, but the son who squarely confronts the implications for them all.

When Jeremiah, Ken Larsen's sternly patriarchal father, hears this conversation, he draws his gun on Peter in classical Western style, but shoots Robert by mistake. Then he fires several times at the agile youngster and misses, only to find himself at the barrel-end of a shotgun grabbed from a saddle-bag and wielded by a determined, skilful, coolly ironic Peter.

Having tried to change the old, racist, sexist code as much as he can in the context of that time, Peter finally extracts his father's shamefaced promise to bring his mother back. He accepts Jeremiah's face-saving lie about "only fooling," thus repudiating the vengeance code of the classic Western. He does not humiliate Larsen senior in front of young Ken as Robert had humiliated him. Instead, by giving Larsen back his bullets and gun, he creates a chance to heal the breach. Thanks to Peter's adult refusal to embarrass Jeremiah, he makes his peace, peer to peer. Robert James, who has created the mess by not telling his bride about his Indian family, is by far the weakest figure in the central narrative, as is his all-white cowboy son in the 1960's framing narrative. In the overlapping triads of "Peter/Jeremiah/Robert" and "Mrs. James/Peter/Robert," Robert James, the strong male entrepreneur and dashingly handsome figure is the only one who does not really change. The stereotyped protagonist of the classic "adult" Western is neither hero nor anti-hero in "The Strong Ones." Instead he is transformed into a subsidiary figure in a time of transition.

The Indian side of Peter James, whose ancestors were there before the ranch and the intruders, is the first to recognize the racist realities, the first to try and change them. Peter is the true mediator of the paradigms. He reprises the major theme of the series as a whole, that is, how the old traditions and the new, the divergent values and perceptions of the people who now live together in the Cariboo, relate to one another. Peter lives to see young Ken Larsen become a successful rancher in the Chilcotin employing his landless, restless, younger half-brother, Robert James's white son.

As a sixteen-year-old in 1910, Peter represented a new generation. But both brothers are repositories of unresolved tensions. Ken Larsen, like his father Jeremiah, represents the old values and the human capacity to adapt within limits to new circumstances. Ken tells the tale with a perspective far less self-consciously ironic than the double narrative structure itself: "So we all grew up together and me that never had any mother. Oh, she was dumb as only an English woman can be dumb, with her classical music and her books." Yet perhaps he does sense the irony, because his tribute to Robert's eastern wife also includes the other mother by inference; throughout the story, he refers to the English bride as the second Mrs. James.

"The Strong Ones" uses some of the conventional themes, characters, plots, and archetypes of the classic American television Westerns, yet it is not derived from the typology of that genre. Rather, it is typical of most episodes in the *Cariboo Country* series.

In Namko, we do not find Marlboro Country or a Paradise Lost or a kingdom of noble savages. There is no hope of recoverable innocence, no urban nostalgia for the sanitized pastoral. We are not encouraged to identify with fast-draw lone wolves who save ungrateful towns or trail bosses or other strong father figures who guard, guide, and occasionally cleanse their communities. The adversaries are not rustlers, outlaws, or gamblers, but rather the climate, the cold, the hard economic facts of ranching in the 1960s, interference from Ottawa and the attempted takeovers of a heritage that is both a burden and a treasure by antique collectors, agribusiness, American deer hunters and "Outsiders" generally.

Unlike the vast majority of the white hat-black hat Westerns already flooding the air waves at that time, the inhabitants of Cariboo Country win some and lose some. *Festival* presented the last of the series in the form of a one-hour drama, *How to Break a Quarter Horse*. A CBC press release (1/1/66) uses dialogue that appears in the film. Smith summarizes the situation thus: "There is a rancher in that country (his name don't matter because he's never goin' to trial anyhow) and Gabriel [an Indian], see, had a piece of land there and this rancher wanted it. So the rancher arranged unlimited credit at the bootlegger—and Gabriel stayed drunk for six months. And one day he happened to have a rifle and he happened to go to Haines' cabin, and finally he happened to shoot Haines. And now the rancher has got the piece of land he wanted and Gabriel is out of the way forever," on the run. He takes refuge on Smith's land, where he is talked into surrendering, and a very reluctant Smith gets drawn into the situation.

Chief Dan George, playing Ol' Antoine, is Gabriel's only witness at his trial. However, instead of giving "relevant" evidence on the witness stand he tells the story of Chief Joseph of the Nez Percé Indians one hundred years before and the peace he made with the American government. Even so, the judge, a long-time resident of the Chilcotin, sees the true relevance of his testimony and releases Gabriel. Before this "happy ending," however, Smith, who got Gabriel to surrender in the first place and thus broke his inviolable rule never to interfere with the Indians, attacks Walter Charlie, the interpreter, because he thinks he has betrayed the old man by conveying his reminiscences. He then defies the judge, who thinks he is a drunken "drug-store cowboy." Gabriel goes free, while Smith gets a fine and three months in jail. Whether or not Ol' Antoine will ever break the quarterhorse is left an open question. Typically, Smith never tells his wife why he is away but instead spins a yarn about cutting timber for three months. However, the Indians do get his hay in for him. This wry blend of comedy and irony with overtones of tragedy was typical of the series and a fitting end for it.

THE BEACHCOMBERS

The Beachcombers (1972-present) had a very different beginning from *Cariboo Country*. By 1971, Vancouver's film unit under the eye of Philip

Keatley had done not only *Cariboo Country* but also *The Manipulators* (1970-71), a series about a pair of social workers which lasted only two seasons. Three needs were apparent to the CBC brass: the Vancouver film unit needed something to do; the corporation needed a "kidult" or family-adventure series to replace the popular independently produced children's series *The Forest Rangers* (1964-66) and *Adventures in Rainbow Country* (1969 and 1970) in an early prime time slot; and the west coast should also be visible in prime time. The CBC described the kind of programme they were looking for to Keatley. Marc and Susan Strange then came up with the idea of a series about beachcombers, featuring a Greek with a zest for life named Nick Adonidas. As an experienced CBC producer, Keatley started to develop the idea into a workable series that would follow the lines specified by the CBC and perhaps retain an imprint of his individual interests and concerns. Unlike *Cariboo Country*, which came out of anthology in the context of and, perhaps, in reaction to the American Western, *The Beachcombers* was intended to be a recognizable variation on a genre which could easily be sold, would fill a gap in the schedule, and might later be exported as *The Forest Rangers* and *Adventures in Rainbow Country* had been.

The Family-Adventure Genre: The Basic Formula from the 1960s to the Present

The formula elements of most family-adventure series are easy to identify. Earlier Canadian examples from the 'sixties, *The Forest Rangers* and *Adventures in Rainbow Country* were followed, after *The Beachcombers*' success, by *Matt and Jenny*, a series about two kids searching for their parents under the protection of a young, handsome bachelor in nineteenth-century Ontario (shown on Global and independent stations), and two CBC co-productions, *Ritter's Cove* (1979-81) and *Danger Bay* (1984-present), both set on the West Coast.

Typical of a formulaic approach to family-adventure series were the characters and setting of *The Forest Rangers*. Here, the "wilderness" is not dense bush, wolves behave like creatures from Russian fairy tales, and the natural world is a backdrop for an adventure for three children who differ in age but are barely characterized at all. The leader of the trio, Pete, is older and more responsible, Chubb is chubby, earnest in a situation and sometimes resourceful, and Cathy, as Arthur Black said in a session of reminiscences with the three actors, "sprained her ankle and had to be rescued a lot."[14] Despite the charm of her infectious giggle, the young actress was caught in a classic stereotype of girlish dependency in contrast to Margaret and later Sara, the young girls in *The Beachcombers*.

The Indian guide in *The Forest Rangers*, who disappeared after a while, was knowledgeable about the woods, monosyllabic, and another responsible adult to get the kids out of difficulties. Pete, the Indian lad in *Adventures in Rainbow Country*, was not given much dialogue or distinctive characterization. In at least one episode, however, "La Chute" (no date available), his knowledge of his own tradition saves a priest who is seemingly lost in the rapids but actually washed up on a ledge. This episode had several nice touches: Etienne Brulé's initials on the ledge proves that Brulé had indeed been swept down the rapids, despite the odd reticence about how he made the passage in his diary. Pete's explains the fact that Brulé's journals did not mention the incident by arguing that he was humiliated that an Indian saved him by shooting the rapids in a smaller canoe just as he and his friend rescue the priest who was swept over (went over deliberately?) the rapids. As well as good action photography and a tight plot, this particular episode adds some texture and nuance to the all-male environment.

Very little has changed in the conventions of the family-adventure genre over the last twenty years. The newest version is *Danger Bay*, a CBC co-production with Disney Studios. A plot summary from *TV Guide* (25/3/85) pinpoints the kind of emphasis on action formulae that weakens the series: "Jonah [the son] is kidnapped by a team of salmon poachers." The woman scientist in *Danger Bay* is largely a beautiful backdrop, and the woman bush pilot performs the traditional twentieth-century version of appropriate female service: chauffeuring. Naturally, she stays back if there are bears or mad hunters. Nicole, the young daughter in the stories, nurtures animals and tags along while her brother gets blinded exploring, finds an otter, or steals a mask. The masculine authority figure, played by Donnelly Rhodes (*Sidestreet*), is Dr. Grant, a veterinarian who specializes in marine life at the Vancouver Aquarium. He is a widower, of course, to allow for transient love interests. Rhodes does give the part a natural authority and charm. He is also comfortable with the blend of action and fathering required by his role, though perhaps more comfortable running, chasing, fixing, rescuing, and trying to get some truthful acting into completely stereotyped, sentimental situations (e.g., father and blind son "face the future" when the audience knows absolutely that the writers and producer will not blind the young male lead) than he is with his role as learned colleague and specialist at the aquarium.

Northwest-Coast Indians do not fare very well in *Danger Bay*. In "The Mask" by Michael Mercer, Jonah suggests to his Kwakiutl friend, Dennis, that they liberate an original mask from the museum as a potlatch gift for Chief David (Dennis') grandfather. Jonah masterminds the theft. The episode could have had the resonance of *Cariboo Country*, or some of *The Beachcombers* episodes, but it does not. Potlatches and masks are explained, though in a somewhat contrived manner, and the culture is treated with respect. When a fragment of a potlatch is shown, the details are close enough to the real thing, though there is no attempt to evoke a fullblooded feeling of ritual or celebration. Unfortunately, the true sacred and social significance of the Thunderbird mask, however, is missing, and the whole issue of the return of native artifacts from museums to their original uses is skimmed over. The potential of the basic tensions arising from a complex situation is also vitiated by spurious hide-and-seek suspense and a ''comic'' subplot about a lizard at the aquarium eating the pennies thrown in his pool. Finally, like many episodes in the series, this one is plagued by dialogue such as: ''Like a hidden conscience, [the mask] will know your heart forever.'' In the end, the two boys repay the museum in cleaning services and, true to formula, a lesson is learned. The contrast with Paul St. Pierre's ''All Indian'' already mentioned is instructive.

In this episode, *Danger Bay* had a chance to inflect the formula using the clash of cultures on the Northwest Coast. In Disney's terms, it may be progress that Jonah does have an educated young Indian friend, and that Dennis's uncle, Charles Mungo (a reference to Mungo Martin, a famous carver?), works as a carver-curator at the museum. Nevertheless, the Kwakiutl heritage is used primarily as a colourful backdrop for a thoroughly predictable adventure yarn rather than to enrich the texture of the series. As will become evident, *The Beachcombers* is much better at handling the clashes and interweavings of the two cultures.

The conventional elements for family-adventure series are: a strong male figure (though *Ritter's Cove* in the more liberated 'seventies featured a woman bush pilot as protagonist with an older male); a motherly female (usually a secondary character, missing altogether from *Matt and Jenny* and *The Forest Rangers*); at least two children (usually one of each sex) with whom younger members of the audience can identify, and an interesting, unusual, or unfamiliar location. There will also be other running characters to interest the older audience and sometimes, but not always, a resident villain with some saving graces.

The plots must involve peril, exploration, and violence against inanimate objects or animals rather than people. Explosions, a favourite and overused device on *The Beachcombers*, fires, floods, storms, accidents, and threatening animals are a few of the formula story elements common to family-adventure series. Domestic problems of various kinds are also a continuing source of dramatic conflict. Usually such a series reflects a clear sense of the sorts of acceptable behaviour society defines and supports. The genre, therefore, often has plots that revolve around deviations from these norms, including serious crimes like burglary, arson and fraud. For other variations on this formula, comedy is essential, either slapstick or the more difficult comedy arising from clearly sketched characters in believable situations—preferably both, for broadly based appeal. The basic characteristics of the formula for successful family-adventure series are wholesomeness and entertainment with lots of action.

How the Beachcombers Developed—A Team Effort

In *The Beachcombers*, we are looking, not at anthology masquerading as series television, but at a series written to fit within and, perhaps, inflect a given formula. The chronology for the development of any half-hour series of twenty to twenty-six episodes in its first season has not changed since the late 'sixties. First, a target audience is identified by the network and a decision made about what kind of programme is needed to capture that audience. In this instance, the CBC decided to look for something fitting the "family-action-adventure" formula. Such a decision, setting up basic parameters, both reflects and reinforces the fact that audiences are conditioned to expect specific types of characters and plots in series television.

The CBC management added its own specific requirements to the general formula which I have extrapolated from the Canadian examples above. They made it clear to Keatley that they were in the market for a programme with a contemporary Canadian setting and characters, a strong father figure who was not biologically the real father of the kids (number and ages unspecified), lots of west-coast scenery, emphasis on action on both land and sea, a strong mother figure (variant on the usual stereotype), picturesque wilderness within reach of the logistics of CBC filming, and violence that was not directed at people. The corporation also specifically wanted a Northwest-Coast Indian boy, a teenager, in the series. With the exception of a grandmother substituted for a widowed mother, all these elements were present in the first and subsequent *Beachcombers'* seasons.[15]

From this general framework came a "treatment" which explored the characters and sketched out some plot ideas. Such a concept is always expected to be "fresh (but recognizable), have the potential for many story lines, and be geared to a specific demographic cross-section defined by age, class, sex, perhaps geographic location, urban or rural, etc. Looking at sixty-one episodes of this series, I find that *The Beachcombers'* intended audience consists primarily of children over five and their parents, together with anyone else looking for light entertainment, beautiful scenery, some action or comedy, sometimes credible plot lines, and likable characters.

When the early scripts had been reworked and filmed, the series was launched in a favourable time slot. While shooting continued on additional episodes, audience reports gave Keatley some early indications of what worked and what did not, and the series was fine-tuned accordingly. Whether it would be renewed or not had to be decided some time in January 1973 if shooting was to begin again in time for the 1973 fall season.

As with more complex, longer dramas, the whole production team contributed to the success or failure of the series. The directors place the actor against an evocative background, or direct a weakly written scene so that it will have depth and interest. The producer casts actors, chooses directors, and perhaps suggests the story line. The actors interpret the lines. The designer adds depth to a character seen only briefly. The episode "Jo-Jo" (7/1/79; wr. Merv Campone) is an example of this. Jo-Jo is a pathologically shy but famous painter; we don't see him—ever, but his character and talent are established by just the right tables and chairs, prints, and colour of paint on the walls. The designer is also responsible for obtaining the superb portraits that have to convince the audience that Jo-Jo is a genius. The editor decides when to give us glimpses of Jo-Jo and when to show Nick's reactions to him, building suspense by intercutting shots just as the writer does with his plot, or the designer with the dingy, dangerous compartments of the entrapping barge. The sound editor may heighten the rapidly mounting tension with a hint of lapping water. Each person makes a crucial contribution to the success of an episode and a series. Each can singlehandedly ruin a good idea, script, performance, shooting, sound mixing, rough cut, or fine cut. But behind it all is the script and its writer on whose dialogue and ability to tell a story the episode depends. In this case it was Merv Campone, one of the series' best writers.

The Beachcombers Evolves: First Season

Some significant changes to the original concept ensured the survival of *The Beachcombers* series, which got off to a rather rocky start in terms of audience acceptance—and management support. Thom Benson, Director of Entertainment Programmes (English), fought for the series until it found its feet.

Over the first season, the slightly raffish, rather lazy, cigar-smoking, woman-chasing Nick, as portrayed in the treatment and in early scripts, was transformed into a hard-working professional, a free spirit who does love the ladies in a discreet but by no means successful way, not lazy, and certainly not a "tomcat." Rather than the loner who thinks of himself as just passing through, originally envisioned in the initial concept, the character of Nick began to take a lively interest in other people's problems and work at solving them. He was still a dreamer, but a reasonable model for the kids of his surrogate family and the ones watching as well. His willingness to respect children for what they can contribute is a key to his character. Thus, in the

first season, Nick was both substantially different from the lovable rascal of
the early discussions and, in some ways, still the same. He evolved further
from the original concept as time went on. In the first season he is also more
clearly defined by contrast with Relic, a very successful foil and antagonist.

The original characteristics of Molly's grandson, Hughie (Bob Park),
who was featured for the first few seasons with his sister, Margaret (Nancy
Chapple), also changed substantially during the first season. What survived
from the initial concept remains a boy with an analytical turn of mind,
though not now bookish or bespectacled, preferring experiments to the
outdoor life. In several early episodes, echoes of the rather bookish kid do
remain. He often supplies polysyllabic words for Nick, who makes gram-
matical mistakes and indulges in a fair amount of flowery rhetoric. But
Hughie's vocabulary soon reverts to a more normal range while Nick's
rhetoric becomes simply one of his tools of persuasion, not an intrusive
habit of speech. The seventeen-year-old sister in the "treatment" was
changed into the kid sister Margaret, eliminating potential romantic
complications for Nick. Their widowed Mother became a grandmother and
the very independent owner of Molly's Reach—a refreshing change from
family-adventure stereotype.

An analysis of some of the early episodes reveals a series in transition,
with certain basic patterns of character and situation conforming to its
"kidult" formula, others abandoned, and new elements introduced.

"The Search" (12/11/72, wr. Merv Campone) demonstrates one of the
archetypes traditional to "kidult" series, the "rite of passage," which shows
us that the character has achieved a new level of maturity or passed from one
significant life phase to another. Hughie gets into trouble for chasing frogs,
as only a young boy might. However, the bulk of the episode demonstrates
that he is quite competent in survival techniques and shows good sense in a
crisis. Humour comes from the interplay of characters: for example, in a
reversal of the usual plot cliché, Hughie blithely rescues his rescuers.
Hughie's character as he grew up in following seasons—good with boats,
athletic, inventive, cool in a crisis, emerged in this episode.

"Evil Eye" (14/1/73, wr. Dennis Donovan) plays coastal Sechelt shaman-
istic magic against Nick's Mediterranean and Relic's Welsh sense of the
"evil eye." Hughie tries ineptly and comically to purify himself by a garbled
version of the ritual because Jesse's disappearance from his new family is
"all his fault." Typical of the series as it developed, the distinctive motif of
the Northwest-Coast Indian's sense of the presence of a spirit world is in the
foreground. Yet the viewer is also offered a rational explanation: Margaret's
resentment is what drives Jesse away from the Reach, not her "evil eye." In
the end, as usual in the mature series, the viewer is free to choose which
explanation fits.

"The Highliners" (19/11/72, wr. Alan Oman) introduces new characters and focuses on the skill and fascination of highline fishing, including the unpleasant task of gutting the catch. The episode demonstrates another characteristic pattern of the series: a documentary photographic style for machinery and work, concentration on accurate detail, and the presentation of various forms of work, not romantically, but with respect.

Of the first cluster of episodes, "Molly's Reach" (8/10/72, wr. Me-v Campone) was the most important in terms of establishing characters and relationships as well as telling the audience how Jesse came to live at the Reach. Changes from the original concept here include Jesse's Uncle Louie handing him over to Nick as a trustworthy partner, despite the fact that no contract has been signed, and to Molly as a caring woman. However, Molly does not "mother" Jesse in any way, agreeing to his living there only "if he pays his way and helps out."

Uncle Louie's anglicized name and understated mistrust of whites, together with his fear that Jesse, with no money and no job will end up on Vancouver's skid row, give the episode a realistic undercurrent. This presence of a different, equally valuable perspective occurs in many other episodes and helps make the series more attractive to its adult audience. A double-edged perspective was also characteristic of and substantially contributed to the success of *Cariboo Country*.

In this episode, Nick's quick anecdote about his "old man" fighting with him yet letting him leave home extends the "rite of passage" theme. In his first appearance, Relic is already a firmly fleshed out character—"an old spider," says Molly. With an intelligent, expressive face and body and an off-hand, often understated delivery, he cheats at checkers and spins a very effective ghost story (which he may actually believe) to distract Jesse from the game. Jesse also loses money to him.

Nick is shown as both ingenious and kindly, talking to Jesse impulsively but warmly. Molly's tough, responsible, independent nature is in strong contrast to Nick's overt sexism. Jesse's determination also becomes clear, as does his sense of humour. There is a wonderfully funny yet wry contrast between the flowery speech Nick composes for Jesse's use in his confrontation with Uncle Louie and Jesse's actual dialogue, which consists of two sentences: "I love you, Uncle Louie, but I have to do things my way. Pass the cream."

The episode "Easy Day" (24/12/72; wr. Marc Strange) focuses on Nick; Relic does not appear at all. This is not an action drama about salvage, but uses a "situation comedy" formula as many other episodes would in seasons to come. Nick, Hughie, and Jesse cope with a busload of stranded ladies on an outing. Molly and Margaret, on their way to Vancouver, appear in two very brief scenes. As a storm rages, Nick fixes shishkebob, lights candles and, with infectious panache, gives everyone including the earnest Hughie, Jesse, the very English Colonel Spranklin, the ladies, and two sceptical locals a lesson in traditional Greek male dancing. The kids do the dishes while Nick sleeps; his theatrics have given the ladies a wonderful time and saved the bus driver's job. Jesse has no real presence in this episode; he watches, serves, cooks, reluctantly dances, and jokes with Hughie a little.

"Easy Day" is a skilfully written piece in a different mode that emphasizes physical humour. It is filled with stock characters among the retired ladies: the officious organizer, the one who makes the tea, the one who worries. . . . However, the jinxed female bus driver provides a good cameo, with the actress filling out the dialogue to turn a simple tale of woe into a glimpse of an insecure but gallant lady.

"Boomsticks" (5/11/72; wr. Merv Campone), shot three months later, concentrates on beachcombing itself, its techniques, laws, dangers, and rivalry with lumbermen whose logs are loose. Relic is established as Nick's professional and thoroughly unethical rival. By then, a perfectly matched camera convention has been developed for him, a head and shoulders close-up that pops up out of nowhere, disrupting the rhythm of shots and the spatial logic. Also, by this point, Jesse has clearly emerged as the practical half of the partnership.

Subsequent seasons of *The Beachcombers* retained some basic characteristics established in the first. We note considerable attention to realism in the portrayal of various kinds of boats, fishing, flying, logging and local customs at Gibson's and the Seahost Reserve. Emphasis continues on action, adventure, beachcombing, Indian vs. white values, situations of peril, and a mixture of slapstick comedy and humour coming from clearly defined characters in usually credible situations. Relic remains Nick's chief rival and the instigator of numerous plots throughout the series' run.

In "Fraser Red" (30/9/73, wrs. Merv Campone/Dennis Donovan), though there is a lot of action with boats and scuba diving, Nick typically reasons and bluffs his way to a kind of poetic justice in his battle with a pirate from the Fraser River. I do not recall ever seeing Nick throw a punch, though fights are common on American family-adventure series.

"The Sea is Our Friend" (4/11/73; wr. Merv Campone) is a classic formula suspense story; Nick's legs are trapped under a log with the tide coming in. However, the plot twist is ingenious, and Nick's love-hate relationship with the sea adds a new touch to his character, as does his inner fight with panic, understated by Gerussi in both dialogue and performance. His deeply felt reply to Hughie's bitter, "The sea is our enemy" is: "This log is our enemy. The sea is our friend." Hughie: "I hate the sea." Nick: "She will help us. . . . Use your mind." This triggers an idea, and Jesse's scuba equipment is used to save Nick from drowning while the tide floats the log free. By now, Jesse and Hughie are fully competent to help think up ways out of dilemmas and attempt rescues.

"Here Comes the Groom" (10/2/74; wr. Dennis Donovan) develops garbage-man McLoskey and shows the viewers a new side of Molly—her impulse to escape being grandmother and proprietor as well as a yearning to end her loneliness. It is an odd mixture of very improbable slapstick (given the circumstances and tone of the episode) and inventive, poetic camera work to convey Molly's subjective point of view. Her decision not to go through with it, made at the altar, and McLoskey's tender handling of her feelings provide the kind of complex emotional balance so often lacking on other action-oriented family-adventure series, with their predictable, sentimental "quiet" scenes.

Early Risk-Taking

"Runt o' the Litter" (3/3/74, wr. Merv Campone), from the second season, is remarkably atypical of the series, a departure from the usual tone, plot pattern, and characterization. The fact that this script was broadcast reveals a willingness on Keatley's part by that stage to risk disappointing his audience. When this happens in series television, and it is very rare, such innovation indicates confidence in audience loyalty and adaptability.

The episode is the key to Relic's character and should probably be rebroadcast every few years for new viewers, particularly since the series is not "stripped" or rebroadcast daily in Canada. Relic's father, a Welsh coalminer, turns up to expose, unwittingly, Relic's lies about success in "America." "Da" intends to collect for all the sacrifices he claims he and his sons made so that "Taffy" could come to the New World. Nick and his friends make an unsuccessful attempt to cover for Relic. The basic situation of egotistical half-truths found out is commonplace in domestic comedy. Yet the image of the father as a sitcom stereotypic lovable rascal is shattered in their first meeting, when "Da," with tremendous energy and viciousness, puts Relic in a headlock, forces him to his knees, snarls, "You always were the runt o' the litter," then kisses him. Further, his hands are all over Molly from the start; not standard behaviour in a "kidult."

The father's devastating effect on Relic's self-confidence and his obsessive determination to break Nick and also to humiliate his son when he discovers the truth about his precarious circumstances as a beachcomber are given truly chilling dimensions. He forces a race with Nick that involves pulling an anchor across a finish line. Spouting invective from miners' strikes in Wales, Da tells Relic, who opposes the race because he knows that it is meant to humiliate him: "Blackleg scum, go stand with your scab friends." Though the old man has pulled coal carts in his day, Nick is thirty years younger.

When Da falls face down in the sand short of the finish line, Nick is already across. The cheering dies away. Relic's wholly ambivalent response to his father's public humiliation, full of hatred, contempt, and yet disappointment at the fall of this giant figure, is written wordlessly in closeups of the actor's face. He stands over his father and looks at him as the camera takes both points of view in turn. Relic grabs the rope and hauls the damned anchor across the line, then drags it all the way back to his "Da" (here the camera takes Da's point-of-view), and says bitterly, "Go on home." The old man weeps. After this episode, we understand why Relic cannot love other people, why he is paranoid, lonely, compulsively acquisitive and fiercely competitive, why his jet boat is sleek and powerful and his houseboat a wreck. Robert Clothier's remarkably expressive face, sideways glance and abrupt body language flesh him out perfectly. The years to come would have a few other episodes of *The Beachcombers* with downbeat endings, but of the ones I viewed, this is by far the most powerful.

Motifs in the Beachcombers

The Beachcombers evolved differently from other family-adventure series because of several factors, one of the most important being the specific characters of Nick and Relic, who interact with Jesse as mediator in the paradigm. Then there is the triad of Molly, Hughie and Margaret that established the fundamental motif of the importance of the "family" in the early years. Jesse, his young sister Sara, introduced in the late 'seventies and later, his new wife, a widow named Laurel, and her son Tommy are essential to the most distinctive pattern of all: exploring Northwest-Coast Indian life in all its variety. The clash of Native values with those of the twentieth-century and the resultant weaving of a new way of life, neither imitative of the white nor fully traditional, have created a continuing theme. In an August 1985 conversation, executive producer Don Williams expressed reservations about continuing to explore this particular theme, primarily out of regard for the sensibilities of the native people in his cast, but in part because he was discouraged about being able to obtain and incorporate accurate information. It would be inappropriate for me to speculate on how this series is perceived by the people who are caught in such conflicts, but I think the subject has been approached with affection, respect, humour, and imagination in the series. Without this particular theme, the series would be much poorer, and certainly less distinctive.

The series began with a focus on adventure at sea, the rugged environment, characters with whom the audience could comfortably identify, romance, characters in peril from tides, storms, accidents, and so on, property crimes like theft and insurance fraud, and dangerous animals.

As characters matured or changed, plots also revolved around various rites of passage. The series used four different kinds of comedy: slapstick, essentially physical comedy; farce—mistaken identities, inadvertent meetings; situation comedy in which small domestic problems are solved, and the more difficult comedy of characters in complex but funny response to situations. A large part of the series' humour is based in credible characters and situations rather than slapstick, situation comedy, or farce.

Content analysis of *The Beachcombers* reveals other motifs and thematic patterns. A motif is a recurring idea of a cluster of related characteristics, ideals, customs, habits, values; or a pattern of verbal and visual imagery. Fully developed and vested with significance, a motif or motifs can be designated as a theme. Motifs function as signifiers, themes as signs.[16]

The major motifs in *The Beachcombers* are respect for outsiders, eccentrics, and particularly older persons, still unusual in this type of television series; respect for the natural environment, animals and the sea; and respect for Nick's Greek heritage and the ethnic heritage of other characters who appear from time to time.

Two or three times a season, the formula of up-beat, action-packed, basically funny plots and characters is modified significantly; "The Hexman" (16/11/75; wr. Arthur Mayse), "Jo-Jo," and "Runt o' the Litter" are three examples. These deviations keep the show and the people who work on it from getting stale, and they extend the awareness and expectations of the audience.

In the first two seasons, the basic elements of the family-action-adventure series were put in place, and the elements differentiating *The Beachcombers* from other series of this type were developed. Following seasons were based on the same motifs, themes, plots and characters that kept the series distinctive and successful. A content analysis of motifs and plots over the period 1972 to 1985 demonstrates how the first two years gave the series its basic shape. These motifs and themes recur, disappear, reappear. The family motifs, for example, include the blood relationships of Hugh and Margaret, brother and sister, and their "Gran." Nick acts as a surrogate uncle, coincidentally (?) an archetypal pattern in the phratries of North American native people, where maternal uncles normally brought up their sister's children. Jesse serves as an older brother by adoption, Sara as little sister and granddaughter; Relic is the obnoxious relative who never quite joins the family circle, but without whom no true gathering of the clan would be complete.

The motif of the rite of passage is focused primarily on various stages of "coming of age," as when Margaret tries a channel swim, Hughie tries to tame a wild dog, or Jesse searches for his particular vision in a ritual fast. One variant of that motif is the problem of coming to terms with fathers after many years; both Nick and Relic confront their fathers, and young Pat O'Gorman has a fantasy father. Then there are many watershed birthdays, courtships too numerous to mention, serious as well as comic, and initiations of all kinds—to work, groups, the oddities of young women, the vagaries of young men, and new responsibilities.

Friendships in the series are mostly male, though of mixed ages and backgrounds. Sometimes, though too rarely, we see women as friends. In one early episode, Margaret goes for help when her young friend is in trouble. The struggle to get help is a conventional plot device, but the nicely sketched friendship departs from the formula. Occasionally men and women form simple friendships, but too rarely now. The obverse motif of rivalry often takes on personal and business overtones at the same time.

Other plot motifs include dreams of fortune, improbable or funny (Nick and Jesse) or rather shady (Relic, McCloskey, and sometimes Pat O'Gorman). As in *Cariboo Country*, the loners, outsiders, and eccentrics are treated sympathetically; for example, Sadie, a middle-aged woman trapper who appears in several episodes, played with great gusto by the ordinarily lady-like and controlled Frances Hyland. We also meet a Christmas mime, "Jo-Jo," a Gypsy King sailor, and many others. As in *Cariboo Country*, old people are treated not as lovable "ol' coots," comic and child-like, but with considerable respect and dignity—for example, Colonel Spranklin, Joe O'Gorman, Chief Moses, and Tommy's grandfather. Plots regularly show pressure on people to conform, sometimes from the community but often from the anonymous "they" of local, provincial, or federal government. The family at the Reach helps individuals either elude or adjust to those social pressures. Molly, and even Nick, have been allowed to age gracefully. Relic is somehow ageless.

The series is enriched by a wide-ranging ethnic mix: Japanese (in several episodes), German, Italian, Dutch, East Indian, Middle Eastern, Swedish. The English gentleman, Colonel Spranklin, eventually moves beyond a Blimpish stereotype in the episode where he is asked to help a young Japanese teenager retrieve his father's fishing boat, which was confiscated in World War II. The Canadian government's wartime confiscation of Japanese property and deportation of Japanese Canadian citizens to the interior are now recognized as unjust. Even so, the colonel lost many men in a Japanese prisoner-of-war camp, and cannot forgive. Needless to say, the lad does get his boat from a remote barn into the water, thus retrieving his father's honour and his own past. The colonel cannot free himself from that past, and we understand why in this example of the occasionally double-edged endings encountered in this series.

The Welsh and Irish are also given historical roots, though McCloskey's character comes perilously close to that of "Stage Irishman." Americans are caricatured several times. Except where I have identified examples, nationalities are not usually stereotyped. Other related character motifs do depend on stereotypes however: the rich, the arrogant, tourists, and land-lubbers are seen as enemies or nuisances requiring rescue.

There is a strong sense of community in the series, owing probably to the fact that it has been shot at Gibson's Landing since 1973. The look and feel of the town have changed with the years—it is more accessible from Vancouver now—and these changes are reflected in the series. Episodes have revolved around festivals, rummage sales, contests, and various forms of community support in crisis, such as making sure the fire engine stays in Gibson's or a successful sit-in organized by Margaret, the champion of lost causes, to save the goats on Goat Island.

The code of the Beachcombers themselves is also communal, yet it stresses the independence of each entrepreneur. Beachcombers settle differences among themselves; they do not bring in outsiders. They do observe certain basic rules of who claims what and where. In *Cariboo Country* there is a similar emphasis on everyone's right to make his or her own decisions, mistakes and choices as long as the person cannot really harm himself, can continue to function, and does not hurt others. The limits of friendly interference are stressed again and again. This is an interesting contrast to the recurrent friendly-but-firm adults, who knows what is best for both children and old people to be found in so many series intended for family viewing.

Peril to central characters is a basic convention in the family-adventure series. The particular inflection in this version is the rivalry between Nick and Relic over their boats, salvage, practical jokes, and occasionally over the loyalty of another character. Another standard plot element in this genre is crime, never really violent but sometimes quite serious: arson, theft, claim-jumping. The formula demands that Relic's questionable get-rich-quick schemes or the harmless but far-fetched notions of Nick or Jesse end in (usually) comic disaster. The implied lesson is that we should stick with what we know best, and not get carried away. Nevertheless, Nick's Greek panache, Jesse's problem with straddling two cultures, and their occupation itself give the show a different flavour from formula "kidult" shows. Another element in most family-adventure series is romance involving both sexes and all ages. *The Beachcombers* stays fairly close to formula in this respect.

Every series that runs this long will recycle its plots once in a while. A few of the more obvious recurrent ones are those in which Nick and Relic get stranded together, Nick runs Molly's Reach or some other enterprise for a friend and gets it into trouble, and, in too many episodes, a character is trapped with water rising, boat sinking, temperature freezing, bridge out, bear coming, boat gone, or whatever.

Continuing Characters Analysed

Working on plot summaries for 221 episodes, content analysis of narrative elements reveals the following patterns:

McCloskey, Pat O'Gorman, Gus, and Constable John are, or have been, continuing characters around whom two or three episodes a season may revolve, but they have not significantly affected the flavour of the series.

Molly drops out of major focus for a significant length of time, although the original intent and the early seasons did establish her as the feisty and yet motherly pivot of the extended family. She was written out of the show in 1985-86 and is missed.[17]

The emphasis on younger children has diminished over the years since Sara has grown into a teenager. For a while, no new major character under age ten was introduced, indicating a significant change in audience identification and thus the audience now sought by the series. However, Jesse's stepson, Tommy, now fills that gap.

McCloskey, the stage-Irishman stereotype, although an amusing character, was no real loss. The young male teenager the formula requires, Pat O'Gorman, is unfortunately a one-dimensional character thus far. Others have come and gone. A more successful continuing character is the good-hearted naive RCMP Constable John Constable. Jackson Davies' persona depends on a gentle, slightly askew brand of humour which adds another dimension to the sardonic quips of Relic, the wry comments of Jesse, the bluff jokes of Nick.

Relic is the other important character focus in the series because he is the continuing antagonist. He is not in the original character list, yet he has been in the foreground for a third of the episodes. Could *The Beachcombers* go on without him? In my judgment, this is unlikely. True, both Relic and Jesse disappear for more than ten episodes at a time. Relic disappeared from the foreground for ten episodes in 1975-76, and is not featured for about ten episodes in 1978 and nine episodes in 1982-83. Was this a matter of contract difficulties? Poor health? Restlessness? A new producer? Different work available? Relic was not as important in the first season as subsequently. However, the series would lose its distinctive character without him. Relic is the only one who serves as a truly well-matched antagonist to Nick's protagonist.

Jesse is also very important to the series and a constant source of plot material. However, he is less essential to the basic structure, fulfilling the role of second lead or best friend-sidekick in the formula. What he adds to the series that is unique (strengthened by Sara, Laurel, and Tommy in recent years) is the distinctive motif of Salish values and customs and the conflicts and synthesis of those customs with modern life. The fact that he still has the inherent freedom of the beachcomber's life is essential to that accommodation. Getting married and acquiring a stepson broadened the range of problems and challenges.

Jesse's character belongs to an archetype going as far back as a "Roland for an Oliver," Little John to Robin Hood, Dumont to the visionary Riel, Spock to Captain Kirk in *Star Trek* (1966-69)—also caught between two worlds—or the loner "Fonz" to Richie in *Happy Days* (1974-83). Jesse is the faithful friend whose perspective is very different by reason of temperament or circumstances (both, in this case) and who is foregrounded in some episodes, comes to the rescue in others, and makes a cameo appearance in many. He does not actively mediate in the stories concentrating on the rivalry between Relic and Nick, yet his presence in the series does mediate some of the conflicting characteristics and values shown by the other two, a standard function of such a character.

Jesse is missing from the last half of 1973-74 to the end of 1974-75. In one episode, he tells his surrogate family that he is restless and he leaves. In that period, no Indian motif appears in the plot summaries. In sixty programmes, this motif is given focus only once. It reappears in the 1976-77 season with the introduction of little Sara. In the second half of 1978-79, there is another gap of about ten episodes. However, the Northwest-Coast Indian motif is still intermittently present. In the mid-'eighties, the motif of the clash of cultures continues to be explored. In one episode (1984), Tommy, though ashamed of his Nootka heritage, saves his tormentors from a bear. Turning the tables on mean kids is a standard situation in this kind of series; suggesting in body language and dialogue that Tommy has not only woodcraft, but also a special totemic power over the bear, is not. In two others (1985 and 1986), Tommy's grandfather breaks free of his intravenous tubes and hospital bed to show his grandson the sacred island where his ancestors are buried, and members of Jesse's band get into a conflict with Gibson's Landing about unsettled land claims that sets the surrogate family of Molly's Reach at odds with one another.

The Northwest Coast Indian Motif

Racist stereotypes of native people have been numerous in North America. On most American television series, when Indians were not massacring innocent settlers they were depicted as either lazy, superstitious people with drinking problems, inexplicable ideas about families and work and no sense of time, or faithful if mysterious guides, hunters, and sidekicks who spoke a tired rhetoric about rivers flowing and forked tongues. Such stereotypes have been with us since the dime novels of the nineteenth century. Contrast with this the CBC record after 1960 which includes *Cariboo Country*; *Wojeck*'s "The Last Man in the World" (13/9/66, wr. Philip Hersch, dir. Ron Kelly); *Dreamspeaker* (1977, wr. Anne Cameron [Cam Hubert as she was then known], dir. Claude Jutra); *A Thousand Moons* (1975, dir. Gilles Carle) and *Gentle Sinners* (1985, adap. Ed Thomason, dir. Eric Till), to name a few.

"Steelhead" (16/10/77; wr. Marc Strange) is fairly typical of the use of this motif in the series. The plot's focus is on Indian custom. Jesse is intent on purifying himself by the ritual four-day fast to seek a vision that will confirm his life path and identity. Unfortunately, there is a parallel and rather distracting comic plot featuring a bear cub which scares off the horses, followed by Nick's attempt to keep the now faint Jesse fed by catching the very elusive "steelhead" fish. However, Jesse's experience itself is presented entirely in visual terms, starting with extreme long-shots of the bleached bones of a logged-over mountain. The irony of the setting is not overemphasized; the series is often allusive in this fashion. In the dead treetrunks and gnarled roots, Jesse glimpses fragments of weathered totems. Mostly in long-shot, we see his wordless figure wandering on the mountainside, half-heartedly and without conviction or accurate knowledge attempting a chant and a dance. Here, photography, editing, and sound play a major role; there is no dialogue. The pace is unhurried, though broken by intercuts wih Nick and the bear. The wind blows, and a gull screams. Jesse's look of calm and inner certainty tell us when something has been achieved. What it is, we learn later in his conversation with Nick.

The Beachcombers series does not perpetuate the stereotype despite the temptations inherent in a week-in-week-out series format. Indeed, in some of its finest programmes, the stereotypes just cited are mercilessly parodied. On the other hand, the "Hexman" episode, which pits a Sechelt shaman against a hexman from Newfoundland via Jesse's talisman, "Soulcatcher," and "Salmon Woman" all illustrate the fact that, as one old Kwakiutl woman put it, on the Northwest Coast "everything has a face." Even the usually literal world of television realism can be made to catch that feeling, though it is difficult and efforts are not always successful. But the series presents the perception of a spirit world integral to the natural world as a truth for native people, part of their experience which we are invited to share, which broadens the scope for plot material in the series, and gives it a very distinctive flavour indeed.

"The Hexman": An Audience Favourite and Why it Worked

I want to conclude this account of *The Beachcombers* with a look at a superficially atypical yet in many ways very typical episode. "The Hexman" by Arthur Mayse was broadcast in the 1975-76 season. A real success with the audiences, it has been rebroadcast eight times and had a less successful hour-long sequel called "The Return of the Hexman." Many pleasures are packed into twenty-six minutes. First of all, it is a cracking good ghost story. The characters are Jesse, Nick, the Newfoundland "Hexman," played with both relish and considerable control by Gordon Pinsent ("John Vincent's me name, sword fishin's me game"), the ghost of "Timmy," only occasionally visible, who helps row the dory, and an off-camera shaman. The story of Timmy and the Hexman is told in a ballad which is the episode's only music, a separate narrative thread, eerie, moving in its simplicity and tragedy and a strange counterpoint to strange happenings.

The episode starts in silence (very rare in series television) with a slow pan past a bare, twisted tree to water, then rock. We hear footsteps from someone out of frame, and the clink of heavy chains on the sound track. The pan continues until the camera finds Nick and Jesse on the *Persephone*, salvaging a log. However, Jesse doesn't like the isolated little cove. There is a petroglyph there, and his great-grandad has told him to stay away from the place. When Nick scoffs, Jesse replies, "it's bad business to bad-mouth a shaman." The sound of wind introduces the Hexman dressed in heavy sweater, overalls, rubber boots and, incongruously, a green plastic sun visor which takes on ominous overtones in closeups as he glances sideways from beneath its translucent peak. He is carrying a lethal looking harpoon, which he plants in the log. Nick is tranced by its gleaming point until he rubs his Greek evil eye bead over his eyes and heart. He then retaliates with logic, that is, with his legitimate salvaging permit. Suddenly, the Hexman has the bead. Nick is torn—"It's all baloney," "hypnotism"—but he is helpless.

In the distance, we hear a raven on the soundtrack. Jesse pulls a carved argillite raven from his shirt pocket and points it at the Hexman, loosening the spell. At this point the Hexman, using a verse of his ballad, tells us that Timmy has done murder and "paid the price." Jesse, who now realizes that the Hexman can overpower them both, uses the totem as a microphone (of all things) to call on his great-grandad. He closes his eyes, half-believing, half-doubting, then says, "I've got him on the line." Jesse (in closeup) becomes the ancient shaman's mouthpiece, and begins an incantation that is also a polemic against "the white men who killed our buffalo." Nick: "Here on the coast?" Jesse (eyes open, in medium shot) " Grandfather is nearly one hundred years old. Sometimes, he slips a clutch." After references to MacKenzie King and Nick's promise of a "4-star arctic sleeping bag," the raven cries and the duel is on. "Is Great-grandad ever mad! A crock of oolachen [fish] grease fell off the shelf." The dory rocks, the wind roars, and the log moves offshore despite the Hexman's spear; so he hexes the boat.

The appealing mixture of horror and comedy, scepticism and the supernatural continues. When the battle is joined again the Hexman demolishes the Shaman's unseen cabin. The shaman, still speaking through Jesse, vows: "I will cross his eyes and turn his feet backward." Finally, in a roll of thunder, the Hexman is defeated. Feeling invisible spiders crawl all over him, he half crawls, half leaps out of the dory into the sea. His last line is, "They beat us fair and square. We'll not trouble them again. Give us a hand up, Timmy."

As the *Persephone* pulls away, Jesse looks back with the binoculars to see if the Hexman is really okay. Nick: "It can all be explained. All we have to do is figure it out. . . . Are they . . . is he okay?" Then Nick discovers that his blue bead is back around his neck. Jesse: "Great-grandad said there would be a delivery charge for your bead." In the closing shot, the camera takes Nick's point of view as he looks through the glasses to see two figures rowing the dory in watery sunlight accompanied by the last echo of the Newfoundland ballad.

The episode is enriched by a first-rate script, as well as with the fabric of Newfoundland speech, an evocative sound track, a very credible ballad sung by Pinsent, crisp editing, lyrical camera style, and really good performances, particularly where Pat John has to play both Jesse and Great-grandad. Both actors also have to balance the anachronisms, humour, and scepticism with real fear, while Pinsent projects menace edged with charm. The episode also gathers up many threads characteristic of this series: the Northwest-Coast Indian motif, beachcombing, the clash of cultures, and friendship. The blend of comedy derived from the incongruous juxtaposition of pragmatism and supernatural power, the comic yet powerful figure of the shaman and the glancing menace and mockery of the Hexman, the physical comedy and the steadily growing tension of the adventure, is particularly satisfying in this episode. So is the fact that no explanation for the haunting is offered except the ancient archetype of the damned sailor doomed to wander the seas forever.

The Longevity of The Beachcombers

It is fashionable among critics, and even around the CBC, to ignore or denigrate this series. There are several identifiable reasons for this. First, it revolves around Bruno Gerussi as Nick; if you don't like his particular acting presence, you don't like the series. Second, the cliché is true. Success, particularly long, hard-won success, is suspect in Canada. Also, the series comes from the west, although unlike *Cariboo Country*, it does not take on the government and the eastern establishment. *The Beachcombers* at its best is very particularly a reflection of some West-Coast values and specific lifestyles and distinctively Canadian as well.

In series television, even if success is defined simply in terms of sales, longevity is a major factor in success. Long life, and thus many episodes, means that the series can be "stripped." It has been sold all over the world. Given the severe budget restrictions placed on the CBC as the 'seventies went on, the fact that about twenty episodes a season have been made for over fifteen years also signifies that the series must be holding its audience figures despite the fact that the young viewers inevitably became adults.

In aesthetic terms, "success" is something else. Many American sitcoms go on for years after their original premise has been exhausted. Eventually, formula is all the programme has to offer. *The Jeffersons* (1975-84) comes to mind. Series that retain their critical reputation grow, evolve and change. *Barney Miller* (1975-83), *M*A*S*H* (1972-83) and *All in the Family* (1971-82) are three American examples that hold up well in reruns, particularly in episodes drawn from their second and subsequent seasons, when the producers, writers, and actors have come to know the characters.

The criteria normally used to measure some degree of aesthetic achievement in series television are: the originality of new characters; consistency and depth in continuing characters, who must surprise us at least once in a while; the originality of the circumstances and fresh variations on the conventions of the genre.

Family-adventure series usually end with the problem introduced in the first five minutes being resolved in the last five. No one ever dies, is badly injured, or ends up deeply unhappy. *The Beachcombers* has legitimately broken that fundamental dramatic convention many times. Ambivalent, open-ended, or even downbeat endings that succeed in persuading the viewer that this unusual event is the honest ending for the storyline attest to the willingness of various executive producers to take risks.[18] Such episodes also demonstrate the skills of the writers and story editor in getting the audience to accept more challenging storylines and character development. The audience's acceptance of this kind of inflection testifies to its sensitivity and flexibility. It is also true, however, that the strength of this kind of series lies in the capacity of the formula to absorb inflection or inversion.

The most striking similarity between *The Beachcombers* and *Cariboo Country* is the way a series can be kept alive when it stretches its conventions towards the more flexible forms found in anthology. *Cariboo Country* had more freedom from the pressures of genre, had one writer and one producer throughout, made fewer episodes in a season and thus maintained a consistently high quality in its episodes. With the changing conditions of television drama after 1968, *The Beachcombers* could not sustain that kind of level of achievement. It had to be market-oriented from the beginning.

Nevertheless, at least one in five episodes contains a spark of originality. More to the point, several episodes over the years have been as good as the very good short plays of the golden age of anthology.

The family-action-adventure genre should permit a range of mood and type of story. Episodes of *The Beachcombers* have ranged widely for their stories; close character study, peril, rollicking slapstick, morality fable, ghost tale, exotic occupation or trip, animal story, explanations of facets of Northwest-Coast Indian life.

Other criteria which I use for excellence would be substance in the issues raised; complex conflicts of character; and what Philip Keatley defined as part of *The Beachcombers*' flavour in his early presentation to the programme planners: humour arising from character, believable motivation, and credible circumstance. The better episodes also display the more difficult ironic, allusive forms of comedy.

By all these criteria, *The Beachcombers* has earned its success. In fact, it is the most successful television series ever made in Canada. When audiences around the world think of Canada, they think of Gibson's Landing and the family at Molly's Reach. It's a good face to present to the world.

NOTES

1. Some of the material on *Cariboo Country* appears with the permission of *The American Review of Canadian Studies* in which an article of mine called "Cariboo Country: A Canadian Response to American Television Westerns," appeared in vol. 14, no. 3 (Fall 1984), 322-32.
2. *The Journal of Popular Culture* (Fall 1969), 254.
3. As before, all dialogue quoted in my analysis of *Cariboo Country* is taken directly from the kinescopes.
4. *TV: The Most Popular Art.* Anchor Press/Doubleday, New York: 1974, 71.
5. "Off-air" observations of late-night reruns. The episodes cited had no titles or decipherable dates.
6. *The Journal of Popular Culture*, 6 (Fall 1972), 339.
7. *The Popular Culture Reader*, ed. Jack Nachbar & John L. Wright. Bowling Green Popular Press, Bowling Green, Ohio: 1972, 271.
8. Bowling Green Popular Press, Bowling Green, Ohio: 1975.
9. *Television Quarterly* (May 1962), 36.
10. Paul St. Pierre in correspondence, and Philip Keatley by telephone and in an interview, supplied me with many of the details which helped me understand the context for the series.
11. "The Higher Meaning of Marlboro Cigarettes," in *The Popular Culture Reader*, 240.

12. Aileen Garland, *Canada Then and Now*, Macmillan of Canada Ltd., Toronto: 1954, 320.
13. McClelland and Stewart, Toronto: 1965, 7-8. Compare too Michael Ondaadtje's *The Collected Works of Billy the Kid: Lefthanded Poems*. Anansi, Toronto: 1970.
14. *Basic Black* CBL Toronto, 4 May 1985.
15. I have had access to correspondence between Philip Keatley and CBC management in the Winter of 1971-72.
16. Dr. J. M. Miller, Dept. of Chemistry, Brock University, assisted me in the formidable job of content analysis of the well over two hundred episode summaries provided for me by the CBC. He also devised a way of plotting the results of our analysis on a graph.
17. In the 1980s, it became fashionable to present characters from old television series in a sort of "reunion" format. *Beachcombers* missed a golden opportunity to reflect on its fourteen-year-run in the fall of 1985, when Molly's wedding (and exit from the show) came and went with no mention whatever of Hugh and Margaret.
18. Philip Keatley, Eli Savoie, Hugh Beard and Don S. Williams

4

SITCOMS AND DOMESTIC COMEDY
"But it's too cold to be
a melting pot, mama!"

Looking at genres like the mystery-copshow, or the family-adventure show, we can see the twenty-year development of television subgenres worked out within the overall "series" concept; episodes with a protagonist, continuing characters, and recognizable if not overly familiar plots, full of either action or laughter. Series television may develop with each episode rooted in a memory of what happened in previous episodes, or the growth and change of characters as in *Seeing Things*. However, like *The Beachcombers*, it may show little or no sign of a memory at all.

Other kinds of television series have had vigorous but more limited appeal. The Western is gone after a long run on radio and television, though there are always rumours of its return. The hospital shows come and go, from Drs. Casey, Kildare, and Welby, plus a variety of paramedical teams, to the hiatus of the 1980s.

One of the most influential critics of American television as popular culture is Horace Newcomb, whose book *TV: The Most Popular Art* (1974)[1] first set up workable definitions of the most familiar genres of series television. He makes a helpful distinction between situation and domestic comedy: the latter depends more on character interplay than what Newcomb calls the "confusion in the plot" characteristic of pure sitcom. The Western as he defines it has now completely disappeared from the air waves, unless we agree with some critics that its characters, situations, and values have been reabsorbed into detective shows. In Chapter 1, I refined Newcomb's mystery genre into the subcategories of copshows and mysteries in order to differentiate between shows featuring police, detectives, and spies in "action-packed adventure" and ones that feature amateur sleuths. The genre that includes doctor and lawyer shows, Newcomb describes more generally as the "professionals," a category which appears in Chapter 5. Finally, he looks at the soap operas, a genre which, since Newcomb wrote his book, has come to include the important subgenre of prime-time soaps like *Dynasty* (1980-present) and *Dallas* (1978-present).

Except for Westerns, the genres he defines have also developed in British formats. Indeed, some of the most distinctive as well as most formulaic American situation comedies have been adapted from British originals—for example, *All in the Family* (1971-77) from the much more biting *'Til Death Us Do Part* (1964-74); *Sanford and Son* (1972-77) from *Steptoe and Son* (1964-73); *Three's Company* (1977-83) from the slightly more subtle sex comedy, *Man about the House* (1973-76). Of *All in the Family*, Halliwell commented that: "The Americans picked up the format rather carefully"— an ironic way of saying that Archie was basically a lovable wisecracker while Alf Garnett assuredly was not. The debate about whether making Archie lovable also made his grosser prejudices more acceptable, or at least appear harmless, continues.

Series and Sponsors—The Bland Leading the Blind?

Series television in North America is intended to build viewing habits, and thus viewer loyalty to the sponsors' products, over many weeks. On both sides of the border, series television lives and, ultimately, dies by ratings. Emmys can persuade network executives to give *Hill Street Blues* the reprieve it deserves, and clamour from the audience can bring back borderline cases like *Cagney and Lacey*, but for every survivor in the 'seventies and 'eighties there have been eight or ten failures in every American network season. The odds of survival for a mid-season replacement are even worse.

Writing his *Prime-Time America: Life On and Behind the Television Screen* for the 'eighties, Robert Sklar has summarized the current relationship between ratings, sponsors, networks and programmes in the vast majority of North American television drama: "Getting someone, anyone, to take responsibility for media power is like forcing him or her to hold the proverbial hot potato. The networks only want to please the vast majority; the advertisers and their agencies only want to get their message across. They all credit the public with make-or-break power over television content. If not enough viewers like a program, out it goes! But this familiar stance completely ignores the fact that the public can only endorse or reject. It has no access to the process of planning, conception or creation. Curiously, when the networks invoke the myth of the powerful viewer, they invite the disgruntled to take extreme remedies—if you don't like what you see, it's no good complaining, turn off the set. They're betting on the belief that most people would rather watch something than nothing."[2] As the audiences for the CBC's *Don Messer's Jubilee* and *Juliette* also found out, the viewer had better be in the demographic group targeted by television sponsors; viewer numbers alone do not guarantee longevity.

"Commercials are accepted by viewers as well as by the networks," Sklar notes, "as part of the basic rules of the television game. Surveys consistently show that more than two thirds of the viewers believe that commercial interruptions are a fair trade-off for the entertainment they get. A majority not only finds the commercials informative, it rates some commercials as more entertaining than the programs they interrupt." This observation does not explain the popularity of channel flipping during commercials, or the fast forward "zapping" of commercials on pre-taped programmes which VCR owners "time shift" to suit themselves, both made possible by technology which came into most homes during the 1980s. It may explain why the strategy of "stacking" commercials, that is, putting five or ten minutes of commercials between programmes, has worked in some countries.

Looking at the "flow" of programmes in 1978, Sklar gives up on his search for meaning and, by inference, audience motivation. Remember, in 1978, home VCRs and even cable were rare in the U.S. ABC, NBC, and CBS were still the triumvirate of a seemingly immortal empire. Sklar: "Now I suppose you want me to tell you what it all means, the sex and violence, the innuendo and circumlocution, the jump cuts and the distancing, the morality and immorality of it all. Sadly I have little to say that you don't already know [about] American commercial television, circa 1978. Just as you have to put out a newspaper every day, even when there is no news, you have to fill so many hours of television every day. Much of prime-time television is like newspaper boilerplate, the stale stuff you use when you have nothing fresh to say."

Martin Esslin defines the issues rather differently in his book, *The Age of Television*. He points out that dramatic entertainment kills time with excitement and suspense, gives us the thrills of empathy with an attractive character in danger and relief when good defeats evil. Other critics would also argue that we feel relief because our central normative values have been reinforced. Esslin claims that television drama satisfies our craving for glamour and fantasy and brings us the cathartic benefits of exercising basic emotions like joy, sadness, patriotism, religious fervour, and anger. Yet he characterizes the United States' television self-image as "violent, vulgar, shallow, hysterical, hard-sell, quite unlike the real America."[3]

Despite his largely British broadcasting experience, Esslin's reference points are chiefly American. He praises Canada at several points for its mixed system of publicly subsidized and commercially sponsored television, for battling the American tidal wave and producing quality alternatives.

Pointing out that "experience in other countries indicates that about 10 percent of the population does want [intellectually demanding television]," he concludes that the preponderance of formulaic imitative television in the United States disenfranchises 23 million people, who can make little use of "the most powerful and efficient medium of communication and information" of them all (174).

Yet it may be unfair to assume such dichotomies for the Canadian experience. As Herschel Hardin points out, Canada is too small to have the luxury of a channel like BBC-2 for the intelligentsia and more specialized audiences, as was lamentably demonstrated by the rapid demise in 1983 of the Canadian cultural pay service, "C CHANNEL." Moreover, the viewer of *Tartuffe* (1985) or *For the Record* (1976-85) may very well also enjoy the power fantasies of *The A Team*. The BBC-2 programmers guess incorrectly once in a while, and find themselves transferring a programme or series which unexpectedly turns out to have broad popular appeal for rebroadcast on BBC-1. Shows that adhere closely to formula, like the "jiggle comedies" modelled on *Three's Company*, the perennial favourite, or imitators of the mindless car chases of *The Dukes of Hazzard*, may not attract more discriminating viewers, but the viewer they do get might enjoy reruns of sitcoms like *Barney Miller* or *M*A*S*H* that stretched the forms and conventions of their genre to the limit. For every twenty versions of series that star the car or helicopter rather than the actors, there will be a hybrid of soap, comedy, and mean-streets police show like *Hill Street Blues*, or an innovation like the until now neglected subject of friendship between women found in *Kate and Allie* (1984-present), a sitcom, and *Cagney and Lacey*, a copshow (1981-present).

The series form got a late start on the CBC, but set a standard of excellence first time out with *Wojeck*. The late 'sixties and mid-'seventies at the CBC, however, were littered with interesting first seasons and increasingly imitative second ones; few lasted for three.

SOAP SUDS

The television forms which are most dependent on sponsors and ratings and which are the most reliable money makers if they are successful are the sitcoms and soap operas. The CBC took a long time to tackle either form. Soap opera, the oldest and most pervasive of the entertainment formulae, eludes all Canadian programme makers, independent or network. The CBC gave up after three widely spaced runs at the extended daytime serial "soap" form. Yet the audience is always there. The British-made Granada production, *Coronation Street*, has been a Canadian favourite for over twenty years. It was started by Canadian producer Harry Elton for Granada's Head of TV Drama, Canadian Sydney Newman. Most American soaps have also been attracting large audiences north of the border since the radio fore-runners hit their stride in the 'thirties. We forget that, in the 'thirties and 'forties, there were popular Canadian soaps on radio, often on private stations, using plots drawn from the area and actors and productions that reflected regional voices.[4] The CBC also followed the BBC model of the Archer family (still going strong in Britain) by having five farm families, one for every region, whose personal problems managed to mix comfortably with the latest tips on farming techniqes. Like most school kids in Ontario, when I was home sick in bed I could tune into the adventures of *The Craigs* on the noon farm broadcast before the air waves filled with the American favourites *Ma Perkins* (1933-60), *One Man's Family* (1932-59), *The Guiding Light* (1938-still running on television), and *Our Gal Sunday* (1937-mid-'fif-ties). We cannot blame lack of audience interest in the form, then, for the fact that we have not had a successful indigenous television soap in English Canada. The gap invites speculation.

The success of contemporary American soaps depended partly on the fact that they were able to deal with controversial subject matter long before the prime-time mini-movies or series would touch such subjects as incest, wife and child abuse, rape, abortion, and racially or religiously mixed marriages. The soaps offered a radical personal and social agenda that was not available in the 'sixties and 'seventies on U.S. evening programming, other than the Norman Lear stable of sitcoms. Canadian prime time was never as bland as the mix in American 'sixties television.

The CBC, from the president down to the second assistant directors, seems to have had a real distaste for this particular form. Instead, to their credit, they spent what money they had on an alternative for women working in the home, like *Take Thirty*, and its replacement, *Midday*. The basic assumption in both cases was that women should have intelligent and stimulating material, tailored for them, and that soaps do not meet these criteria. Moreover, by the 'eighties far more women were out in the work force than in the 'fifties and 'sixties.

Throughout its history the CBC cast a suspicious eye on the U.S. afternoon's never-never-land round of adulteries, murders, amnesia, life-threatening (but not disfiguring) illnesses, sweet patient heroines and vixenish, seductive young villains of both sexes, and the endless doctors, nurses, lawyers, and businessmen. The perfect hostesses and mothers, their adventures broken up by equally improbable rescues from the disgrace of dirty floors and dull dinners by Mr. Clean and the Pillsbury dough boy were definitely not classy enough for the English network of the CBC—though, as evident in sitcoms like *Delilah* (1972-73) and *Flappers* (1979-81), the corporation's drama department was not free of blatant sexism when it came to getting laughs.

Soap Opera Formula and How It Works, Year after Year

Though the soaps' plots move much more quickly than they used to, their structure has remained the same for half a century: exposition and recapitulation on Monday; several narrative threads in each hour segment, with suspense building to an emotional climax over the week, and concluding with a cliff-hanger in at least one plot on Friday. If a dilemma is resolved, another has to be in full crisis. The quintessential line of soap dialogue is ''Joshua is Bill's son, not Clay's. How can we keep the truth from him?'' Nor has the soap's loss of innocence in recent years done away with its classic shot, the stricken closeup. The wives still deal with the seducers, and the tempted regularly join the fallen. Passive, inept, subservient or psychotic women prevail. Women are still regularly attacked, charged with murder, impregnated, diseased, or abandoned. With their all-purpose blocking, missed lines, paraphrases, slang, cliché, predictable action-reaction camera rhythms, ''dream sequences'' glimpsed through vaselined lenses, the soaps continue their remarkably durable lives.

Irma Philips, creator of some of the most successful soaps on television, has defined their essence as ''a home, a child and a man to kiss.''[5] Despite radical changes in society, this continues to be an accurate description. Note that the narrative perspective of soaps is overwhelmingly from the woman's point of view. In my sampling of soaps across the dial over the last four years I have yet to see a story line which shows the circumstances from the male viewpoint. For the most part certain basics have not changed, since soaps went to television—the emphasis is on the middle- and upper-class professionals and the focus is on personal problems not on the problems of society at large, even if those societal problems impinge directly on the home. There are few specifically topical references, though the range of acceptable subject matter has widened dramatically. The values of the more conservative segment of society are titillated, even challenged, but in the end reinforced.

In an interview with Jim Bawden (*Toronto Star*, 28/12/82), Paul Denis, once television critic for the *New York Post* and founder of the magazine *Daytime TV* talked about how addictive TV soaps are. ''It's a close personal bonding'' which he attributes to the constant close-ups, the continuity for viewers who are uprooted, or travelling, the release for the house-bound and disabled, the very moral point of view. The first commandment of soaps is ''thou shalt be found out and punished,'' a distinct contrast to the night-time soaps. Most important, the sense of identification is paramount.

Yet, as John Hartley pointed out,[6] soap opera lovers in Britain (and in the USA) can exert control over these programmes in ways closed to them in prime time. Audience research often determines if a character goes or stays. When the producers make a mistake, as they did in cancelling the contract of the leading actress, Noele Gordon, on ITV's *Crossroads* (1964-present), the public simply would not put up with it. Having written her out, they had to write her back in.

Halliwell on *Crossroads*: "a perfectly dreadful soap opera"; Purser: "the lowest, feeblest and laziest form of drama ever invented as if they'd simply gone home and left the tap running." Their distaste for soap opera is at variance with their more general sympathy with popular television forms. My general observation is that series or serials based on power and money are treated more seriously by both academic and newspaper critics than series based on human relationships. When the soap operas shifted their timeslot to prime time and their focus to include money and power, critical attention also shifted into high gear.

According to Canadian actor Donnelly Rhodes (*Toronto Star*, 30/12/78), his character in the American soap *The Young and the Restless* (1973-present) was killed off because he had had a fight with the writers. He demanded an instant demise rather than a lingering illness and got it. However, the audience was so taken with him that there were flashbacks from previous episodes and references to the character for over a year after his "accident."

Esslin theorizes that the fascination of soaps rests on the fact that we get to know the characters far better than our next-door neighbours. Moreover, as we have just seen, the longer a soap runs, the more fine-tuned it is to the audience. Producers find out what the audience wants or expects through the usual ratings, more detailed audience research, and through the volume of mail received. Is Esslin then right to worry that, through the quantities of soaps and long running series now seen by millions, their "disbelief may be permanently suspended"? (33) Despite the incontrovertible evidence from fans over the years who send flowers to "Mrs. Archer's" funeral (BBC), who write to soap heroines to tell them that their husbands are cheating on them, or who confront actors in supermarkets with the news that their soap opera girlfriend is pregnant, I think not—at least not for the average citizen of either sex whose viewing is interrupted by a cake in the oven, a baby waking up from a nap, or who fixes a lamp socket with one ear on the story line.

The perennial love of gossip, the joy of seeing others who are worse off than oneself (particularly if they are rich or male authority figures like doctors and lawyers), and the intimacy and continuity of daily viewing are enduring attractions. In these respects Canadians are no different from their American or British counterparts, except that, unlike the rest of the world which makes indigenous soaps, Canadians are addicted to foreign ones.

Soaps are the life-blood of the American television industry. With their studio sets, three or four camera set-ups, smaller (but growing) casts, they are relatively cheap to make. Two hour-long episodes are taped in a day, five in three days. They are immensely popular with large audiences of both men and women, many of whom now own VCRs with a specific button which makes it easy to tape their favourite soap five days a week. Soaps generate enough advertising revenue to subsidise money-losing ventures like prime time documentaries. They are popular with all ages and classes. Although the majority of the audience is 18- to 49-year old housewives, a wider audience of working women, shift workers, professional athletes and now the general male population is growing fast. Even so, their ability to generate revenue does not appeal to our independent television producers or networks. They are even cheaper to buy.

Since there had been very popular Canadian radio soaps, the reason that there are no Canadian television soap operas on the CBC or CTV appears to be that the private networks find it much cheaper to import the problems of Middle America than to invent new ones in North Bay or Battleford. In 1985-86 all EST CBC stations carried *All My Children* (1970-present) at 1 p.m., *Dallas* at 2 p.m. [some affiliates carry the CBC news-features package *Midday* at that hour], Granada's *Coronation Street* (1960-present) or other programming on the affiliates at 3 p.m. and a sitcom at 3:30 p.m. CTV's *Lifetime* [1-2 p.m.] makes a real effort to have Canadian content for an hour while Global runs a very light package of news and light information and entertainment items with thoroughly pleasant people at noon. But the fact is that afternoon across Canada on all stations belongs to American or British soaps and Americanized game and talk shows with a scattering of cartoons and fitness shows. There are quality reruns on PBS. The Canadian educational networks broadcast programmes like *Math Patrol*. We are still waiting for an English Canadian soap opera.

Round One: Scarlett Hill A.K.A. Room to Let

The CBC did make one short-lived attempt, in the early 'sixties, to broadcast a rather odd mixture of soap and traditional serial. Based on recycled radio scripts, *Scarlett Hill* was an independent production by Taylor Television Broadcasting. The widow of Robert Howard Lindsay, who had written the radio pieces on which the new serial was loosely based, helped Canadian writers to adapt the old scripts.

In the first season, 1962-63, the aim was not a traditional soap, but a weekly serial in five episodes, daily at 4 p.m., with a new story every week. This heartwarming round of stories with changing locales and characters was to range from light comedy and human interest to suspense drama (*Globe and Mail*, 24/8/62).

In its second season, the programme was retitled *Room to Let* and changed from self contained week-long serials to a serial set in a boarding house, with continuing characters like an orphan keeping her family together, a young idealistic reporter, a ten-year-old girl, a lonely old gambler, and a successful, bad-guy lawyer. The programme began to develop more obvious resemblances to soap opera. Most of Canada's best actors appeared on *Scarlett Hill/Room to Let* in those two years, giving the rather thin and often slow-paced plots the benefit of top acting skills, unusual in afternoon soaps, and some sense of timing and subtextual nuance. Two-and-a-half hours were taped live over a four-day period.

I saw episode one of "The Luck of Amos Curry" (1962-63). As Amos, John Drainie gives the audience a gritty integrity which is overlaid on the dramatic clichés of his loneliness and boredom as he tries to settle into the Bella Vista Rest Home. Joe Austin as his confidante, Mac, shares the first scene, his children and their kids take focus in the second. Except for the fact that there is only one plot line, visual and verbal styles are reminiscent of soap opera. For example, although there is no standard voice-over explaining the little town as a setting, *Scarlett Hill* opens with a film insert with recognizable soap music as a main title. The *CBC Times* even published a map of Scarlett Hill for its fans. Predictably, directors continued to use many closeups and tight two-shots. In fact, for much of the episode, all we see are talking heads. In the only plot summary surviving in CBC files I later found that Amos wins $20,000, goes to stay with his children, and finds that he is happier away from their lives on a daily basis.

"Return from Night" was a concluding episode. The set for the general store was fairly scant, but the dialogue in this episode had a more colloquial Canadian flavour than others I saw. I had no difficulty picking up the plot threads from the previous four episodes because there was plenty of recapitulation, another basic soap convention. The acting was crisp and quite genuine compared with many American soaps, possibly because they had already worked with each other many, many times in CBC anthologies. Unlike most other soaps and serials of the time, this one had no music between scenes. The plot was entertaining, a little slow to contemporary eyes but not by the standards of 'sixties soaps, as a more mature man and woman find each other after many years.

Generally, the first season seems to have featured good, wholesome stories with bits of sharp social and psychological observation and the odd flash of humour. It was handled by one of the most skilful freelance directors for the CBC, John Trent.

The second season of *Scarlett Hill*, transformed into *Room to Let*, also shown from 4-4:30 p.m., but saw a new set of characters. Kate Russell (Beth Lockerbie) is the all-purpose mother figure who lets rooms with her daughter, awkward brother (Ed McNamara), curious Pearl Tolliver (Cosette Lee), impetuous Walter Pendleton (Ivor Barry), and a student nurse. The episode of *Room to Let* I watched (wr. Paul Wayne, dir. George McGowan—at about the same time he was doing *Mother Courage* for *Festival*) was taped in week four of the season and set, like the others, in a large, ornate, boarding-house. Guilt and its penalties seem to have been the focus of several plots—a very conventional motif in soaps. The interlocking, unfinished plots, the stereotypical characters, the dialogue, and the intrusive music over confrontations also belonged to the soap opera formula. Unlike *Scarlett Hill, Room to Let* did have a sponsor. There was no third season. Since most soaps are over before 4 p.m., I wonder if the fact that the kids were home and keeping mothers busy by then contributed to its demise.

The House of Pride

Eleven years later, the CBC English network, in one of its drives for ratings and in a self-conscious attempt to find Canadian versions of popular entertainment forms, tried another soap under John Hirsch as Head of Drama. *The House of Pride* (1974) was ahead of its time in that it was scheduled in prime time, a shift that, together with a major speed-up in the usual slow moving pace of daytime soap opera narrative, helped put 'eighties programmes like *Dallas* and *Dynasty* on top of the ratings a few years later. However, *House of Pride*, their forerunner, had several problems. In the first place, the idea was to broadcast interconnected stories about a sprawling family based in various cities across the country, thus giving regional centres a share of production and reflecting the country's diversity. Unfortunately, the viewer was expected to tune in three times a week, a time commitment no television series has ever coaxed from its public over a season in prime time. Worse still, according to Audience Research reports (Tor. 75-67), viewers became very confused by the plethora of characters, plots, and settings. The plots simply did not catch and hold their attention. In defiance of serial convention, there was no single star like J.R. (*Dallas*, 1978-present), or Alexis (*Dynasty*, 1981-present), both glamorous villains, or even Joanne, a mother figure-heroine for at least thirty-five years on *Search for Tomorrow* (1951-present). Finally, the logistical problems of taping in so many locations were immense.

The audience enjoyment index averaged 60 for the first twenty half-hour episodes, with much of the audience female, older and less educated. This was not high enough, and older, less educated females rarely have money to spend on the kinds of products prime-time sponsors sell. Regrettably, the enterprise failed.

Frank Penn, the reliable critic of the Ottawa *Citizen* (25/10/74), articulated a widely held perception about *House of Pride*, which, like the successful American series that followed it, combined the characteristics of the serial and of soap: "It's difficult to point an accurate finger at any flaws in any level of the production. The acting is, at best, excellent, in Murray Westgate's totally convincing portrayal of the dour and bigoted—yet curiously likable Dan Pride [the clan's patriarch] and, at worst, merely competent. . . . There's nothing at all wrong with the direction and only occasional lapses in the dramatic conception by Jack Nixon-Browne and Herbert Roland [both very experienced producers]. Moreover, within the high quality soap opera framework of the basic story line, George Robertson's writing and dialogue is plausibly human [Penn compares it favourably with *Jalna* of then recent memory]. But something is missing, and I suspect it is less a matter of substance than simply being out of the right television time. *The Plouffe Family*, of almost sacred memory to some millions of Canadians on both sides of the language fence, was not nearly so slickly produced nor, could we but sit through a few reruns, any better written or acted than *House of Pride*." He went on to ascribe the success of the Plouffes, "an attempt to bring soap opera serial to prime-time," mainly to the fact that there was only one channel available in much of the country in the early 'fifties. Thus the audience, intrigued with the new medium, "got caught up in the most trifling details of the comings and goings and felt as comfortably familiar in their kitchens as our own. Yet I doubt if the Plouffe magic would work in today's multi-channel world. We would find the pace too leisurely [quite true] and the commitment required for keeping up with a serial story too binding, against the instant appeal of shows which offer a beginning and end with each episode." *Dallas* et al. proved Penn wrong, but *Dallas* appeared once a week and concentrated on the exotic, the world of oil and high finance set in larger-than-life Texas.

The Plouffes

I do not completely agree with Penn's analysis of the success of the Plouffes. Only one episode in English seems to have survived, but I did see clips of some French-language episodes at an ASCRT dinner honouring the series' creator, Roger Lemelin. It was evident from the very eloquent tributes offered at that dinner that *La Famille Plouffe* (1953-59; English version, 1954-59) meant something else altogether to Québec audiences; a distinctive voice and precursor of nationalism, a reflection of working-class dilemmas and aspirations, a slice of a life long vanished, but fondly remembered. Though the actors were identical, English Canada saw something rather different. For them *Les Plouffes* was a stylized look into the kitchen window of a culture almost unknown to Anglophone viewers, a wonderful mixture of the exotic world of Québec in the 'fifties and the universal wrangling of family life. It went off the air in English Canada only because it was cancelled in Québec, not because interest waned. Lemelin's replacement for both networks (*The Town Above* 1959-60 on the English network, 1959-61 on Radio-Canada) did not work out.[8]

In Québec, however, from the 'fifties on, Radio-Canada was commissioning fine scriptwriters like Marcel Dubé to boost the ratings with more teleromans—hybrid blends of soap opera and serial conventions with topical subject matter, often working-class settings, easily recognizable characters drawn from a distinctive culture, realistic issues, lots of plot twists, some comedy and, above all, entertainment value.

The CBC made one last try at a daytime soap in 1979, according to Marilyn Kneller who has been with CBC Edmonton since 1961. I rely on her memory here, as the thirty-five tapes have been wiped and I have found no other record. *Country Joy* was a summer replacement produced in Edmonton by Mark Schonberg. Writers included playwrights Sharon Pollock, Warren Graves, and newcomer Bob Tessier. The cast of characters included a new wife, the mother of two teenagers (one of each sex), her husband's mother, a family doctor cum friend, and a (standard) rich bitch, all living middle-class lives in a small town not specifically located in Alberta. According to Ms. Kneller, the soap was killed, not by low ratings, but by lack of funds to continue.

No traces have come to light from the 'sixties of other independent soaps bought by the English network of the CBC. In the 'seventies, CBC stations struggled along with what might be termed "one-set semi-soaps"; for example, *Paul Bernard: Psychiatrist*, a 1971-72 prime time co-production shown on the CBC, and rerun extensively on private stations. Chris Wiggins played the comfortable, detached, authority figure of the psychiatrist. The "hook" was that an actor with a well-researched problem "talked it out" with the "doctor" over the course of a week, while he made comments and suggestions. I did not see many episodes, although the series is still shown on channels unavailable to me. From those I did see, the series did not appear to fall into traps of sensationalism or gross oversimplification—at least, not when compared with the current wave of pop psychiatrists who answer complex questions in five minutes on radio phone-ins and television sex therapy shows.

About the same time, OECA devised a hybrid soap-sitcom called *The Real Magees* (1973, produced by Brian McKeown) which combined sketch-comedy, interviews, improvized situations and discussions, triggered by but not necessarily connected with the guest. Michael Magee played a man who was articulate, sardonic and stubborn; unfortunately, the persona of his wife was the stereotypical, shrill nag. Both sounded uncomfortable in his/her role. It was not a very satisfactory experiment in indigenous daytime television. In fact, it reminded me of *55 North Maple*, Max Ferguson's uneasy foray into television. Neither show captured the special flavour of the talent of these two very funny men.

Family Court (exec. prod. Dan Enright, prod. Herbert Fielden) re-enacted real court cases at 4 p.m. using the same enclosed acoustic environment and live quality enforced by a restricted set characteristic of soap operas. Again, well-briefed professional actors presented testimony. The cast included judge, doctor and social worker. A series of small dramas, it had overtones of soaps, sitcoms, and crime shows. All these programmes are located on the generic grid somewhere between soap opera and sitcom.

SITUATION COMEDY: THE HARDIEST PERENNIAL IN THE GARDEN

Despite a temporary loss of popularity in the early 1980s (which *The Cosby Show* reversed in 1985), sitcoms have been as popular a staple in prime-time television as soaps are in the afternoons. Like the copshow and the soap opera, situation comedy and domestic comedy have been entertaining broadcast audiences since the 'thirties. Most early television sitcoms were adapted from radio. Some, like the very popular *Amos'n'Andy* (1926-50), did not survive the transition for very long, dying on exposure to lights, cameras, and scenery. Some, like the Allen's Alley segment of the *Fred Allen Show* (1932-49), were truly inventive radio and therefore too flexible, intimate, full of fantasy and verbal humour to survive a shift to early television. Others were stuck in the 'forties while the world was rapidly changing. Radio sitcoms of the period that did make the transition included *The Life of Riley* (1949, 1953-58), with its working class background, and *The Goldbergs* (1949-51), using second generation Jewish New Yorkers. Still others were adapted from successful Broadway plays, examples being the pleasant period sitcoms *Life With Father* (1953-55) and *Mama* (1949-57). Perhaps because of expense, few successful sitcoms since the 'fifties have been given period settings except *Happy Days* (1974-83, and some of its spinoffs) and some of Radio Canada's teleromans, which are serials, soaps and sometimes part sitcom and, above all, themselves.

Numerous articles have been written about sitcoms since television studies finally began to concentrate on the programmes themselves. By contrast, the premise behind much of the average television journalist's criticism is that sitcoms are somehow not really worthy of attention. As Stephen Neale points out,[9] in film and television as well as in the theatre, some genres are considered more "fictional" than others, and some have a higher status than others. Such hierarchical assumptions are not useful critical tools, and I will set them aside. After all, Mike Eaton's paraphrase of the BBC attitude to comedy obtains for our own television: "You can get away with more in comedy."[10]

"Sitcom Domesticus: a species endangered by social change"[11] is the provocative title of an article by Susan Horowitz, written before *The Cosby Show* and *Kate and Allie* began the resurgence of a form almost dormant by the early 'eighties. In it, Horowitz reminds us of the remarkable success of the progenitor of the television sitcom, Lucy [Lucille Ball] as "clown/child beset by familiar problems, surrounded by husband and friends." *I Love Lucy* was pure television, not radio at all; full of visual gags, mugging and reaction shots.

The viewing context was the 'fifties love affair with the novelty of television, the stress on the nuclear family, and the values of Dwight D. Eisenhower. Fathers knew best. Moms supplied comfort and apple pie to little Beavers and their older brothers, or were deviously inept as they tried to get round their bemused husbands. Audience research at this time showed that women "controlled the dial" and loved sitcoms.[12] In the vast majority of houses with only one set, the family-based nature of such comedy made it acceptable all-family viewing, snuggled as it was in a night's scheduling heavy with other sitcoms and stand-up comics like Bob Hope or Jack Benny. The salt and pepper came from variety shows featuring some of the most durable, witty and trenchant comedy sketches yet seen on television by such inventive comedians as Ernie Kovacs, Sid Caesar and Imogene Coca, Jackie Gleason, Audrey Meadows, and Art Carney. The last three appeared in a series of sketches called *The Honeymooners* (1951-56) which has never been off the air since. Neither has its obverse, *I Love Lucy* (1951-57).

Whether it was squeaky-clean, all-white, mostly middle-class suburbia, its manageable domestic problems solved by all-knowing fathers or patient, managing mothers, or the manic slapstick and occasional sharp wit of *Our Miss Brooks* (1952-56) or *Private Secretary* (1953-54, 1957), the form of the situation comedy of the 'fifties deteriorated into "endless rows . . . spawned by expediency and nourished by apathy . . . the sand dunes of Newton Minow's vast wasteland . . . believability all but vanished."[13] Shulman and Youman go on to chastise "contemporary sitcoms [for consisting of] artificial characters in artificial situations, being egged on by artificial laughter" on laugh tracks: "Creativity seemed to end once an occupation had been chosen for the major character in the series." There could not be a more precise description of some of what went wrong with the CBC's first prime-time sitcom, *Delilah* (1973).

In the early 'seventies, spurred by the new subject matter of gritty, working-class British sitcoms, Norman Lear and the MTM production company revitalized the form in the United States. *Mary Tyler Moore* (1970-77) could not be a divorcée in 1970, but she could be a single middle-class woman who slowly learned to stand up for her rights and whose professional family had realistic, occasionally insoluble problems. There were several spin-offs; *Rhoda* (1974-78), starring Valerie Harper, was the most successful. *All in the Family*, first broadcast in 1971, was videotaped before a live audience. It broke new ground with believable working-class characters, first-rate idiomatic writing, and some recognition of previously taboo topics: homosexuality, racism, menopause. It improved significantly over the years as the comedy of situation came to prevail over the battery of one-liners. Characters were allowed to grow and change. *The Jeffersons* (1975-84) (with upwardly mobile blacks) was one successful, if formulaic, spin-off; another, *Maude* (1972-78), gave us a loud mouthed, obnoxious and "liberal" female protagonist with the same mixture of punchy energy and occasionally touching story line as its original. In 1985, Bea Arthur returned to series television to play her usual wisecracking and confident woman, and Betty White, reversing her role as the acidic and selfish Sue Anne on the *Mary Tyler Moore Show*, played a lovable innocent on the other major hit of the 'eighties sitcom revival, *Golden Girls*, a largely formulaic sitcom whose single break-through is to focus on three late middle-aged and one old women. A more interesting variant is *Kate and Allie*, where two very different divorced women and their children progress through the stages of single parenting, working, dating, and so forth. This sitcom is confident enough to have offered in 1987 a stylish, funny and affectionate tribute to *I Love Lucy* as an implicit yardstick of how women—and sitcoms—have changed.

We must not forget that the 'seventies also saw the wildly successful *Happy Days*, a barely updated version of the 'fifties family sitcom, whose only bow to the intervening decades was the audience identification with a mildly rebellious character, the Fonz, rather than with the clean-cut teenager intended to be the star, and the spinoff variant of 'forties working-class slapstick, *Laverne and Shirley*. The obverse of *Happy Days*, the remarkably popular yet shopworn farcical comedy of *Three's Company*, did not do well in the 'eighties when it was reborn as *Three's a Crowd*.

Not all critics saw these shows as unadulterated blessings. Some worried that Archie Bunker's rampant prejudices reinforced rather than altered audience misconceptions. Sklar has this to say on the 'seventies sitcoms: "The morality lesson [was] one of the staples of the 'social relevance' sitcoms of the early 'seventies. Don't be a racist, keep your money in a bank, respect the elderly, don't abuse your credit card. From the lofty ethics to everyday advice, the sitcom plot seemed to point a moral finger at the audience, saying (with a smile and a laugh), Hey, you, watch out, shape up.''[14] As he points out, sometimes the lesson was a psychological rather than a moral one.

Sklar is even more skeptical of late 'seventies sitcoms. "Television seems more and more to be emphasizing verbal styles favored by adolescents of all ages—the cheeky impertinence that makes it better to be witty than to be right; indeed, that makes being witty being right. . . . Sitcom characters spend so much of their time these days trading one-liners that when they actually hold conversations, one might mistake it for Shakespeare. It has become harder to understand what the arguments are about anymore, other than argument for the sake of argument: in other words for the convenience of plot construction." The new generation of successful sitcoms, centred on veterans like Bill Cosby, Bea Arthur and Bob Newhart, has returned to an emphasis on dialogue and situation. Situation comedy is the most prolific form in American television. It helps the longevity of the form that is relativey inexpensive to make, once the series is established.

Sitcoms in any period do eventually reflect some of the new popular values of their times. Very little of the social turbulence of the 'sixties appeared in truly vapid sitcoms like *I Dream of Jeannie* (1965-70), *The Flying Nun* (1967-70), *Green Acres* (1965-71), *Petticoat Junction* (1963-70), and the notorious *My Mother The Car* (1965-66). By the 'seventies, however, perhaps conditioned more by what had happened in the 'sixties streets rather than by television's old conventions, audiences valued the new socially relevant sitcoms from the Norman Lear stable: *All In the Family*, *Maude*, *The Jeffersons*, and the likable characters of the surrogate work families in MTM's *Mary Tyler Moore* and *Rhoda*, as well as *WKRP in Cincinnati* (1978-83). At the 1979 Edinborough International Television Festival, Norman Lear proudly proclaimed his list of breakthroughs in subject matter fit for the sitcom: "death, infidelity, black family life, homosexuality, abortion, criticism of economic and foreign policy, racial prejudice, problems of the elderly, alcoholism, drug abuse, menopause, the male mid-life crisis."[15] This is material found regularly in CBC anthologies and series from the early 'sixties on, but seldom in prime time on American TV until the late 'seventies. In the beginning, the CBC saw sitcoms as light entertainment filled with slapstick stereotypes and general confusion. It may have been when Lear and MTM pushed back the boundaries of sitcom, and *M*A*S*H* started wide-ranging experiments in form as well as content, that the corporation's programme planners finally saw a way to blend more serious, or at least more interesting, characters and plots into the redefined sitcom formulae.

By the mid-'seventies, the stereotyped characters were also taken for a ride. *WKRP*'s luscious Jennifer had more brains than all the lechers who pursued her and a sense of humour to boot. *Barney Miller*'s Sgt. Wojo was a non-psychotic Vietnam vet, and Sgt. Harris, the superbly tailored, money-smart, urbane black bore no resemblance to the shufflers or the saints of previous American portrayals of black life. Above all, Americans dealt with Vietnam itself (in the guise of Korea) in one of television's major pop icons, *M*A*S*H*. As Horowitz points out, the serio-comic tone and multiple focus of characters in *M*A*S*H* influenced programmes as diverse as *Hill Street Blues* and 'eighties action-adventure series with comedic overtones.

In Britain, superbly eccentric comedies of social manners like *Solo* (1983) or *To the Manor Born* (1979-81), and the biting political satire of *Yes, Minister* (1980) and *Yes, Prime Minister* (1985), share the channels with pants-falling-down humour like *On the Buses* (1970-75) and *The Two Ronnies* (1970s and 'eighties). The British have a comic tradition entirely their own, stretching from the radio comedy of *The Goon Show* (1950s) through *Round the Horne* (1960s) and on into the wonderful television lunacies of *Monty Python's Flying Circus* (1970s) and *Not the 9 o'Clock News* (1980s). In 1966, Frank Muir, the urbane and witty man about words, delivered a *BBC Lunch Time Lecture* on "Comedy in Television" in which he remarked that "there is a significant difference between what the viewers will tolerate without complaint and what they really want."[16] What they got in Britain in the vigorous 'sixties were some sitcoms presenting a "comic attitude which is a product of one writer's mind and talent and cannot be written satisfactorily by anybody else." Muir's view of comedy ripe for export was formula comedy from which "you take out the nerve, iron out the idiosyncratic, immediacy, reality, topicality and make everybody lovable," a fair enough description of what happened to many British series after being adapted to American network demands.

But then, television writers do not enjoy the prestige in Hollywood or Toronto that they do in Britain. Muir identified this as one factor in the individuality, high quality, and success of British television comedy. He also pointed out that comedy series were (and still are) "hand-made," seven or ten at a time. Later, writing in *The Listener*,[17] Muir claimed that good shows like *Steptoe and Son* and *'Til Death Us Do Part* "leave no wake behind them . . . nor does this . . . way of creating comedy lead to easily recognizable trends or fashions." He saw the basic success of these two series as depending on actors and not comedians, on crisply executed realism, working-class characters, and the breaking of subject taboos. He also underlined the importance of stretching such traditional television conventions as Alf Garnett's "great slabs of uninterrupted monologue," and plots that are "either miniscule or pretty awful." Critic D.A.N. Jones thought that Alf, unlike his counterpart, Archie Bunker, was sometimes sick and vicious, with a perverse charm like Parolles or Falstaff.[18] Only in Britain is there likely to be an unselfconscious comparison of television characters within a comic tradition stretching back to Shakespeare, and obscure Shakespeare (*All's Well That Ends Well*) at that.

The range of material and tone in sitcom is enormous, but the narrative structure used in Britain and the U.S. follows a pretty rigid formula. Sitcoms can revolve around comedy of characters and situations that are improbable but somehow credible; or the ancient and well-loved comedy of physical humour, slapstick, mistaken identities, coincidence, inopportune meetings, dives behind convenient doors or under beds; or pure, silly fantasy; or slices of manageable and familiar life, as in Newcomb's "Domestic Comedy." Britain still has forms of situation comedy in which stars, like Benny Hill and Frankie Howerd, who often have a music hall background, do "turns." The American variant of this has vanished, and with it Jack Benny's deadpan monologues followed by sustained sketches using recurring characters. Breaking the usual fourth-wall convention of situation comedy, Benny would "include" the audience in an intimate little conspiracy with his famous look straight into the camera just before he delivered his perfectly timed, "wellll!"

George Burns and Gracie Allen (1950-58) shared Benny's vaudeville background. Mike Eaton points out that they did two parts sitcom to one part vaudeville. Their most interesting formal innovation was George's habit of watching Gracie's serene illogic on a television set in another room, resulting in what Eaton calls "the inscription into the text of the television viewer,"[19]—another way of intimately addressing a living room audience which, like Benny's asides to camera, broke through the fourth wall. Eaton argues that this piece of play, developed in the mid-'fifties (1950-58), "far from being an alienation device serves to establish George as the index of our identification." Such asides did take us into his confidence, but when Burns flipped the channel and caught Benny taking a crack at him, surely the convention had been flipped inside out. Such a device seems to be self-reflexive, derived from a special kind of "in-joke." No one in North America plays with the form like that any more.

C.K. Wolfe sees sitcom divided into two categories, the traditional sitcom as mimetic and the less traditional as deliberately stressing artifice with charades, skits, gags, plays within plays.[20] All the CBC sitcoms fall into the mimetic form. In either case, however, my observation is that the most important ingredient for success is characters with whom the audience can identify. A few sitcoms have a memory, and their characters are allowed to grow and change. Most do not; it is harder to sell a non-formula series for reruns, since stations might have to keep to a chronological order or use episodes they do not like. The best example of the latter is the famous *M*A*S*H* episode in which the unit's popular commander, Henry Blake, is written out of the show. *M*A*S*H* undertook many experiments in its testing to the limit the sitcom form. One episode presented surrealistic and revealing nightmares, one was shot completely from the point of view of a mute wounded soldier, another was presented as a black and white news documentary, and one was simply a monologue by Hawkeye.

Another characteristic that seems to distinguish the better sitcoms is a very specific sense of time, place, and mores. It may be realistic or more overtly "fictional," but it is a detailed, convincing, and minutely observed world, with its own rituals, geography, recurring and highly individualized characters, motifs, unresolved conflicts, and a wide range of emotional tone. Obviously, neither the mildly amusing *Gilligan's Island* (1964-67), nor the somewhat manic *Mork and Mindy* (1978-81) would fit that description. On the other hand, some of the socially relevant sitcoms can date rather quickly, where the daffy antics and titillation of *Three's Company* may not. Some sitcoms are self-reflexive: Horace Newcomb maintained that *Happy Days* was a fictional commentary on *Father Knows Best* of twenty years before, just as *The Flintstones* (1960-66, a prime time cartoon series) originally parodied *The Honeymooners*[21] He was writing before the success of SCTV, though not before the successful parodies of television sitcom on the *Carol Burnett Show* (1967-78).

The fundamental conventions established in the 'fifties of American sitcoms remain: the studio audience overhears what is going on behind the fourth wall of the single set, while viewers at home are made aware of the studio audience by its recorded laughter, "enhanced" by a laugh track. There are some exceptions to the live audience convention, but very few, aside from two or three episodes of *M*A*S*H*, to the idea that the viewer requires the companionship and prompting of laughter. The actors do not require it, of course, although it is fun to observe the odd times when laughs have come in an unexpected place during both tapings, or have not come, or when an audience whoops in delight at a plot twist. Even with two complete run-throughs to work with, editors cannot always iron out the unexpected. As a rule, Canadian sitcoms do not have live audiences, just laugh tracks. In both cases, the standard narrative structure staged with a three- or four-minute exposition of the week's misunderstanding; twelve-minutes' complication, broken in half by a commercial, with, finally, a three- or four-minute resolution and a little lesson either articulated, implied, or simply a sorting-out of the confusion of the past twenty-two minutes. To persuade people to sit through the last set of commercials, this is often followed by a one minute follow-on joke or a restatement of the "lesson learned." The credits usually run over stills of the episode as a nice little reminder of the chuckles. This has been the formula available for imitation, inflection, or innovation in Canada.

The CBC got off to a very slow start as far as situation comedy was concerned. Until 1969, in fact, the only glance in the direction of the sitcom wasn't really a sitcom at all. Three months after Canadian television finally opened up shop, the embryonic CBC drama department took on the task of presenting *Sunshine Sketches* (1952-53), thus bringing a popular Canadian myth to the new medium. The series was live, produced and directed by Henry Kaplan and written by Andrew Halmay, Rita Greer, Don Harron and others. Many of the kines survive; I watched "The Mariposa Light, Power and Census Situation."[22] Even at this date, we can see ingenious technical solutions to live television demands on real time and space. Borrowing from radio, Kaplan relies on sound effects to do a lot of the work, such as creating a complex effect, or a large crowd, or other characters; the audience, used to radio, fills in the blanks. The sets are often two flats forming "corners" in which actors perform scenes using restricted but not necessarily predictable blocking. The producer cross-fades from scene to scene. There is music in the background to brighten scenes, and voice-over while characters mime a scene which condenses the action. A graphic states boldly, "three weeks later." The wigs are rather bad, the make-up good, the sound sometimes rocky, and the picture quite clear with good definition, showing the designer's sense of detail in period and place—for example, the sign on the wall of the hotel, reading, "British Beer at all times." The general look of the sets is theatrical but the performances, blocking and camera work are scaled to television, not theatre. There is no sense of its being under-rehearsed, despite the quantity of drama put out that first year from tiny Studio 1. The most important thing is that the dialogue is a competent pastiche of Leacock's lines and lines in the style of the master.

Despite the running characters, small-town setting, and concentration on domestic problems, however, this is not sitcom manqué, but comedy derived from famous, familiar short stories adapted to the new medium. With the gentle social satire, period flavour, and rounded characters, *Sunshine Sketches* is very far from its contemporary, *I Love Lucy*.

In the 'sixties, CTV was the first to try indigenous situation comedy. *Trouble with Tracy* was a pale imitation of the child-clown-wife prototype, laced with painfully sexist character clichés and unredeemed by good writing. (It must be acknowledged that the series still had a weekday run on CHEX Peterborough, a CBC affiliate in the 'eighties.) In this same period, *Pardon My French* deserves attention as a fairly competent and relevant sitcom based on the promising idea of a francophone wife and an anglophone husband; here, because I was unable to track down much information on this production, I am depending on my own memory from some years ago. In the 'eighties, CTV bought and broadcast a new Canadian sitcom called *Snow Job* (1981-83). It was indeed a snow job! Basically it updated the worst sitcom clichés, set the whole thing in a ski resort, and added a dollop of trendy prurience. That trend continued with the even more offensive *Check It Out* (1985-present), a co-production with American pay TV of truly dreadful proportions, badly written and featuring a faded American television star, Don Adams of *Get Smart* (1965-69). Abetted by predictable scripts and sloppy execution, it is a crime against two gifted comedians: Dinah Christie, his sexy, blonde mistress, and Barbara Hamilton, the booted, tyrannical store owner, both continuing characters in this imitation of a derivation of an American sitcom. Products, idioms, stereotypes, jokes, even the cash in this supermarket are all American. In my view, it is an embarrassment but not to Canada, since nothing in it betrays its country of origin.

CBC—LATE BUT NOT NEVER

The first of the prime time CBC sitcoms, *Delilah* (1972-73), set in a small town, was not much better than *Trouble with Tracy*, though innocent of the leering that characterizes *Snow Job* and *Check It Out*. It was an inept version of folksy sitcoms of the 1960s. *King of Kensington* (1975-80) began as topical and ethnic and ended as a rather better written and more completely characterized version of the familiar "genus domesticus." *Custard Pie* (1976-77) was a satire that lost its nerve. *Nellie, Emma, Ben and Daniel* (1979) took real potential for gentle social comment and buried it in bad writing, and *Flappers* (1978-81) was both sexist and close to racist, as well as dull. *Hangin' In* (1981-87) was an odd mix of the 'eighties multi-cultural perspective and 'eighties problems, with 'seventies social commentary and 'fifties stereotyped slapstick. To the present the CBC has not really reached the innovation in form and content in sitcom that it reached in copshows. Before the corporation really even got down to work on sitcom, however, it made one false start in the genre.

Toby

The CBC's first sitcom was a show called *Toby* (1968). Perhaps because the CBC drama department seems to have distrusted the genre, *Toby* was developed by the Schools and Youth departments. The target audience for the Friday late-afternoon offering was teenagers, who would be interested in what the Corporation announced as "an exciting, lovable, teenaged girl, her family and friends." At sixteen, Toby had "the IQ of Einstein and the imagination of Salvador Dali." The latter, however accurate (and it was not), was very odd publicity indeed aimed at adolescent audiences. In a more conventional vein, interviews and press releases indicated a middle-class family and teenagers who were "lovable, amusing and intelligent," with "basically kind and well-intentioned parents." Plots were to be derived from "the pleasant, little amusing things which happen." Other characters included a "glory-hunting big-wig, a tormented chess master [very topical then], a globe-trotting uncle and a ponderous, middle-aged lawyer who loves birthday parties," whatever that meant (CBC press release, 9/9/68).

Toby also featured "JJ," Jean-Jacques Roberge (Robert Du Parc), a French-Canadian exchange student who lived down the street, as a foil, antagonist and friend for Toby (Susan Petrie). This generated such plots as: "JJ's presence leads to hilarious complications. When Toby, after a spat with him, vows to speak nothing but French, she is mistaken by her father's business associate for the family maid." I do not know whether or not that episode actually explored the potential for sharp social satire implied in the plot summary, but I rather think not.

After all, the plots came from the writers of *The Tommy Hunter Show.* One of them was quoted as saying: "We are writing to get laughs. Sure, kids spend time talking about Vietnam and drugs and sex—we just don't write about those times." Another press release felt constrained to point out that "Toby's ambition is to be queen of the universe and her destiny is to be a mother of three." Perhaps the last word on the series, which survived until June 1969, should come from actress Susan Petrie (age seventeen), who said of her character that Toby was the "type of girl who will go to college, marry the guy she meets in English 345, arrange flowers, be a good hostess and raise more Tobys." The love and peace generation then carrying anti-war signs on the streets, dropping tabs of acid, and occupying boardrooms —or even just reading about it all—failed to identify with that sort of character. At 4:30 p.m. on Friday, no one else was interested.

Delilah

The next try at sitcom was made by the drama department. In the four years between *Toby* and the pilot for the first prime time anthology, *To See Ourselves* (1971-76), the department went through an acute crisis of leadership and morale. It was not an auspicious time for trying any genre new to the CBC. What came out of the hopper and disappeared without a trace, thirteen weeks later, was *Delilah*. The basic premise was simple: Delilah is left a barbershop in trust for younger brother Vincent so he can finish school. Even at the time, the basic premise, that a "lady barber" has to be intrinsically funny, seemed a trifle out of date. Yet, the inherent sexism and coy title were not the real problem.

Both internal correspondence and press releases indicate that, when the CBC finally acknowledged the legitimacy and ratings lure of situation comedy, there was no thought of developing a low-level satire like the yokels-in-Hollywood, *Beverly Hillbillies* or city-folk-on-a-farm, *Green Acres* of the 'sixties. Certainly no one had any intention of patronizing rural and small-town life when *Delilah* was dreamed up, and yet that was the ultimate effect. The programme planners did understand, on paper, that "the small-town today is no backwoods Mariposa of comic yokels, but is very much a part of mainstream life in Canada. A little slower perhaps than the big city, it still reflects a total picture of people living in Canada in the 'seventies, but with a comic awareness. The trend of TV comedy today, in general, is towards a more realistic kind of situation." The plan was to forgo Hollywood's canned laugh tracks and tape in front of an audience: "a live audience provides energy and a greater challenge to artists."[23] Preparation for the series was meticulous; to find out about the logistics and how to set up a production schedule for this kind of programme, producer David Peddie and corporate official Robert Lawson went to California to talk to the producers of Los Angeles based situation comedies.

Bryan Barney wrote six of the scripts, giving the series some continuity, and the very able Jean Templeton was the script editor. Since thirteen episodes were planned to be interchangeable, one may conclude that no provision was made for development of character or plot. Still, the cast included some of our best comic talent: Terry Tweed as Delilah, Barbara Hamilton as Aunt Peggy, Eric House as T. J., Drew Thompson as Franny Tree, plus young Miles McNamara as Delilah's brother, Vincent, and one or two guest principals and extras.

No series should be held accountable for the sins of inept promotion, but this CBC press release of 21 September 1973 provides accurate clues to what went wrong: "Can a young attractive city girl find true happiness as a . . . yes a lady barber in a small Ontario town? [How many readers would recognize the arch parody of the daily introduction to the American soap *Our Gal Sunday*?] Surround her with a super hip teenage brother, some rather untypical [sic] friends and relatives and a haphazard assortment of local townspeople [too true] with their varied idiosyncracies and if what you have isn't exactly true happiness, well it's light-hearted and amusing T.V. fare."

Well, it wasn't. The dialogue was limp, the scenes often apparently under-rehearsed, given the elaborate slapstick gags, and, worst of all, the chance to make loving, detailed, funny observations about life in a small town was sacrificed to predictable formula comedy. For one thing, the writing revealed no familiarity with life in small communities: the condescension was almost certainly unintentional, but it was pervasive. If a viewer happened to remember the early production of *Sunshine Sketches*, the contrast was painful. Audience research reported that *Delilah* was perceived to be overacted, slow-going, not amusing, and uninvolving (Tor. 75/21). No one at the CBC defends the series; in fact, no one wants to talk about it. *Delilah* lasted one season, 1973-74.

In 1974, with John Hirsch as the new Head of Drama, the department set out determined not to commit another *Delilah*: the aim was to pull a winning sitcom out of the hat, even though expertise was still lacking. As usual, there was no opportunity to make a dozen pilots, not even a realistic chance to junk a series that was not working out once it got into production. Also as usual, the policy makers expected a hit sitcom without the backup of hundreds of script treatments, dozens of pilots, and the ten or fifteen series that make it to air on a network in the United States before one or two American hits emerge. Yet the CBC did manage to achieve a bona-fide hit on its next attempt.

King of Kensington

If one of the criteria for success in sitcoms is that a recognizable and well-liked star (in this case, Al Waxman) develops along with the show, then *King of Kensington* has been the most successful sitcom to date. On 26 November 1974, a pilot went to air that had been written by Arthur Samuels and produced and directed by Perry Rosemond, a Hollywood producer-director-writer (*Good Times* and CTV's *Pardon My French*) whose backgound included a stint as assistant stage manager at the Manitoba Theatre Centre.

We learn from a 28 November 1974 CBC press release that Sandra O'Neill, not Fiona Reid, played the "pure-WASP" wife, Cathy, and that Paul Hecht was the king of his multilingual neighbourhood, Toronto's Kensington Market. At this early stage of development, King was unemployed and living in a derelict house. Helene Winston, the only lead who survived from the pilot, played his stereotypical Jewish mother who "continues to remind him of his dead father's hope that Larry might one day become Canada's first Jewish Governor General."

After some basic changes in the concept, *King of Kensington* was given the go-ahead for the 1975-76 season. Many articles at the time repeated the unfortunate CBC release phrase, "ethnic zoo," used to describe the series' social and economic context. Luckily, the scriptwriters were not that insensitive. They did start with stereotypes—voluble Italians, and so on—but some of the continuing characters were rounded out as the series progressed. The show did use predictable gags of "the curry is too hot for King and Cathy" variety, but the clash of cultures and the unrealistic expectations on all sides also gave humorous situations arising from character and circumstance rather than outdated reflexes. Early episodes introduced an unsuccessful bookie, a Ukrainian alderman, a francophone gambler who is so successful they call him the "Bank of Montreal," the "Duke of Milan," a cabbie, Nestor, a black postman by day, aspiring comic by night, and other regulars of the "club" where they all play poker and kvetch. The protagonist, Larry King, helps anyone with anything. Yet he is another of Joyce Nelson's lovable losers,[24] dependent on a variety store to make ends meet and living above the store with his wife and mother.

Appropriately enough, *King of Kensington* replaced the American minority sitcom, *Chico and the Man* (1974-78), in the broadcast schedule. Jack Humphrey, subsequently producer for other CBC sitcoms, and Louie Del Grande wrote several early scripts as a team, as did Aubrey Tadman and Garry Ferrier. By episodes fifteen to eighteen, Humphrey and Del Grande were producing the show. Herb Roland and Gary Plaxton were the regular directors with Perry Rosemond.

King of Kensington had a very interesting premise based on the rich potential of fresh insights into Canada's ethnic mix. The original concept also depended on a unique commitment to topical humour. In every episode, the writers tried to incorporate jokes and comments about issues in the news during that week. The show would be taped immediately, and six days later it was on the air. The logistics were difficult, the concept imaginative; for example, writers managed to incorporate lines about John Turner's resignation as finance minister into the taping that same day. The actors and director adapted their delivery and timing to fit, and, six days later, the audience saw it.

At this point, the series was not being carried by the affiliates, but by "Metronet," the CBC owned stations that cover several of the largest population centres. After a short time, five more episodes were planned in addition to the thirteen already under way, and the show was renewed for the next season. The last five were each done in a different style, including one in a broad satire, to test the market for the direction the show should take. The one constant that remained was that "King" was to be a likable guy. As Sklar points out, U.S. sitcom at this time was in a very aggressive phase, with characters often on the attack. Hirsch and Rosemond correctly saw that tone as not characteristic of the Canadian self-image or what we would watch, so they did not include the subgenre among their test models.

Confirming this intuition, viewers did not really take to the other comedy show sharing *King of Kensington*'s time slot. For the second half of the season, the spot was filled by a "Canadian" (that is, made at the CBC) series of truly dismal inanity starring Britain's leering, pants-falling-down master of the double-take, Frankie Howerd. It "bombed," as they still say in the trade.

The given circumstances of *King of Kensington* are summed up in the lyrics of the song and the images under the titles. King is the "people's champion." He is a "king without a buck . . . his wife says helping people brings him luck / His mother tells a slightly different story," say the lyrics, while we see him strolling the colourful streets of Kensington Market "with a smile for everyone." At the end, there is a sendup. King is making a telephone call to help a buddy, but what we hear in a voice-over is, "That line is still busy." As he hangs up, a deep, male voice-over drawls in exaggerated bonhomie, "Whatta guy!" Throughout the run of *King of Kensington*, Waxman's basic persona remained the same—an easygoing, friendly guy, a little overweight, centred and at home in his place in the community, his store and his streets, surrounded by frantic, often shrill women and other lovable male losers. Yet, to use a very old Yiddish expression of respect and praise, Larry King is also a "Mensch."

In the second season, 1976-77, the immediately topical references disappeared because research (Tor. 75-21) had indicated that the fast one-liners were too obscure for the audience. Note, however, that the same kinds of one-liners were treasured by the audience of *Seeing Things* and often added to the script by co-producers Barlow and Del Grande or ad-libbed during the shoot. Perhaps the audience for comedy has changed somewhat.

By the second season, Rosemond and writers Ferrier and Tadman had left. Writers Tom Hendry, co-founder of the *Toronto Free Theatre*, and Anna Sandor, who went on to *Home Fires* (1980), episodes of *Seeing Things*, and the impressive drama special, *Charley Grant's War* (1985), occasionally teamed with Martha Gibson and George Allen to write scripts. Meanwhile, during the twenty-four episodes of that season, the comedy's focus shifted from the "club" to King's family.

This move may have been in response to positive aspects of earlier audience research reports indicating interest in the series' distinctly Canadian humour, enjoyment of its fast pace, and favourable reaction to Waxman as well as to the negative finding that the audience felt that the series was "too demanding." A later report (Tor. 76/04) found that, as the first season progressed, the sitcom appealed more to women than to men, more to teenagers and rural people, and that there was a rising enjoyment index and a warming response to the three leads as the show found its audience. Martin Knelman put his finger on the quality that did survive all the changes over the four seasons: "It belongs to the world of corny old Frank Capra movies . . . but with an ethnic touch . . . [a world] where the threatening qualities of life can be turned with gags" ("The Little Sitcom that Could," in *The Canadian*, 18/12/76). There was one dissenting voice. John Herbert, the author of the world-wide successful play and film, *Fortune and Men's Eyes*, in a stinging, to some degree accurate review, charged that the show was racist, condescending, and full of caricatures of

Jews, East Indians and blacks "with dumb wife, interfering mother, cronies in a card game from a beer commercial and Caucasians choking on curry" (*Onion*, 15/10/75). Some representatives of these groups were also bothered by the swiftly stereotyped characters, but others were pleased to be visible on television in roles that did not necessarily portray them as victims, or as decorative and quaint.

In its third season, 1977-78, the series was syndicated and sold to nine U.S. markets with a potential audience of three million viewers. The Canadian audience had also built to 1.5-1.8 million a week, and Waxman had become a familiar household name. The series underwent further changes when producer Jack Humphrey decided that the continuing characters "Duke of Naples," Nestor the black postman, and Max, the Jewish owner of the "club," should disappear. Since Fiona Reid wanted to return to the stage, leaving the character of Larry King in effect without a wife, these changes were not intended to focus more narrowly on the family.

Such crucial transitions can make or break a series, as the producers of *All in the Family* discovered when Archie's wife, Edith, was written out: the follow-on series, *Archie's Place* (1983), did not survive. (For an analysis of the difficult transition episode where Cathy leaves King, see below.) By the 1977-78 season, David Barlow, unit manager and then assistant producer, and Del Grande, who had been a creative force since the show was first discussed also left, to write Hollywood sitcoms. They returned to write a very funny satire of a sitcom as part of the plot for one of the three pilot episodes of *Seeing Things*.

In the revamped *King of Kensington*, a slimmer, sexier King was the protagonist in plots featuring his new job at the community centre and his girl friend Tina, played by Rosemary Radcliffe, a delightful comic actor. The series now widened its focus to include the serio-comic problems of teenagers. The complicated sexual tensions as King relearned his dating skills gave these scripts a little more depth. Meanwhile, Gladys had found a boy friend, Jack Soble the druggist. Aided by some good scriptwriting, Peter Boretski brought Jack's superbly salty character to life. Situations were now less prone to the plot and character clichés of "Jewish mama and only son" that had too often vitiated the comedy of previous seasons.

In the fourth and last season, Jack Humphrey became executive producer while trying to get *Flappers*, another sitcom, aloft, and Joe Partington took over day-to-day production. More time was spent at the community centre, a big new set, where King acquired a female boss—played by Jayne Eastwood, who is both a gifted comedienne and a fine serious actor. Also in this season, Gladys married Jack. Anna Sandor wrote many of these transitional scripts, including the final episode; other writers were Harvey Patterson, George Allen, writer/performer Alex Barris, and David Mayerovitch. The regular directors were Alan Erlich and Ari Dikijian. On 11 December 1979, after 111 episodes, the show wound up, largely because both Humphrey and Waxman wanted to move on. In the final episode, Gladys and Jack retire to Florida and Gwen and Larry agree to marry. Through all these fairly radical shifts in emphasis, most critical comment continued to be favourable and the audience stayed loyal to the show.

Joyce Nelson[25] quotes John Leonard, TV critic for the *New York Times*: "What is allowed in U.S. sitcom, even according to the people who do the best ones, is quite constricting. The characters have to be vulnerable, and to a certain extent they have to lose every week. You have to have somehow stumbled through to the right solution so that you are not presented as being better than anyone else. I sat in on discussions with the best writers in the business and they said the public doesn't like to feel that people on TV are better than they are"—a curious premise for writers of comedy. Admittedly, Aristotle and critics for 2,300 years since have argued that comedy is distinguished from the other major modes of epic, romance, and tragedy by its concern with ordinary people in recognizable situations. It is a definition, however, that excludes comedies as diverse as many of those by Shakespeare, Jonson, Congreve, Shaw, and Sam Shepard. Since the 'fifties, the viewers have loved television characters who are larger than life, especially the fathers or mothers or maids or butlers in comedy who can solve every comic family crisis. King belongs to that archetypal role, average enough yet a problem-solver; vulnerable, but a fixer; a smart-alec outsider (though little was ever made of his Jewish heritage) who is widely loved by other misfits.

In addition to watching the series during its run, I reviewed twelve episodes of *King of Kensington*. Taking them chronologically, "The Real King" (11/12/75, wrs. J. Humphrey/L. DelGrande, dir. P. Rosemond) explored a genuine dilemma: who is "Mrs. King," Gladys or Cathy. The course it took, however, was predictable. As a spoiled WASP, Cathy can't cook and Gladys claims the kitchen. Gladys makes a transparent and not very convincing (or interesting) attempt to change her ways. In the end, the women unite to shield King from endless calls for help. Thus, the women learn to co-exist. In this early episode, the sets and costumes were very good, with a real feel for the contrasting tastes of the two women.

"Delma's Decision" (5/2/76, wrs. J. Humphrey/L. DelGrande, dir. H. Roland) had a few unexpected moments such as when Delma bursts into tears, and the Kings, oversensitized to cultural mishaps, automatically wonder what they have done to offend their West Indian friends. Delma explains that it is not that her husband wants to see *Ilsa, Harem Keeper of the Oil Sheiks* while she wants to see *Barry Lyndon* but that she hates Canada. When Nestor says, "She gets like this every winter—but Daddy's going to take care of his sugar pie," she ploughs the anniversary cake squarely into his face. Both the male chauvinism and her display of temper are given a specific cultural context, and there is even a fleeting reference to prejudice as a fact of life in Canada. Unfortunately, much of the tension is dissipated by a scene at the all-male club which contains a few topical jokes and a lot of digression. Predictably, Delma goes home to the Islands, Nestor appears with another woman at King's to make her jealous (long-distance), and King and Cathy quarrel about her. In the end, Nestor flies to Jamaica just as Delma flies to Canada. Of course, all are reconciled before the last commercial. What is striking about this episode is both the hint of a comedy of character arising out of real circumstances and the fact that it is smothered in predictable plotting and digression.

In "Cathy's Hobby" (22/1/76, wrs. A. Sandor/M. Gibson, dir. G. Plaxton), the problem is stated very plainly by Gladys: "Artist? You're a wife!" Macramé is Cathy's rebellion against the fact that the household revolves around King. Nestor: "Some choice you got, King. Max's hot chili or your wife's cold rope. She's got you tied up in knots." A trendy "tourist" who is a gallery owner visits the store where Cathy's masterpiece is hanging and puts it on display. Cathy is vindicated when someone buys it; her independent identity is established at last. The buyer, however, is the ever-protective Larry. By implication, her search is unresolved, her restlessness unquenched. Unfortunately, the inevitable reconciliation does not really allow for any real ambivalence through subtext or coda. The macramé hangs in their bedroom and that's good enough.

These three episodes from the first season are probably intended to reflect the topical issues brought temporarily nearer the top of the social agenda by the International Year of the Woman (1975) both directly, through a character wearing a button saying, "WHY NOT"—many of us wore them despite the ambiguous and easily misconstrued message—and indirectly, through situations. The first episode explored a permanent source of comic tension (mother-in-law, daughter-in-law in same house) in very predictable ways. The second started to examine a pervasive sexist attitude, but was undercut by displacing the basic conflict into the safer environment of a clash between two people from a different culture. For the most part the episode followed formula, including the obligatory compromise—with Delma doing the compromising. In the third, Cathy's "hobby," that is, Cathy's rebellion, comes to nothing, thanks to misplaced chivalry. One wonders if Fiona Reed would have stayed with the show if her character had taken other directions. It is true that these three episodes "told it like it was, baby" for most women in 1975. Nevertheless, the writers could have implied ironic intention, or used open-ended or down-beat endings. But complex characterization, unexpected plot twists, and both dramatic and conscious irony are largely missing from the show. In contrast to early *Beachcombers'* episodes, the situations did not really develop new facets of character.

In this first season, most of the topical humour was not connected with the apparent subject of a given episode. Worse, the topical jokes were not really a crisp commentary on the news of the day. In *Seeing Things*, topical humour is largely a function of Louie's persona as newsman and struggling outsider. Most lines are improvised and very funny. Equally important, comments on trends and fads are worked into the plots as part of the texture of people's lives. Perhaps because a copshow is perceived by audiences as demanding, not relaxing, the topical humour works on *Seeing Things* despite the time lags and reruns. Morris Wolfe in *Jolts* cites weaknesses in *King of Kensington* as partly caused by the fact that the show had few experienced staff, "especially writers and editors," and contrasts this situation with that of American sitcoms, where nine people are devoted to various stages of writing an episode of a show like *Alice* (1976-83). Wolfe also gives an example of where in *King of Kensington* "the writers and producers get cold feet and resorted to smart-ass humour," something that happened too often to scenes of character interplay.

In the second season, Cathy continued the search for her separate identity, at one point boring guests at an anniversary party with psychiatric jargon ("Gestalt of Kensington," 16/11/76, wrs. A. Sandor/J. Cunningham, dir. S. Larry). In "The Check-up" (5/10/76, wrs. A. Sandor/M. Gibson, dir. G. Plaxton) the doctor rushing to get to the golf course cuts short King's hypochondriac recital of his ills and puts him on a diet. King goes jogging and orders a pizza. Cathy nags, then goes on his diet with him and faints from her 900 calories plus exercise. He decides to remain chubby (that is, himself), and she resigns as his coach. When Cathy flirts with her sexy, muscular, blonde dance teacher in "The Teacher" (20/11/77, wr. A. Sandor, dir. R. Arsenault), King is predictably jealous. In this episode, the scenes between them have the distinct flavour of a comfortable marriage in shock. His vulnerability and her comic honesty do strike some sparks. She goes to Montreal with the class. He hires gorgeous Linda Thorson (in a broadly funny performance) as his "cello" teacher. Cathy comes back. Of course, since this is a formula sitcom, both are acting out innocent fantasies with the "teachers"; eventually they must, and do, return to the status quo.

Fiona Reid's exit took place in an episode that was very different from the usual run of the show. Although written by three people, Anna Sandor, George Allen and Harry Patterson—all veterans of *The King of Kensington,* the script seems very much of a piece. "Cathy's Last Stand" (3/26/78, dir. A. Erlich) verbally reprised Cathy's various abortive searches for herself from other shows but, in an interesting twist, took a closeup look at the basic premise of the show itself. Cathy is leaving King, not because she doesn't love him, or he her, but because he is too helpful: "I don't know what I can do on my own . . . I love you too much." The opening scene in the kitchen, where King welcomes Cathy back with Chinese food, champagne and candles, is intercut with her rehearsing her speech: "Look him straight in the eyes and say, I'm leaving you." The effect is wry, yet verging on pathos.

A very good scene concludes this episode. Cathy is packing. Both remember how they met at a peace rally where, typically, King shielded her from a police charge. She remembers that when she had asked about his politics, in a typical King reply he had evaded the question with, "Peace." As he recalls the day, he wryly makes the peace sign. A long kiss and a hug. Cathy: "I'm going to chicken out." King: "Sounds good to me." (*car horn*) Cathy: "The car's here." She clings to him; gently, he extricates himself, goes to take her bag, then stops. She takes it and runs out. He turns out the light and follows. The camera pans around the room, by now familiar to the regular viewer, then back to the empty bed. After the commercial, the closing credits show him with another friend, still interfering and interrupting. The show itself is atypically understated, deeply felt and emotionally truthful. Reid and particularly Waxman show their mettle as performers in this episode.

It is quite remarkable when a situation comedy uses the exit of a character to question its most basic dramatic premise. Here, we see Larry King's "helpfulness," not as comic bumbling, or even as an effective Mr. Fixit role, but as a "cage" which drives those nearest him either into continuous nagging criticism, as in Gladys's case, or away altogether. This kind of implied critique is particularly striking in a sitcom, because a character who is helpful to everybody and sorts out their problems each week has been the fundamental premise of many, many sitcoms for close to fifty years.

The next two seasons of *King of Kensington* took a rather different direction. Now the dipsy female was Tina, played by the touching and funny Rosemary Radcliffe as less brittle in her vulnerability than Fiona Reid. Gladys was allowed to mellow a little with Jack on the scene, and his tartness made her less the overprotective mother figure. Moreover, she no longer had a daughter-in-law to bully. Nevertheless, many of the scripts continued to present her as a mugging cartoon character.

Christmas shows are compulsory on sitcoms. *King of Kensington*'s 1978-79 Christmas show (wr. A. Sandor, dir. A. Erlich) developed the rather hackneyed idea of the characters being snowbound in a highway restaurant with a pregnant woman about to deliver her baby (a son, of course), helped reluctantly by a spoiled, mink-and-migraine female; there was even a Santa Claus. The reader will be able to fill in the rest. Two other episodes, "School Daze" (10/20/78, wr. D. Mayerovitch, dir. A. Dikijian) and "Diabolical Plots" (11/18/78, wr. A. Sandor, dir. A. Erlich), had better scripts. In a neat twist, King, trying to better himself, also helps Guido, a drop-out taking a night course exam, which both of them pass. A caricature of a middle-aged school-marm unjustly accuses Guido of cheating. King, though still frightened of her after twenty years, defends Guido. Unexpectedly, she believes "Lawrence." Thus, behind the caricature, there is a lovely, quite funny glimpse of the delight of the good teacher when a student really has done the work. Coincidentally, King's bête noir is slain. "Diabolical Plots" is enlivened by a running gag, some zippy lines on various forms of sexual and racial prejudice and a funny but credible fight between King and Gwen, his new boss.

In "Down but not Out" (6/12/79, wr. G. Allen, dir. A. Erlich), times have changed enough on family shows for Gwen to come to work in the morning in evening dress and earrings and to clean her teeth with coffee. The major plotline centres on an old man who comes to the community centre every day, but will not take part in the activities. He has been forced to retire, is now mysteriously impotent, and too much the proud Scot to tell his wife he has been retired. Ruth Springford, as the wife King summons to take him home, colours her few lines with just the no-nonsense tenderness needed to resolve the situation credibly. Meanwhile, there is also a crisis about a planned gay dance, a bunch of women are doing an art class using a "life" (i.e., nude) model and Alderman MacReady, inspecting the centre, is having apoplexy. The satire on the politician is quite funny, particularly when he decides that the whole centre is controlled by a "deviant ethnic conspiracy." In fact, I found this episode the funniest of the ones I saw.

The series ended with an episode titled, "Moving On" (13/3/80, dir. A. Erlich). Anna Sandor was given this difficult job. For once, reversing roles, King makes a decision for Gladys: she should sell the building and move to Florida with Jack. The agent's cold eye roams around the familiar set and creates a nice tension for the audience, who do not necessarily want a "stranger" to see its imperfections. Meanwhile, a self-made, struggling young Italian couple fall in love with both store and apartment and can't fake it, as shrewd buyers are supposed to. There is even a reprise of the immigrant theme through a reference to King senior's self-made success with the store before he died. For once, a little Yiddish creeps into the script. Then the Cortinas are turned down by the bank because they have no credit line. Finally, King gets a mortgage to buy the building from his mother, rents the store to the Cortinas, and decides to stay on in the flat above. As people from previous episodes gather for a farewell party for Gladys and Jack, Gwen, having turned King down previously, finally consents to marry him. Realizing that his new wife should have a fresh start, King floats a second mortgage and the Cortinas own the whole building after all. Throughout the episode, Waxman runs a subtextual history of the series on his face and in his body language as first the agent, then the Cortinas, and lastly Gwen go over the apartment. Altogether, the episode made a very satisfactory conclusion to *King of Kensington*'s four successful years.

As light entertainment, *King of Kensington* worked. Much of the series was uninflected but skilfully performed formula sitcom. Some of the writing was predictable, thin, even unfunny. The richness of Kensington Market and King's own Jewish heritage were never fully explored, the topical humour never integrated into the general comedy. Often, episodes were almost pure slapstick. Rarely did actors get a chance to stretch their characters in new ways. Had the episode about Cathy and King parting been more typical of the standard of writing and complex humour arising from the characters themselves, the first two seasons would have been more distinctive, and more fun. Once the eternal triangle of Gladys, Cathy, and King was broken, however, the series did take on new life. Writing was still uneven, but the plots and jokes were less predictable and the new ambience gave relationships and situations more energy, more credibility, with a broader range of emotions than the first two seasons had managed to achieve.

Custand Pie in the CBC Eye

Custard Pie was an atypical sitcom concept in its early stages. It began life as *Rimshots* (1976), a pilot devised, like *King of Kensington*, by Perry Rosemond. The characters were conceived for Andrea Martin, Catharine O'Hara, and Dave Thomas, at that time members of Toronto's Second City improvisational troupe (later of *SCTV*), and Saul Rubinek, then an actor in Toronto's alternate theatres. In October 1976, the pilot was made on film, directed by CBC veteran George Bloomfield. Apparently, it was never broadcast, but Martin Knelman, through what he calls devious routes, saw the show and enjoyed both "its desperate punchiness" and the interplay of rounded characters: "What came as a shock was how good it was" (*The Canadian*, 24/9/77). He much preferred it to the watered-down version that became *Custard Pie*.

Between the pilot and the series, several decisions were made by the Head of the Drama Department, John Hirsch, and his immediate supervisor, Jack Craine. The show was retitled. Film actor Peter Kastner (*Nobody Waved Goodbye*, 1964) was cast as the lead, changing the balance. The show was not to be filmed, but taped twice a week, using the studios. The original four actors, who had no contracts, had demanded that the series be shot on film and also that they be given script control. What the three Second City actors (and others) could accomplish with full freedom to devise their skits became clear with their well-deserved success on *SCTV*; but the sharp, self-reflexive, adult parodies of *SCTV* are not sitcom. The original four were replaced.

The new series situation revolved around a "four-member comedy team in a boarding house. . . . [The plots arise] from the clashes between their show-biz aspirations and their boarding house realities as the four fight together about staying together. [The series is centered on Kastner's character], hooked on variety, stardom, power and keeping the group alive, using his charm, stubbornness and entrepreneurial skills" (Bruce Kirkland, *Toronto Star,* 3/9/77). Kate Lynch played "a serious actress," Doug McGrath, "a naive and goofy but lovable clown," and a singer played "Maggie . . . the hip, heavy rock chick who plays off the others' more straight lives." The emphasis had shifted from wit and confrontation, from socially significant humour to a fairly typical, late 'seventies "work" family. My off-air notes from that era corroborate both producer Perry Rosemond's claim that the series was primarily farce with a "more staccato rhythm than most sitcoms have . . . zanier than anything done in the U.S.,"[26] and Blaik Kirby's judgment that it was "one of those hyperthyroid shows about four young people who do silly things that are supposed to be funny, always at the highest pitch of hysteria" (*Globe and Mail,* 24/9/77). The show was clearly not to his taste, though he liked Rick Moranis (then D. J. on CHUM, later of *SCTV*) who played an empty-headed American D. J. In fact, both British and American sitcoms with this kind of stress on physical comedy often work for audiences who enjoy good slapstick. *Custard Pie* did not. The bite of the original concept was gone. The viewers did not warm to these particular four actors as characters, or their "aspirations" or their jokes. *Custard Pie* died after one season.

Flappers

In 1978, the drama department, now under John Kennedy, decided to try again with a pilot for a longer-lived sitcom called *Flappers* (1978-81), set in a Montreal speakeasy in 1927. The team assembled for this sitcom were alumni of *King of Kensington*, then still going strong: the executive producer Jack Humphrey; writers Sandor, Allen, Patterson, and Mayerovitch; producer Joe Partington, and director Alan Erlich. Quoted in a CBC press release, Humphrey called it "unique . . . a comedy created for its stars." These were Susan Roman, a pretty and personable actress, Victor Dey, Robert Lalonde, and classical actor Edward Atienza, with Denise Proulx as Francine the cook. "It is structured around several stories rolled into one half-hour supported by exciting original music." Like *King of Kensington* in its settling-in season, 1979-80, this series was carried only on Metronet and not really promoted. The audience-enjoyment index was variable, but the numbers climbed slowly, indicating acceptance of the French-English mix. In the second season, twenty-two episodes went out over the full network, where it found an enthusiastic audience. Humphrey thought of it as "not formula sitcom, but a half-hour comedy play." The continuing characters were May Lamb, the lovely owner of the club; the sexy singer Bunny; Andy, the sardonic band leader; Francine, surrogate Mother Earth to one and all; Oscar, the gruff, domineering chef; Yvonne-Marie, the flirtatious cigarette girl, and Robert, the macho boy friend.

I found the characters undeveloped, the plots predictable, the francophone stereotypes offensive (particularly in the context of the 1980 Quebec Referendum), May Lamb singularly vacuous (repetitiously the "comic" victim), and the laugh track particularly intrusive—as did many of the newspaper critics. In fact, despite the visual delights of 'twenties costumes and the display of shapely legs, *Flappers* reminded me of *Delilah*. Nevertheless, the series enjoyed modest international sales in Italy, West Germany, and Australia, and went into a third season before being superseded by *Hangin' In*.

Meanwhile, in the late 'seventies, the CBC had begun a search for other sitcoms with the help of Stephen Hickcock, an American from a New York audience-research firm. Hickcock was given $500,000 to produce a flock of pilots which were to be broadcast, rated, and studied. A self-professed admirer of Fred Silverman, ABC's sitcom king, he came up with five. Within a year, however, Hickcock resigned his post charging that the CBC had thrown the pilots down the drain with no promotion, no star interviews, no press stories, and that a clique of about twenty people made all the network's comedy shows. ("All" equalled two at that point!)

Available information about these pilots is sketchy. Of the five, "Fit to Print" was about a big-city editor gone home to run a Nova Scotia weekly. "Any Number Can Win" was about lottery winners. "Midweek," "a parody of a public affairs show," was never broadcast. "North East Passage" was to be an innovative revue, a comedy send-up of Canadian history. Jack Miller caught it, and later called it "Pythonesque" (Toronto *Star*, 4/2/78).

For good or ill—and in the aftermath of a fight between Light Entertainment and the Drama department which was not resolved until John Kennedy took over as Head—the middle management trusted their intuition, not the "science" of conceiving, promoting, and testing sitcoms as their new American experts thought they should. Without promotion and evaluation, a pilot hasn't a prayer. With all the promotion and audience evaluation in the world, however, no one can guarantee that a sitcom will survive its first season. The executives may well have been right to trust their intuition and cut their losses.

Nellie, Daniel, Emma and Ben

The only survivor of the flurry of sitcom pilots was "the one about the old people:" *Nellie, Daniel, Emma and Ben* (1977, 1980). Despite the lack of publicity, three hundred favourable phone calls about this Vancouver-made pilot were received after its telecast in late 1977 (Peter Wilson, *Vancouver Sun*, 4/1/80). Critics found it "sprightly"—an unfortunate word, roughly equivalent to, "how amazing that the old dears can do it," but perhaps I am being unduly harsh here. Two years later, in 1980, the show finally went to air from Vancouver on tape with a low budget, in front of live audiences.

The actors were indeed senior citizens, significantly older than the cast of *Golden Girls*: Alicia Ammon was ninety-three; her son, Jack was seventy, Barbara Tremaine was eighty-three; and Roy Brinson, sixty-four. This kind of casting gave the series an authenticity such portrayals often lack, but also imposed some physical limitations on writers and directors. The four actors played a "rebellious group of oldsters who set up a communal house after escaping from a prison-like home for the aged."

The difficulty was that the genuine problems and triumphs of aging Canadians were not the focus of the series. Despite the freshness of the basic idea and the "realism" of casting actual old people, the series was an average sitcom, no worse and no better than dozens of American sitcoms that come and go with the seasons. Plots included one senior having a "torrid affair with an Avon Lady"; another coaching a baseball team to prove that he is insurable; a group protest against a bus-fare increase; and discrimination against the old people's commune by neighbours. Not much was developed during the season from "Nellie's malapropisms," "Dan's hard-working blue-collar background," Emma's "irascible grey-panther image," or Ben's "high energy con-man," to quote the press release. The writers did not reflect the lively and varied ways senior citizens see themselves. Instead, these actors were trapped in a sitcom which did not even seem to be making use of what they themselves knew about being old. The pity of it is that, in Canada, when a promising idea does not work out, it cannot be given another chance. *Nellie, Daniel, Emma and Ben* simply vanished after thirteen weeks.

Hangin' In Does

In 1981, Lally Cadeau appeared in a highly rated, emotionally charged three-part mini-series called, *You've Come a Long Way Katie*. A few days later, she opened in a weekly sitcom called *Hangin' In*, as a calm, together, wise-cracking social worker who exudes warmth and professional wisdom.

The audience research report wryly points out that the series' low audience enjoyment index in its early days could have been the result of resentment at "their heroine [Katie] now exchanging cheap one-liners" (Tor 81/12). "Cheap" is an undeserved pejorative, but the audience was already alert to one substantial difference between this sitcom and its predecessors; they were enjoying the story more than the humour.

Under John Kennedy, the CBC Drama Department had decided to create light drama with comedy rather than a sitcom where the story was sacrificed to a set number of laughs per minute. In its early years, *Hangin' In*'s scriptwriters often gave us a range of comedy arising from both credible situations and characters. Like any promising series, it grew on the audience who came to accept and then enjoy the balance of humour and seriousness in each episode. In *King of Kensington*, King is the neighborhood problem-solver who does not always see how to resolve his own domestic problems. In this double sense, Kate, the social-worker protagonist of *Hangin' In*, resembles her popular precursor. She also tackles, with varying success, the problems of adolescence and ethnic background among inner-city kids of all classes, while grappling less successfully with her own problems as a single, professional woman. In some ways, she resembles Gwen, King's boss at the community centre, except that she had no developing private relationships until the series' last few episodes. As time passed, Kate got involved in the private lives of her work family: Mike, the younger, male worker and Webster, the broadly comic receptionist. Unfortunately, their private lives were usually rife with cliché: Mike was intimidated by his father-in-law, Webster wanted to win a bake-off, and so forth. The dilemmas of the teenagers are what spark the better episodes.

Again, Joe Partington was producer and Jack Humphrey, executive producer. Anna Sandor is credited as the creator of the series. In my view, it had two major strengths. First, week after week we met new young actors in their teens who were fresh, full of energy, usually well cast, touching or awkwardly and plausibly adolescent, and often fun. Secondly, the problems faced by these kids, leavened by the relationships with the adult continuing characters, were often presented credibly, and sometimes with imagination and sensitivity. Once in a while, Kate and her younger colleague Mike didn't have all the answers.

In this series, episodes are usually structured so that there are two problems or situations of varying degrees of seriousness and one short skit-length interlude. The stories could be about which twin to cast in *Annie*; a Greek father refusing to let his daughter be a cheerleader because of the "costume"; a voluble client with agoraphobia who refuses to leave the centre; a grandfather forbidden access to his grandson because his son walked out on his volatile Latin American daughter-in-law; three children who live in a van because their father can't find affordable housing; or a young prostitute trying to find a job. Often it is a simple vignette:

Pre-teen: Excuse me. How do you French kiss?
Webster: How old are you?
Teen: Twelve.
Webster: You kiss first on one cheek and then on the other.
Teen:—and Diane tried to tell us it was such a good deal. . . .

Too much of the time in later seasons the scripts of *Hangin' In* were reduced to a series of one-liners rather than credible dialogue, with Ruth Springford's Webster handed many of the truly inane lines. Characters appeared without even a brief sketch of their background to place them and make us care about them. Too often, we endured a frenetic pace, stereotyped characters, and worse, a set of glib responses to trivial problems.

In a sitcom, the lines and the "business," that is the actions of the characters, have to be funny. By 1986-87 *Hangin' In* was getting tired, primarily because the Kate/Mike/Webster triad was going nowhere. The three recall an archetypal family of older sister-mentor, brash younger brother, and gruff aunt who is something of a poor relation, often the butt of the comedy, and usually devious in getting her own way. None were presented as developing, changing characters with complex working or personal relationships. Mike's new wife and in-laws did not really open up new opportunities for the show as Cathy's exit did for *King of Kensington*. When the problems to be solved were presented with warmth and humour by good young actors, the permanent members of the cast also shone, but when the plots were stale, there was not a lot of shared experience and character interplay to fall back on. Too often, flippancy and predictable physical humour replaced real wit, irony, or good humour. I did not ever see a tenth of the dramatic range I know from other programmes that Lally Cadeau and Ruth Springford were capable of showing. I don't know what David Eisner can do, since the episodes I saw did not make many demands on him. Indeed, after several seasons, I am hard put to characterize "Mike" at all.

The programme touched on many serious problems as well as various cultural backgrounds, yet there were few hints of racial or religious prejudice or any real poverty or their long-term effects on young lives. The converted old house that serves as the drop-in centre provided an interesting visual background for the action, yet the blocking was often flatly horizontal, an impression intensified by predictable shots from the usual three-camera set-up, and the uniform lighting. The sitcom rhythm of two-shot, closeup, reaction shot, two-shot was wearying. Unlike *The Beachcombers*, the longevity champion in CBC series television, it did not add new, continuing characters in the foreground or evolve into a broader range of tones or try formal innovation. *Hangin' In* was markedly better than its rival *Snow Job*, and beyond the reach of *Check it Out*. However, it ran out of energy long before Lally Cadeau decided to move on in 1987.

COMIC PERSONAE

Two other areas of comedy that overlap with sitcom, in the sense that sustained characterization and some narrative form are part of their structure: sketch comedy, which uses a revolving set of characters from *The Big Revue* (1950s) to *SCTV* (1970s); and CTV's *Bizarre* (1980s), and the more difficult to define genre created by "Charlie Farquharson," "Sergeant Renfrew," "Rawhide," and "Fred Dobbs." I omit any detailed consideration of that round-the-world favourite, *Wayne and Shuster*, (1952-) because Paul Rutherford in his forthcoming book on Primetime in Canada can and will do a better analysis of their particular mixture of broad slapstick and literate parody. Morris Wolfe has it right in his characteristically short and pithy comments on the show in *Jolts*, particularly when he recalls the impact of their early television sketches: "Their literate slapstick was a sheer delight," and he goes on to cite "The Brown Pumpernickel." I would add the "Julius Caesar Caper" and "Hamlet at the Ballgame." Wolfe speculates that their comedy was weakened by the demands of being in a sitcom for CBS in 1961 and the move to videotape, which permitted laugh tracks: "There's a much more self-conscious quality to their work now." Yet, as Wolfe says, they can still mount wicked parodies of TV commercials (and blockbuster movies).

In fact, except for *Wayne and Shuster* and *SCTV,* both of which show the Canadian prediliction for revue sketches and parody—particularly of the successful forms of foreign cultures—and *Seeing Things*, Canadian television is short of pervasive, jokey humour. Yet, as *The Canadian Connection* (1986) demonstrated in its parody of CBC's exposé documentary style, American television is riddled with expatriate Canadians writing, acting and producing comedy. This probably was not a result of Mackenzie King's secret revenge on the U.S., as the programme mockingly alleges, but it may be a form of retribution on the CBC, because Robert Goulet, Lorne Greene, Anne Murray, William Shatner, Michael J. Fox, and Howie Mandel did not find enough opportunities in Canadian film, or Canadian record labels, or on the CBC.

Canada has, however, produced two of the best political satirists in the world in Max Ferguson (Rawhide and Company) and Don Harron (Charlie Farquharson). We also rejoice in the political wit and wisdom of the acidly cantankerous Fred Dobbs, invented by Michael Magee, and the Member from Kicking Horse Pass, Bobby Clobber, and Sergeant Renfrew, who are the creations of Dave Broadfoot. Presumably their humour is not exportable, or maybe their creators just prefer Canadan audiences. In any case, unlike *Wayne and Shuster* and *SCTV*, these four masters of topical humour rooted in specific, continuing characters have stayed in Canada.

The character of "Charlie Farquharson" developed in part in versions of *Spring Thaw* and appeared, unnamed but unmistakeable, in the CBC TV's *'55 in Revue*. Both Canadian theatre and television have since shown a remarkable gift for this form of entertainment. Its essential characteristics are parody, satire and sketch, sometimes loosely organized around a book or story line, as in the 'fifties hit, *My Fur Lady*, or around familiar characters, as happens on *SCTV*, *Bizarre* (1980-, which unlike *SCTV* has no running thread of Canadian parody and allusion) and *The Royal Canadian Air Farce* (CBC radio and television, 1973-). In other instances, the comedy gathers around the actors, as in most *Wayne and Shuster* specials, or one comic actor and his mask, or masks.

Throughout the 'sixties, satirical sketches were part of the new format of current-affairs programmes like *This Hour Has Seven Days* and the purely comic and more uneven *Nightcap*. Annual satirical revues, late-night comedy, and pointed topical sketches attached to more serious fare have disappeared altogether on television—regrettably, since one cannot help but think that the considerable accomplishments of *The Journal* could be leavened with a little investigative satire. On CBC radio, with *Dr. Bundolo's Flying Circus Medicine Show*, *The Frantics* (renamed *Four on the Floor* for CBC television, 1985), and *The S and M Comic Book* specials (1986, 1987), *The Royal Canadian Air Farce*, and on some private stations, as well, these forms still flourish.

Harron, Magee, Ferguson, and Broadfoot

When we look at our four perennially popular comic actors—Harron, Magee, Ferguson, and Broadfoot—the most obvious common thread is that all are members of the radio generation and still comfortable in that medium. Yet they are also part of our television history. Oddly, they have no apparent successors in the rising generation. When they retire, there will be no new comedians interested in creating durable characters, their wry comments on the passing scene rooted in a particular place—Beamsville, a farm outside Parry Sound, a tacky CBC studio, a log cabin on the fifteenth floor of RCMP headquarters—or a set of Canadian icons: country and western disc jockeys, CBC announcers, hockey players, mounties, Members of Parliament.

Harron, Magee, and Broadfoot are basically affectionate towards their characters, while Ferguson seems more detached from his personae. None "play" themselves, as stand-up comedians often do. They can be quite merciless in their satirical thrust. All four take special delight in skewering politicians but, with the exception of Magee's, their humour is not overly aggressive. None of the actors would be recognized on the street without their comic masks of voice, carriage, accent, and idiom. These characters do not depend primarily on costumes or make-up. They depend first on their voices, and then face and body language for characterization. All four use regional accents, idioms and even dialects with skill and exactness that are not "stage" rural or "jocktalk." Of course, with Ferguson's Rawhide voice, accent and idiom were deliberately parodic.

None of these comedians is honing a persona that has an off-stage public life of its own, as Jack Benny, Bob Hope, or George Burns did. The masks really do hide the men, perhaps a distinctively Canadian characteristic. Harron and Ferguson have both said that the trap for them is anger about a given subject. With anger, they lose the humour and then the audience. Magee treads the line with hard boots, Harron and Ferguson with more subtlety; Broadfoot is the least likely to cross it.

In contrast to American comedians, behind the masks of these four men are other careers: Magee is a well-known sportscaster. Don Harron has been a classical actor, radio and television host, starred in very good drama specials including *Reddick* (1968) and *Reddick II* (1969), and directed the excellent, much underrated TV drama *Once*. Max Ferguson is a writer of eerie radio plays, broadcaster of a music programme drawn from his own collection, and a cook. Only Broadfoot sticks with being a comedian. However, he is at home as writer and actor in radio, TV, theatre, and industrial shows.

Each has built a specific world that is unique: Fred's sleazy, now long-gone Ford Hotel (Magee), Charlie's barn (Harron), Sgt. Renfrew's log cabin on the umpteenth floor of the concrete tower of RCMP headquarters (Broadfoot), and a rundown CBC studio full of crazies (Ferguson). A strong sense of fantasy characterizes all four in their sketches, particularly in yarns spun to live audiences where anything can, and does, happen. When Magee's Fred appears as a guest on a show, he often simply blathers on until the producer pulls the plug.

Broadfoot's three main comic characters are really Canadian sacred cows: a mountie, a hockey player, and a Member of Parliament. Harron's two, old Charlie and rather-come-lately Valerie Rosedale, are very consciously drawn from opposites—the farm and the city, male and female, declassé and upperclass. Ferguson's most beloved country-and-western D.J. character, Rawhide, Charlie and Fred all take the time-honoured persona of old age and its privileges, Lear and Fool in one. Their rural flavour is not the nostalgic pastoral that usually surfaces in advanced, urbane societies, but the sharp, wry, rather conservative and wholly self-confident stance of older rural people with long memories and no personal stake in power games.

All improvise brilliantly in character, but it was Max Ferguson who took on the task of writing and performing a satirical radio sketch every single morning. The pressure was great indeed, and he refuses to resume that kind of broadcasting. All four write all or part of their own material, a significant contrast to most American comedians.

Like many other comic talents, these actors are definite loners. Fred Dobbs has no on-stage soul-mates. Broadfoot does belong to a comedy company, but "the Member" and "Renfrew," who tells his tale of woe in a monologue, were his creations long before *The Royal Canadian Air Farce* existed. Max Ferguson invented his own company, with Max playing all the parts and the indispensable Allan McFee as straight man. Harron can act with others, but Charlie for the most part kept his "wife and former sweetheart" Valeda (and son Orville) in the wings—though Catherine McKinnon, his wife, was introduced as Valeda to theatre audiences in the summer of 1983.

Though Ferguson is not now seen on television, the fact is that these four men have performed in character in every medium including print. American television consigned Jackie Gleason to reruns, dumped Red Skelton, and retired Carol Burnett's wonderful, blackly comic and absurd "family." But even in an age when the television variety show is dormant, Charlie, the Sergeant and Fred are still around. The man behind Max Ferguson's masks retired from comedy of his own volition, though he may be back one day with a new stacked deck of characters. On one wonderfully loony occasion in 1970 on the CBC, "the Member," "Fred," "Charlie," and Max as various characters got together in a revue "book" show involving a trip to Ottawa, civil servants and live chickens loose in the corridors of power. Would that the CBC could get them together for another.

The francophone equivalents, Gratien Gélinas's *Fridolin*, Viola Léger's performance of *La Sagouine*, and Yvon Duchamps's large cast of characters do not often appear on anglophone television or play with their anglophone counterparts. A superbly simple, unobtrusive taping of a live performance of *La Sagouine* (1978, pr. Mark Blandford) did introduce the character to much of English Canada, to our great delight. Gélinas has acted in adaptations of his serious plays, and Yvon Duchamps appeared once with Don Harron in a television special which both evidently enjoyed as much as the audience. Yet the cross-fertilization of talent is rare in either direction.

"Up here" we have enjoyed the uniquely flavoured comedy of Magee,

Harron, Broadfoot, Ferguson, and Wayne and Shuster for many years. We are also faithful fans of American comedians, sitcoms and soaps. As with American series television, when Canadian soaps and sitcoms are analyzed there is no automatic correlation between quality—inventiveness, range of tone, depth of characterization, fresh subject material—and ratings. *Flappers* was as successful as *Hangin' In*. *The Trouble with Tracy* has run for years in syndication on CTV affiliates as cheap Canadian content. Nevertheless, when we think about the slow, even reluctant entry of CTV and CBC into these evergreen forms of series television, the basic viewer context is no different from that of copshows or family-adventure shows. Most Canadians can flip to American versions at any time. Once again, the shows most fondly remembered and successfully sold abroad are the ones that are distinctive variants on the familiar formulas.

NOTES

1. Anchor Books, Garden City, New York: 1974.
2. Oxford University Press, Toronto: 1982, 35.
3. W. H. Freedman and Co., San Francisco, 1982, 74.
4. See "Nazaire et Barnabe: Learning to Live in an Americanized World," Joan E. Pavelich, in *Canadian Drama*, vol. 9, no. 1, 1983, 16-23, and Renée LeGris on World War II serials in *Propagande de Guerre et Nationalisme dans le Radio-feuilleton (1939-1955)*, Fides, Montreal: 1981.
5. 10 November 1982, on a panel on TV Ontario.
6. In a paper at the Fourth Annual International Conference on Television Drama, May 1985.
7. PAC Acquisition identification; *Scarlett Hill*—Crane films ltd., 84-344, no telecast date.
8. See Renée LeGris, *Dictionnaire des auteurs de radio-Feuilleton québécois*, Fides, Montreal: 1981. Her entry on Lemelin points out that *Liberty* magazine called him the best television writer in the country.
9. Genre and Cinema, 19.
10. *Popular Television and Film Reader*, eds. Tony Bennett, Susan Boyd-Bowman, Colin Mercer and Janet Woollacott; Open University, British Film Institute in association with Open University Press, London: 1981, 31.
11. *Channels*, Sept.-Oct. 1984.
12. Ibid, 22-3. Patricia Mellencamp's stimulating discussion of two 'fifties classics in "Situation Comedy, Television and Freud: Discourses of Gracie and Lucy," in *Studies in Entertainment: Critical Approaches to Mass Culture*, ed. Tania Modleski. Indiana University Press, Bloomington: 1986, 80-99.
13. *How Sweet It Was: Television: A Pictorial Commentary*, Arthur Shulman and Roger Youman. Bonanza Books, New York: 1966, 122-23. Though this is

essentially a picture book, the comments are both trenchant and knowledgeable. Both writers were long-time *TV Guide* staff members.

14. Op. cit., 67.
15. *Popular Television and Film Reader*, 42.
16. Fifth series -3-BBC, London: 1966, 17.
17. 10 August 1967, 161-62.
18. *The Listener*, 10 August 1967.
19. *Popular Television and Film Reader*, 31.
20. Paper at The Second International Conference on Television Drama, May 1981, "Bilko's plots and the *Bilko* Plot: Toward a Structural Definition of TV Sitcom."
21. *University Film Association Journal* 30, no. 2 (Spring 1978), "Toward Television History: The Growth of Style," 9-14.
22. Continuing characters were William Needles as the banker, Paul Kligman as the proprietor of the hotel, Eric House as Jeff Thorpe, Alex McKee as Judge Pepperleigh, Jack Marigold as hotel owner Jack, and, in this episode, Hugh Webster as Shorty, the census taker (with John Drainie as narrator).
23. Gloria Lyndon, planning memo, 5 February 1973.
24. "TV Formulas: Prime-Time Glue," *In Search*, Fall 1979.
25. Ibid.
26. *TV Guide*, 19 November 1977.

5

OTHER SERIES AND MINISERIES
*"The best of tries and
the worst of tries!!"*

Generic analysis does not always provide helpful insights into the nature of series television. Nevertheless, a critic looking for the distinctive has to separate the variant or hybrid which is different from or inflects formula from the failure which differs not because it provides fresh insight, or even because it is a good example of a successful formula, but because it isn't very good.

In *TV: The Most Popular Art* (1974) Horace Newcomb first identified a popular subgenre derived from the television of the 'fifties, 'sixties and early 'seventies. He called it "The Professionals," television drama about the lives of doctors, lawyers, social workers, teachers, editors and reporters. Nine years later, together with Robert S. Alley, Newcomb interviewed David Victor, producer of *Dr. Kildare* (1961-66), *Marcus Welby, M.D.* (1969-76), *Owen Marshall, Counselor at Law* (1971-74), and *Lucas Tanner* (1974-75), all of them series he would classify as "professionals." Newcomb and Alley describe Victor's work and, incidentally, much of this subgenre as mainstream television. When we apply the template signified by the comments in square brackets to Canadian variants of this genre, we find similarities—and differences. "Contemporary in setting, [yes] unrestricted by the tightly defined codes of action-adventure formulas [yes], focussed on tense, life-crisis moments [yes], these works always return for their core to the emotional responses of realistic, threatened, perplexed central characters [no]."[1] A collection of CBC series in the early 'seventies did bear superficial resemblance to this sub-genre. Victor's distinctive touch was to add "specific information regarding incidence, treatment, legal implications and social services available [yes]" and touch such topics as autism, homosexuality, mental retardation, senility, and mastectomy in the early 'sixties when the word "breast" was forbidden on American television. The CBC approached topical issues with even more range and honesty—see *Wojeck, Quentin Durgens, M.P.*, and *The Manipulators*. "Plots are infused with emotion [yes]. Character interplay is the main focus [yes]." Critics argue that these protagonists sketch the ideal man in his particular professional role creating, as some real doctors and lawyers say, unrealistic expectations on the part of clients. This is not true in CBC drama. Newcomb and Alley argue that "Victor's particular stylistic emphasis is on a skilful blending of ideal and social reality that points simultaneously to failure and the possibility. Indeed, he measures our failures precisely in terms of our dreams." They admit, however, that Victor "does not challenge implicitly or explicitly the deeper forms of society that create and maintain these institutions." Wojeck, Louis Ciccone, Rick the parole officer in *The Manipulators*, many characters in the *For the Record* topical dramas and film specials do that. McQueen, the reporter and Corwin, the doctor really did not.

The first Canadian series, *Wojeck*, was a hybrid that might have simply been an imitation in the David Victor mould of ideal "professional." But as we have seen, *Wojeck* did not imitate anything on television at the time. Neither protagonist nor plots bore any resemblance to the tidily structured melodramas of *Dr. Kildare* and *Marcus Welby, M.D.* Indeed, it more closely resembled the intricacies of the copshow-mystery genre. The CBC did not try to follow the 'sixties American trend toward "the professionals" subgenre. Instead a handful of series about a variety of people whose work puts them in touch with people in crisis, some successful, some less so, appeared on our screens largely because Ronald Weyman with his Toronto film and tape units and Philip Keatley in Vancouver saw them as a way of looking at topical issues while supplying the continuing characters who build audience loyalty demanded by the programmers. Ratings were becoming more important. The CBC Drama Department was also entering a phase of crisis in leadership. In Toronto skilled directors were resigning and morale was low. But before that period of creative burnout occurred *Wojeck* and *Quentin Durgens, M.P.* pointed CBC drama in new directions.

Quentin Durgens, M.P.

Ron Weyman devised *Quentin Durgens, M.P.* (1966-69) as the "other half" of the same time slot, presumably intended for the same audience. David Gardner produced and directed most of the first season episodes with Peter Boretski directing others; Ron Weyman was executive producer. Like *Wojeck*, *Quentin Durgens, M.P.* was tested in a pilot, but unlike its partner, it had a trial run of six half-hour episodes, written by George Robertson, in *The Serial* (1965). Durgens, elected for the first time in his father's former riding, tries to get a dam built to prevent spring floods from damaging Mennonite property. Unfortunately, the Mennonites see the flooding as God's will, and worse still, they do not vote, so there is no political advantage in helping them. David Gardner was quoted in the (*CBC Times*, 9-15/10/65): "It's basically an ironic drama. Our M.P. wins the battle but loses the war. George Robertson's scripts show the relativity of values in political life—the issues that aren't just a matter of right and wrong." The series also explored that sense of irony in the powerbroking of "The Road to Chaldea." Some episodes were written to be scheduled for showing during the time of a real election or with a leadership convention. In this sense, it paralleled the complexities of human behaviour found in the best *Wojeck* episodes.

Yet *Quentin Durgens, M.P.* did not have the unexpected and exhilarating impact of *Wojeck*, partly because the scripts were not as consistently strong. Another difficulty was that the show was taped (on location and in the studio), not filmed, with the inevitable limitations this implied, particularly in the days of clumsy editing and flattened images. Certainly, as the rebellious but very decent and inexperienced young member, Gordon Pinsent had as much charisma as John Vernon in the role of the weary, yet determined coroner with an acute conscience and a jaundiced eye. Both series looked at current issues. Quentin Durgens found himself tangling with the definition of obscenity, whether the country was looking after its Korean veterans, violence in minor-league hockey, and how women Members of Parliament were systematically excluded from the power structure.

How power is exercised behind closed doors is a matter of endless fascination to the average viewer when it is mixed with glimpses into the personal lives of powerbrokers, as we have seen in *Dallas*, *Dynasty*, *Empire Inc.* (1982), *Kate Morris V.P.* (1984), and the documentary series *The Canadian Establishment*. On its own however, the subject is not fascinating enough to be habit forming as *Vanderberg* (1983) demonstrated. In *Quentin Durgen*'s case, viewers may have cared about such limited topics as the conflict between home responsibilities and the demands of being an M.P., or the problems of caucus leaks, or conflict between Anglo and Québec Members because they could see analogies in their own lives. Nevertheless, the dramatic impact of such issues was less than the subjects of *Wojeck*.

In the episodes I saw, there was usually a strong opening, clear exposition, a fairly balanced case made for and against the "good-guy" position (if there was one) on an issue, a subplot with some humour, a sense of what small towns look and feel like, and an ear for educated rural speech with a regional accent—something that is still very rare in CBC television drama. There was also some good precise detail; in "You Take the High Road" (24/1/67), for example, the ex-con's release from prison into a flat, frozen countryside as bleak as the walls he had left, or the hockey action and coach's dialogue in "A Case For the Defence" (28/1/69).

Camera and script worked well together in "It's a Wise Father" (10/1/67). Its main subject, pornography, is balanced by a subplot about Durgens not spending enough time with his adolescent son. There is some carefully observed social tension in the episode. In the scene in which the all-party, apparently all-male, parliamentary committee meets to consider the issue, the magazine distributor defends his business with the familiar excuse that "no one's forcing anyone to buy the stuff." In fact, he supports an "adult" designation as a booster for sales. When a woman committee member arrives, Quentin is suddenly unable to continue to display the evidence. We hear him flounder as the camera shows, in clear but anonymous closeup, a male hand rapidly doodling a naked female rump. Meanwhile, the M.P. refuses the suggestion that she withdraw to examine the material separately, pointing out dryly that she is over twenty-one. Abruptly, the issue takes on a different coloration.

The move to draft a law to censor this material falls though. Not only is Durgens defeated, but Pinsent's portrayal also suggests that he is partly persuaded of the futility of his cause. The camera follows him as he takes the material to an incinerator to protect his own son from exposure to it. A closeup reveals a picture (creased and folded to shield the viewer and the CBC), of little girls, six or seven years old. Dialogue in the programme's debate is balanced by conflicting statements like, "You might as well try to ban loneliness," or, "We have to dry up the demand for their product." The conclusion is that pornography is not controllable by legislation. Perhaps so, but the programme began with a violent near-rape in which a bruised, terrified girl is left on a floor amid battered copies of pornographic magazines, and it is this image and that of the children which linger uneasily in the mind. The tension between reason and argument and the potency of image making, whether on television or in print, gives the episode a complex irony beyond its surface resolution.

McQueen

Quentin Durgens, M.P., Corwin (1969-71), and *McQueen* (1969-70) were all closer to the successful *Wojeck* model than to David Victor's prototypes. Topical issues often inspire plots with documentary, behind-the-scenes flavour—more successfully in *Durgens* than *McQueen* or *Corwin* because the character of Quentin Durgens offered much more to work with. He had to deal with complex, occasionally funny constituency problems and, at the same time, learn the caucus system and find his way over the tortuous paths between the back rooms and the floor of the House of Commons.

Playing a newspaper ombudsman cum investigative reporter, Ted Follows as McQueen was never given a distinctive character rooted in a recognizable place. The situation should have provided plenty of variety in the story lines, but it did not. A half-hour format does not compel a series to be superficial, but the whole enterprise had a tired feel to it; predictable plot twists, flat dialogue, and rather uninteresting subsidiary characters, subsiding too often into stereotype.

One episode I chose to review tackled the highly charged issue of Americans who dodged the draft by fleeing to Canada during the war in Vietnam. Despite the crowded frames, rough sound, jump cuts and 'sixties music, the episode simplified the complex question, and the divisive passions it aroused, into clichés such as Jenny's, "Maybe Gary's fighting a different kind of war." When McQueen does come on side, it's with this kind of rhetoric: " I'll get him through that customs shed if I have to use a crowbar." Even the title is self-consciously hip: "Home Is Where The Heart Ain't" (18/11/69). Immigration officials are 'sixties "pigs," and Margot Kidder's Jenny is the beautiful young reporter with her heart on her flowered sleeve. Unlike most episodes, this one ends on a "downer" as Jenny's tears intercut with Gary's walk back across the border into the arms of the FBI. We are not even spared a last glimpse of the WELCOME TO CANADA sign. By this time, however, even the legitimate sub-issue of Canadian immigration policy depending on the individuals who actually man the border has been undercut by the conventions of the American professionals genre: self-important, closeup-ridden, sentimentalized "topical" television that offers condensed and oversimplified versions of life-and-death issues. Followes, a competent actor, but without the star quality which kept many American professional shows alive, could not compensate for the scripts. The series lasted only one season.

Corwin

Corwin, a psychiatrist turned idealistic G.P., played by John Horton (also produced in 1969), gave its professionals a full hour to unravel their crises. By its second season (1970), however, writer Sandy Stern had gone to Hollywood and René Bonnière had been replaced as director, which may explain why the series showed little continuity. It was scheduled in the *Sunday at Nine* (1970-73) slot, where "single dramas alternate with light entertainment and special documentaries" (*CBC Times*, 8-14/11/69). Viewers must have wondered what this series had to do with the other dramas in the slot: *Volpone*, *Noises of Paradise*, and *Power Trip*, "a futuristic teleplay by Antony Lee Flanders"; or with Wayne and Shuster, a documentary on Bonaparte, a revue called "What's New," and two specials featuring "Our Pet" Juliette and folksingers Simon and Garfunkel. Perhaps that is why the press release identified *Corwin* as a "miniseries." It was not a miniseries as the term came to be known in the 'seventies but the label was needed to fit the performance somehow into the eclectic grab-bag which had by then replaced the much better and more prestigious anthology, *Festival*.

The episode, "What Do You See When the Lights Go Out?" (2/11/69) dealt with a case of what one character calls "psychological murder." Ron Hartman supplied one of his sinister, blonde-fanatics as Cunningham, the obsessed minister of the "Church of Spiritual Science." Unfortunately, John Horton's Dr. Corwin faded beside his intensity, and the obligatory mentor, Dr. James (Alan King), was trivialized by bromides like, "You don't treat a patient with large doses of indignation." Admittedly, the episode does try to evoke the minister's nightmarish visions with some desolate location shooting, smoke, surreal children shot out of focus or in slow motion—even skull masks. It's a good try but, like so many of the drama's moments, far too calculated in its theatrics.

As Cunningham's growing paranoia unfolds, the plot is full of suspense, reasonable pacing, subplots, and good subsidiary characters. Cunningham's death by vision and heart failure provides a suitably melodramatic ending. However, as a psychiatric case study the piece is overwrought. Worse still, from an aesthetic point of view, the laboured, surrealistic images do not augment our sense of the minister's inner torments, which are far more vividly portrayed on the actor's face.

Corwin should have been able to count on audience fascination with the human mind and its terrors, but the protagonist Dr. Corwin was neither written nor played as a strong presence—by contrast to John Vernon's Wojeck, who made his presence felt even when he was not in the foreground of an episode. The two parts of "Any Body Here Know Danny?" and the episode "Who Killed the Fat Cat" suffered even more heavily from stereotyped characters and heavy symbolism. The series symbolized to me the loss of skilled workers, of the imaginative energy and leadership Weyman had shown in the 'sixties.

The Manipulators

When Newcomb looked at "the professionals" as a subgenre, he found a plethora of doctors and lawyers, often in pairs: usually one older, wiser and more cautious, and the other, younger, more intemperate, sometimes father and son but more often, mentor and protégé. All were male. Together, they would tackle societal problems and questions of responsibility. The issues might be new to the audience and even a touch controversial, but the basic formula prevailed: two stars heal, protect, or solve the problems of guest actors every week. "Increasingly we have also come to expect some sort of knowledge, some information that is directed toward the audience. Such advice embodies traditional American values and in structuring these shows television is openly and explicitly didactic."[2]

Two of the American series Newcomb considered strike me as connecting to *The Manipulators*. The first was the wildly popular *Ben Casey*, first aired in the fall of 1961. Played by Vincent Edwards with dark brooding charm and an undercurrent of barely controlled violence—a refreshing change for audiences tired of Richard Chamberlain's bland, blond Dr. Kildare—Casey and his mentor, played with a raffish air by Sam Jaffe, opened up a "masterfully contrived medicine cabinet of torn emotions and mended nerves. . . . The appeal of this show lies in our thirst for violence—one aspect of our hypochondria."[3] As a performer, Marc Strange, star of *The Manipulators* (and in 1987, one of the writers of another "professionals" series, *Street Legal*), had the same kind of dark magnetism, often projecting pent-up rage. Like Casey, he was in constant tangles with bureaucrats, the champion of his losers, parolees in this case, and sceptical without being cynical.

The other American series I compared with *The Manipulators* was *The Defenders*. Produced by Herbert Brodkin (*Playhouse 90*) and shot on location in New York, the series was script-edited by the superb anthology writer Reginald Rose, with the result, according to Edith Efron, that all the episodes have his hallmark, "a logically constructed drama built around a severe ethical conflict, intelligently complicated by a few more brain-cracking ethical conflicts and building to a sharp climax. Characterization too bears the Rose brand of moral paradox."[4] The producer called it the most liberal show on the air. Nevertheless, when all the paradoxes are uncovered and admired, Efron concludes that "the viewers remain secure in their knowledge that, before the hour is up, justice will triumph over the procession of high-minded lawbreakers and murderers. In the universe of Reginald Rose, good intentions still pave the road—if not to hell—to jail."

Yet in many ways *The Manipulators* had no real connections with American models. The best episodes were like one-act plays, full of texture and distinctive both visually and verbally, with the story and issue not the star in the foreground. Even the obligatory pair of leading characters swiftly became one. Strange started out as Rick Nicholson, a wrongly convicted ex-con, with Roxanne Erwin as Maggie, a widow with a child. However, the older woman parole officer did not appear in the first programme and had no real presence in the others. The ambivalent title that replaced the initial working title, *The Double Bind*, was intended to convey the point of view identified by Patrick Scott in *Star Week* (30/1/71) that the two parole officers had "to manipulate the precarious checks and balances between society and its crime-scarred outcasts in an attempt to redeem the latter and educate the former." In fact, it was not that simple-minded.

This was the first full-scale regional drama series filmed in Vancouver. It was created not by a script factory but (typically of the CBC) by Ed McGibbon, an investigative reporter from Toronto, in consultation with many real parole officers. Scripts came from experienced drama writers like Merv Campone as well as McGibbon. Several were by Ben Maartman, who had not only written documentaries but also had been a parole officer. After a set of regional pilots ironed out some of the difficulties and a change in cast [I saw one of the pilots and it was not as bad as Keatley seems to think it was; in fact it was quite powerful] this "miniseries" opened to instantaneous applause and a storm of controversy. Under the experienced hand of Philip Keatley as executive producer, the first and perhaps best episode, "Spike in the Wall," was directed by Daryl Duke (*Q for Quest*). The images stayed with me for fifteen years, until I saw the episode again. Others were directed by Vancouver producers Don Eccleston and Len Lauk. McGibbon called it "documented drama," which again brought *Wojeck* to mind.

I take issue with the newspaper critics fairly often in this book, but this time I think they charted the course of the series with intelligence and sensitivity. Nearly everyone recognized, that "Spike in the Wall" (28/1/70) was good. In fact, it was memorable. It also seemed daring—not because Linda Goranson appeared "topless," as the phrase then was, but because it was unsparing in its detail of the personal agonies of a junkie, here portrayed in a remarkable performance by Jace Van der Veen.

The shy ex-con is still on a small "maintenance dose" of heroin, tolerated by his parole officer but leaving him open to brutal hassling by police. He meets, woos, and marries a small town girl who will not consummate their marriage until he swears absolutely never to use heroin again. He is too honest and strong-minded to make such a promise; no addict can promise to give up the substance forever, even if he is addicted only to sugar or tobacco. As further proof that he is not simply a weak-willed victim, he kicks his habit cold in a series of short scenes that are harrowing without tipping into melodrama. The young wife still holds out for the absolute promise, and the crucial dramatic scene finds him standing indecisively in the middle of their small apartment with the "spike"—the needle—in his hand. The threat of losing him to his old despair suddenly moves his young wife to new and touching wisdom: she removes her nightgown and offers him a choice, though by then, the choice is not an easy one to make. Convulsively, he drives the "spike" into the wall and embraces her. Not all their problems are solved, but both have left the dream world of black-and-white ecstasy and damnation for the contingencies of human warmth.

"I'd like to review the play without even mentioning the business of nudity but I guess I wouldn't have been doing my job if I'd done it that way," wrote Bob Blackburn in the Toronto *Telegram* (29/1/70): "It was something new for North American television and the fact that it was a perfectly normal and intrinsic part of the drama may just have escaped some viewers." Patrick Scott in the *Daily Star* (10/2/70): "In revolt and disgust I have witnessed as much as I could stomach of the first episode of *The Manipulators*. Is this the Canadian artistry and talent for which we are striving? is this Mr. Juneau's [then head of the CRTC] reason for protecting us from infiltration of American programming?" That letter to the editor of a Toronto newspaper reflects the sentiments of roughly three quarters of the correspondence and phone calls the CBC has received . . . out of a total of 977, the largest public response the corporation has had to any drama this season. The answers to the two questions raised are both yes, emphatically." Scott had already called it (29/1/70) "one of the dismayingly few dramas on the CBC (or anywhere else for that matter) this season that attempted to deal honestly with real people and real problems. Where *Corwin* is glossy soap opera and *McQueen* inadvertent comedy, *The Manipulators* promises to be at least a stab at real drama." Blaik Kirby in the *Globe and Mail*, also on 29 January, was more ambivalent but approving of its production quality, noting that it was "the first major fruit of the CBC's decentralization campaign."

Other episodes that season revolved around a runaway boy and an old man who befriends him, a convict obsessed with getting parole who is refused twice, an alcoholic teacher convicted of drunk-driving and now back on the street, and a good kid convicted and jailed on a charge of marijuana possession. In all of them, the emphasis remained on the difficult decisions people have to make for themselves, rather than the quick fix that so often characterized the "professionals" genre.

The second-season scripts were weaker. Another problem was that the writers did not really find any character development for Rick. The problems were still topical: a controversial treatment for narcotics; a spoiled rich girl caught up in an anarchic terrorist movement; the generation gap between a logger and his son—an episode full of pop music and soft drugs, which focused on the question, "Will he rat on his friends?"; the problems of a man running a shelter for transient youth; and Indians protesting exploitation. There is not much to say about the ones I saw—"Turn to the Wind," "X-kalay," and "The Flock"—except that they were confusing, padded with trendy anecdotal scenes and fragmented. Bob Blackburn, Patrick Scott, and Urjo Kareda all agreed. There were some good performances, notably those of Susan Hogan as the spoiled kid and Margot Kidder as the temptress of an ex-parolee, but *The Manipulators* still represented an opportunity lost. First and foremost, it badly needed stronger scripts: the basic idea was promising but, like most series, it also called for an interesting set of continuing characters to give it texture, as in Keatley's successful Vancouver series, *Cariboo Country*. Instead of developing into a long running series with a range of tone and a highly regional flavour, *The Manipulators* disappeared, to be followed next season by the more traditional copshow formula of the Toronto-based *Collaborators*. *The Beachcombers* was the only regular dramatic series to come out of Vancouver after this.

Judge

From 1975-82, shows about doctors, lawyers, teachers and social workers largely disappeared. In 1983, the CBC turned once more to such a series in *Judge* (1983-84). However, the choice of a judge doesn't fit the formula of helper or guide. A lawyer is a protagonist or antagonist. A judge is the *deus ex machina* of the comedy or tragedy, the person who, like the audience, must choose between complex rights and wrongs. By definition then, *Judge* was unlikely to fit any existing TV formula. When the judge is the protagonist, his role parallels that of the viewer—to assess the evidence presented in a dramatic conflict.

A summer replacement series in 1983, it was given a winter berth in prime time, 8:30 p.m. Thursdays, in January of 1984; it was also used to train new directors in the use of tape under experienced producer Herb Roland. In the next season, fall 1984, it suffered badly from being placed in the 7 p.m. Saturday night slot, "the children's hour," even though its best scripts had dealt intelligently with very specifically adult material. Thanks to this misjudgment on the part of programme schedulers, it disappeared.[5]

Judge presented an interesting mix of predictable and quite gripping drama. The half-hour time constraint meant that scripts hit only the high spots, which often led to a surfeit of intense emotion and gave the series a melodramatic cast which accentuated the pattern imposed by commercials, dictating that viewers be left on a note of suspense at the end of each "act." The series also had some very strong actors as defendants and in its second season, a woman judge was added as Tony Van Bridge's colleague and confidante.

Judge was very set bound, focusing almost entirely on the courtroom. The angles of the shots, the framing, and the over reliance on extreme close-ups often created a sense of crowding and visual monotony. Yet the writing was often first-class, using idiomatic, truthful, individualized dialogue that was sometimes eloquent, a quality too rare in 'eighties television. Tony Van Bridge, the star, held it together with his trustworthy, tough, yet avuncular persona. Repeated motifs included his agonizing over difficult decisions and being angered at the laws or the judicial system. Combined with a sense of irony and wry humour, these elements gave the series continuity. Occasionally, some issues were treated superficially or sensationally. Quite often, however, *Judge* avoided the temptation to approach loaded issues and then, by dispensing an artificial poetic justice, to satisfy the viewer that the issue had been dealt with and resolved.

Two episodes strike me as particularly worthy of comment. In "A Man or a Child" (wr. Douglas Rodger, dir. Randy Bradshaw, 19/1/84), a boy from an alcoholic environment, failing to appear in court, is arrested and held overnight. The judge's wife, a young lawyer, acts as the boy's counsel at a bail hearing. The presiding judge (not Van Bridge) is soured, brutally sarcastic, and indifferent to the endless, soul-destroying task of setting bail year after year. The kid is not too bright, loves cars, and has been befriended by a garage owner and his wife (Jayne Eastwood), who uses the idiom and grammar of her class to stand up for him and watches, appalled, as the judge denies bail. Throughout the episode, it becomes more and more plain that the kid is terrified of jail and has already been threatened with homosexual rape. His agony is wrenching as he is dragged off. In the next scene, the judge and his wife are at their cottage, talking about the case. They receive a phone call. The kid has hanged himself. Freeze frame on her face; end. Her shock is ours.

In the second episode, "Flesh and Blood" (wr. Gordon Rutton, dir. Danielle J. Suissa, 19/1/84), two black-garbed Jehovah's witnesses are defendants in a case where they have denied their child a transfusion. The case for the doctor and prosecution is very strongly presented in act one. The father's (David Fox) attempts to justify his actions with a pamphlet on alternates to blood is demolished. Both lawyers grandstand and are chastised by the judge. When the mother is examined (Jackie Burroughs), however, her strong convictions and unassailable faith, her suffering and loss balance the scales in a very surprising *coup de théâtre*. The judge finds them guilty, in law, but the Crown has not proven that their actions did kill the boy. More importantly, he recognizes in his verdict that society and its lawmakers will not face the issues of parental and juvenile rights, or religious freedom and its limits, expecting an insensitive system and judiciary to lock this problem away. He finds them innocent of the charges and, to our surprise, we agree.

Judge did not survive after its third season, for several reasons. No appropriate time slot after 9 p.m. could be found for its adult material, and so sensitive episodes were dropped into unadvertised, late-night half-hours. Also, training became a luxury the CBC could not afford. Perhaps, too, serious half-hour drama was suspect. There was no money available to get the series out of the courtroom once in a while. The result was that a series with considerable potential was unceremoniously deleted from the schedule.[5]

Miniseries Take Off in the 'Seventies: The British Show The Way

A Gift to Last (1978-79) (three seasons), *Homefires* (1980-82) (three seasons), *Backstretch* (1983) (two seasons), *Empire Inc.* and *Vanderberg* (1983) (six hours in one season) and shorter miniseries like *You've Come A Long Way, Katie* (1981) (three hours), *I Married the Klondike* (1983) (three hours), *Anne of Green Gables* (1986) (four hours)—have built a loyal audience for variants of the miniseries from scratch. "Miniseries" is an often misused catchall term. I define a miniseries more precisely than common usage denotes. First of all, it is not a serial; the stories are interlinked, characters do grow up, change and die. Yet each episode is a complete dramatic unit in its own right. What miniseries have in common is, theoretically, that each one has a story worth telling which requires more than an hour or two to do it justice and that the last episode should provide closure to the whole narrative. In that respect, *Backstretch* and *Vanderberg* are atypical because they were cancelled. Miniseries are justified when the viewer is invited into a richer, more detailed world than is available in a one-hour drama special or a two-hour film.

Miniseries are sustained television narratives with a memory, plot and character development, and a wide range of tone built into them exploiting television's strengths. What television drama lacks in spectacle, it can capture in the potential offered by an extended time scale, filled with emotional detail projected into the intimate setting of a living room. If scripts, performances, and production values are good, the audience will either return for several successive nights or set aside a few of its Mondays to follow the adventures of a usually sprawling cast, with many plot twists, richly detailed sets, and exotic locations or distant times. Often prompted by nostalgia or the feeling that such miniseries are somehow "good for them," huge audiences have tuned into American and British miniseries like *Holocaust* (1978), *A Man Called Intrepid* (1979), *Shogun* (1980), *Masada* (1981), *Brideshead Revisited* (1981), *The Winds of War* (1983) and *North and South* (1984). With *Roots* (1977), Fred Silverman gambled on throwing away a chancy miniseries and dumping it all into one week: he won a huge and unexpected ratings success and many imitators in presenting "miniseries" over several successive nights. A few experimented in the range of effects and narrative conventions available to television. Those which enjoyed a *succès d'estime* were dependent on excellence in writing, direction, editing, design and performances; two such were the masterful adaptations of Le Carré's spy novels, *Tinker Tailor Soldier Spy* (1979) and *Smiley's People* (1983).

The British first explored this form in the late 'sixties and early 'seventies with historical productions on *The First Churchills* (1969), *The Six Wives of Henry the VIII* (1970), *Elizabeth R* (1978), and *Edward and Mrs. Simpson* (1978). All were accessible for Canadians on the CBC or on PBS. Their common ancestor was *The Forsyte Saga*, a blend of series and serial (1967; twenty-six episodes), based on the John Galsworthy novels, which were a hit around the world. As Philip Purser noted in Halliwell: "For the British audience a hidden significance was its function as a kind of communal family history, complete with births, deaths, and juicy scandals, but also a nostalgia for what were imagined to be better days." The same comment might also be made of the otherwise very different *A Gift to Last* (1978-79), longer than a miniseries, yet sharing some of the form's characteristics (twenty episodes, wrs. Gordon Pinsent and Peter Wildeblood).

Miniseries present in a new form old questions of television ethics. The danger of popularizing emotionally loaded issues like slavery or the holocaust has been succinctly identified by the British television playwright Dennis Potter, quoted in Halliwell: "The case against *Holocaust* is not that it is bad soap opera but more—much worse, that it is very good soap opera." Fictions present no major problems; historically based dramas can telegraph how close they are to the basic facts. Some events are sacrosanct: the South must lose; Hitler's Reich must fall; the Romans will find that the Jews at Masada have committed mass suicide; World War II starts on schedule. Coincidence will bring husband and wife together in *Holocaust*, and the protagonist in *Winds of War* will be in London, Moscow, and Berlin at just the right moment for events to unfold.

Our successful miniseries do not quite look like those; for example, the short "problem" drama of the CBC's *You've Come A Long Way, Katie*, heavy on research into the problems of alcohol and drug dependency, with good writing, and an excellent lead performance by Lally Cadeau—though weakened by a didactic and rather dull middle third—or the expansive, nostalgic, carefully crafted, and immensely successful family saga of *Empire Inc.* told over six episodes, each set in a different time period. Both of them achieved national and international success.

The fact that VCRs can now timeshift such programmes makes it easier to win audience loyalty, though viewers with fast forward VCR buttons may not be willing to sit through the commercials that pay for such programming. For the CBC, the huge VCR sales in Canada also mean that the corporation is less likely to suffer in the battle of the ratings. *Winds of War*, presold by saturation publicity worth several million dollars, and the fact that it was based on a best-selling novel, was programmed against the last episode of *Empire Inc.*, in what turned out to be a real test of viewer loyalty, which

Empire Inc. won. By the time the sequel to *Winds of War* is ready, the audience will have a choice. Nielsen, the ratings bible, now asks its householders to report programmes recorded on VCR; since viewers may well "zap" through commercials, however, the industry is still uncertain whether or not to count them.

Sponsors take a chance with expensive miniseries. They can build huge audiences or be a complete bust, like non-formula movies of the week or the old anthologies, miniseries present risks. We must also remember that on American television and increasingly on Canadian television the idea is not to present a programme. The idea is to deliver to sponsors X numbers of viewers of a specific age, income and set of interests. You can be sure that many research dollars are being spent on the impact of VCRs on the sales of products advertised on television and, indeed, *Channels* (January 1987) features articles on the death of the form in the U.S., arguing that they were coopted by their own offspring, "the continuing plot series."

As one might expect, Audience Research (Tor 81-17) found that what they define as "daily miniseries," (programmes on successive nights) get a higher number of viewers (without VCRs) than "weekly" miniseries. VCR viewers include those who trouble to record the scheduled epsiodes of weekly miniseries. *You've Come a Long Way, Katie* (wr. & pr. Jeannine Locke, dir. Vic Sarin, 4-6/1/81) ran three successive nights on the full network starting on a Sunday with a strong lead-in. Over the three nights, its enjoyment index climbed from 69 to 75, and over three million viewers saw the last episode. Better still, 90 per cent of the audience were "return" viewers. As usual, the largest audience segment consisted of women over fifty, the most stable group for serious, demanding television drama. As demographics change, their numbers are growing. Their spending power may also rise over the next decade, as the issue of pay equity is addressed.

The story of Katie as a successful daytime talk show hostess fell naturally into three acts. She is given a chance at a more prestigious television job, but the pressures of the job lead her to drugs and alcohol abuse. The second episode—and the least successful because it was overloaded with information rather than dramatic conflict—chronicles her rehabilitation in a clinic; the third shows her relapse and death. As far as the subject was concerned, the script pulled no punches, did not sentimentalize or pop-psychologize Kate, and did not provide the expected easy out of a happy ending which her struggle should earn, according to the formula of "message" miniseries. Instead, the problems had not gone away, nor were her coping mechanisms up to the test. But the programme made no value judgments; it merely showed a sympathetic protagonist defeated in a serious battle for her life.

The audience report summarized viewer reaction in this way: "For many viewers the story provided a novel insight on how easily this sort of cross-addiction can arise." It went on to say that the high audience figures and enjoyment index indicated that the miniseries was perceived to be "insightful . . . not unrealistic," even though 74 per cent of the audience thought Katie would triumph. Surprisingly, only 36 per cent found the ending too depressing. This suggests several things to me as a critic: that the older women who made up the largest audience segment have a fairly high tolerance for tragic irony when the context presents such an outcome as truthful to the character; that Canadian viewers do not necessarily demand endings that fit the usual TV melodrama, and that the programme's refusal to moralize or simplify the issues did not alienate the viewers. These factors, however, did not seem to mitigate the more hostile audience reaction to another documentary drama ending of *Turning to Stone*, which I deal with in Chapter 8). Critics and audiences abroad reacted with equal admiration for Lally Cadeau as Katie and for the series itself.

The scope presented by three nights of television does not always create momentum and depth, as evidenced by the anecdotal and rather lifeless *I Married the Klondike*. The on-location scenery was beautifully photographed in this production, and some of the anecdotes were amusing. Unfortunately, however, neither writers nor actors got below the surface of what could have been a fascinating character story about a most unusual woman. We knew no more about her after three nights than we did in the first ten minutes, and we had not been shown much more than details of harmless local colour about life in the Klondike.

Empire Inc.

Empire Inc. (exec. prod. Mark Blandford, prod. Paul Risacher, wr. D. Bowie, dir. Denis Heroux, 1982) may have been the finest example of pure storytelling to come along in several years at the CBC. Despite its focus on the secrets of the self-made rich, it was not typified by delicious villains like J.R. from *Dallas* or Alexis from *Dynasty*, to which the show was inaccurately, but surprisingly, compared by television critics. The character of James Munroe had a magnetism that was closer to the repulsion-attraction exerted on viewers over so many hours by Soames Forsyte. The plots' pace and drive alone were enough to hold the viewers' attention. Audience Research findings (Tor 83-24) were echoed in my own family when my father, who seldom enjoys television's fictional modes, gleefully acted out to friends the hilarious episode in which a Scottish laird, prospective father-in-law to Munroe's weak son, goes to a brothel to enjoy Munroe's special "reserve" whore—only to find himself in the middle of a classic farce. Later, when Munroe is so financially desperate that he has to turn to the old Montreal money barons, such as his wife's father, she agrees to intercede on his behalf only if he gives up his weekly rendezvous. At this point, Munroe and the viewer discover that his wife is indeed his match, as tough at striking a bargain as he is when her opponent is down. To the laird, the young son-in-law-to-be in the brothel is simply undergoing a rite of manhood; to her, however, Munroe's special whore is a deadly insult to her pride. This mixture of comedy, misogyny, upper-class assumptions, steamy sex and casual possession—with the lovely detail of the whore's amused, languid and rather bored demeanour, forestalling any oversentimental response to these particular facts of life—is typical of the better moments in the series.

Audience research indicated that viewers relished the strong characters, especially Munroe (Kenneth Welsh), the robber baron yearning for respectability and dreaming superbly hubristic dreams; his strong-willed, spoiled, clever daughter, Cleo (Jennifer Dale); his steely wife (Martha Henry); as well as Larry, his left-wing son, and his late-in-life feisty mistress, but they were less enchanted with gentle, malleable Amy and the son who yearned to be a musician, but failed in the family business.

The series held other attractions for a larger audience. Many viewers were fascinated to see the chasm between rich and poor Montreal in the 'twenties, the Great Depression so accurately conveyed as a backdrop to the francophone/anglophone tensions, the resurgence of labour movements, and the growing fascism in Duplessis's Québec. Viewers were also offered glimpses of a few of the machinations of Mackenzie King, C. D. Howe, and others in the clique that ran Canada just after the war. The wealth of period detail created a sense of nostalgia and discovery for older viewers. For younger viewers, there was surprise and quickened curiosity about a period and place seldom seen on English-language television. There was also the unmistakable enjoyment of the predictable: when a character is making a mistake, we know it and he does not, as when Munroe suggests that his inept son bring a menacing young Communist union organizer in to keep his aircraft plant free of strikes; naturally, the son is outsmarted in every way, and Munroe has to invoke Duplessis's infamous Padlock Act. There is a nice touch of irony in the outbursts of patriotic gratitude from his duped workers. Meanwhile, viewers are also finding out something about a little known chapter of labour history in which all sides manipulated the ordinary working person.

What remains in memory of this series after a three- or four-year interval, reinforced by subsequent viewing, is a set of powerful images and well-written vignettes coming from a fully realized social milieu, a comprehensible history, a specific place and time, and a particular set of central characters. Each of the major figures took focus and held it for one or two episodes, but it was all tied together by Sir James Munroe. Munroe shares some of the energy and driving ambition of the dispossessed, the sense of irony (though not the wit) and the well-buried, aching conscience of Richard III. This powerful archetype is fleshed out by Kenneth Welsh as a protagonist who ages without mellowing, falls hopelessly in love too late, and loves and exploits each of his children, discovering that he may even have loved his wife. He is also a man who needs and gets the forgiveness of them all, even his patrician wife. Martha Henry's performance especially demonstrates her ability to convey pages of emotion without dialogue. She gave many scenes remarkable texture, serving as a counterpoint to the direct speech and devious actions of Munroe.

The more successful longer series or miniseries has a natural lifespan and satisfying ending. *Empire Inc.* concludes with its businessman antihero off to the jungles of Africa to confront a new challenge after a period of disastrous retirement. *Homefires* ends with the war. *A Gift to Last* ends when Sergeant Sturgess and his wife Sheila move from Tamarack to take up an appointment at the Royal Military College. Each provided a comfortable sense of closure to its devoted audience, yet left viewers sorry to see the close. Miniseries like *Vanderberg* and *Backstretch*, which had continuing story lines but failed to build large enough audiences or simply did not work well, do not end with an episode that completes a phase in the characters' lives. They simply stop leaving the viewer either relieved or with an unsatisfied itch to see the series revamped and given a proper farewell. Unceremonious cancellation of the programmes that are neither series nor miniseries is usual in the United States. Second chances and persistence are also becoming a rare commodity in Canada.

Vanderberg

Vanderberg (1983) was an attempt to show the excitement, tension, and conflict backstage in the Calgary oil patch, the acquisition of oil and delivery systems, and worldwide energy deals. Like Munroe, Vanderberg (Michael Hogan) is a self-made man in the sense that he builds his father's business into an empire, flying in his own jet to all corners of the world. The series was filmed on location at great expense: unfortunately, all the European locations looked much the same since, after a while, hotels bear a strong family resemblance. The Mafia at home in Italy spiced up the action with a little local colour, but not enough.

Here was a protagonist who was upstaged by his treacherous mistress— Jennifer Dale again, and doing quite well in the type of role in which she excels. He was also upstaged by his father (Jan Rubes) and by his cowboy-rough, rancher-smart business rival (Stephen Markle). His wife, on screen and off screen (played by the badly miscast Susan Hogan), whined a great deal and gazed woefully into space or at her vapid lovers. Vanderberg's kids were not only spoiled but also dull. His glamorous life was lived at the level of deals, with no human emotions beyond greed and anger. The plots were convoluted and too often predictable, with few fresh ideas or emotional nuances. One of the most touching and interesting moments occurs when his superbly macho cowboy rival tries to cover up the fact that he cannot read contracts; he is illiterate.

How does one explain that Michael Hogan and his offstage wife Susan, both quite good television actors, could play scene after scene with no spark reaching the screen? How could the skilled and sensitive executive producer, Sam Levene, responsible for several successful seasons of *For the Record* and overseer of progressively successful adaptations of Stratford's Shakespeare productions—or head writer Rob Forsythe, who has done such good work, or the devoted crew, who gave it more than 100 per cent—end up with a loser? I don't know; they don't know, nor did the network's management, though they cut their losses as soon as it was evident that the series was not working and cancelled plans for another season of episodes. A regional producer may have been right when he attributed the series' problem to the fact that both the concept and personnel were "parachuted into Calgary from Toronto" rather than rising out of a first-hand, long-term feeling for the place and time. *Empire Inc.* was excellent popular storytelling. *Vanderberg* was incoherent. Television drama is not a science or a business—it is an art.

Backstretch

Backstretch did not make the home stretch for a different reason. The early episodes were quite good; it was refreshing to see a contemporary small town portrayed with its rich variety of characters, enforced intimacy, and strong sense of community. The central characters at this run-down racing track can be seen any night in the barns or in the stands at Elmira, Hanover, Orangeville, or dozens of other B-tracks in towns across Canada. Florence Patterson gave Madge the tough edge and warm heart of the traditional earth mother. She also lent directness and real credibility to her struggles, her beer in the fridge, and man in the bed without promises or guilt. The writers also had a good ear for colloquial rural speech and, for once, the actors gave the regional accents of southwestern Ontario true value. The Winnipeg *Free Press* television guide put it this way (10-17/12/83): "We know people just like that . . . a goldmine of hometown characters each loaded with real-life quirks." The Ottawa *Citizen* (30/11/83) agreed, calling it "home-spun, earthy . . . vintage Canadian television." Even *Maclean's* greeted it as a welcome surprise, though contrasting it with *Dallas* for some unfathomable reason: "The locales and conditions of conventional life turn the series into a feast as satisfying as a strong hot mug of tea and a plateful of homemade butter tarts."

The designers caught the grubby yet glamorous ambience of a race-track where the owner actually lives in her office and the stables are a world of their own. The tension between town and track, the sexism of the racing world, the hand-to-mouth existence, the complex lore of breeding, and the legends and traditions of the sport were lovingly explored. At the same time, the series also developed stories around Ronnie, the hot-headed son of an alcoholic father, who dreams of owning and professionally driving his own horse. His scrapes and his on-again, off-again love affair with a waitress provided lots of entertaining plot material throughout the first season. It must have seemed to writer-producer Peter Wildeblood and director Jack Nixon-Browne (both of whom had done good work on *A Gift to Last*) that the second season was off and pacing.

Not so: somehow, the gritty realism, the dry wit, and the original, eccentric characters gave way to a focus on Ronnie's pseudo-Byronic temper tantrums, shallow and predictable "relationships," improbable plots, and sentimental endings—in other words, to melodrama. Melodrama is common on television; a dollop can be just the seasoning for a series, as *Empire Inc.* and some of its honourable predecessors attest. But the memorable miniseries gives viewers much more; its segments approach good anthology in the range of tone and variety of plot.

In its first season, *Backstretch* was not formula television. When I observed a studio taping of an episode, expectations were high and the atmosphere good. After the focus shifted from Madge and her buddies to Ronnie and his troubles, however, *Backstretch* began to "be on a break," lose the disciplined gait required of a successful trotter or pacer; from a strong start, the series started to fade as it rounded the turn. Given the fact that rural and small town life, the lower-middle and working classes are seldom seen on the CBC (or any other North American network), this was a pity. For whatever reason, the series was not or could not be given the chance to get back on track in the next season.

A Gift to Last

Polymath actor-writer-director Gordon Pinsent created *A Gift to Last* and wrote many of the scripts and starred in the series, which was produced by Herb Roland. The series had been developed from a successful Christmas special and went to air for three seasons, fall 1978, winter, 1978, and fall 1979, for a total of twenty-one "chapters." *A Gift to Last* might appear to conform to Purser's "communal family history . . . [set in] what were imagined to be better days'' (in Halliwell, *The Forsyte Saga*), but there its resemblance to that series or to *Upstairs, Downstairs* ends. In fact it was one of the most distinctively bred-in-the-bone Canadian television dramas in the CBC's long history.

Terming episodes "chapters" brings to mind a three-volume nineteenth-century novel. *A Gift to Last* was a blend of series and serial, a continuing narrative about people whose lives change, made up of episodes linked by continuity and recurring characters in each of which the main action is completed. The tone varied enormously both within programmes and from programme to programme. One episode juxtaposes the simple fun of the children's first car ride with Edgar's quiet despair as he deals with the loss of Sheila through his own inability to make a commitment and Lizzie's painful humiliation when not one of the Parish Committee ladies turns up for tea with the exhilaration of Uncle Edgar, spinning yet another yarn about the big world outside Tamarack. Every episode moves from the nostalgia of the theme song, evoking a pastoral world long vanished except in memory, to pragmatic detail about life with wood stoves, wash basins, and gas lighting.

A Gift to Last had many strengths. Ruth Springford played Lizzie, the puritanical old matriarch whose rebellious blue eyes match those of her favourite unconventional son, Edgar. The scope provided by three seasons gave space to present many facets of character, and Springford was challenged to stretch herself instead of playing the tartly wisecracking stock character of so many of her other roles. Gerard Parkes as James gave a strong performance as the good son and successful shoe manufacturerer in the small town, satisfied to be at the top of the social hierarchy, conscientious, loving in a humourless way, and never admitting to himself that he envies the scapegrace favourite. Dixie Seatle as Sheila made the transition from ignorant yet proudly Irish Catholic servant to wife, widow, and wife again as a Sturgess. She made a fine contrast to Janet Amos as Clara, a conventional wife and widow who needs Edgar's help to find the courage to marry John, the entrepreneur. Clement (Mark Polley) and Jane (Kate Parr) as the children were less developed as characters, but still essential to the plots of several episodes.

Best of all was Pinsent's Sergeant Edgar Sturgess. Through this character, Pinsent, the writer, grapples with his own most popular television and film persona, the "rowdy man." Magnetic to women yet unable to face the responsibility of marriage and children, yearning for foreign parts yet disillusioned by the Boer War, rootless when demobbed until he buys an old hotel and renames it the Mafeking, Sturgess continues to scandalize the town by drinking with the town alcoholic and taking in stranded and penniless Slavic immigrants. This rowdy man does eventually mature, though retaining his passion, his gift for fun and for improvisation, never fully settling down in one place.

Parallel to the changes in the protagonist, Edgar Sturgess, the series' strongly regional voice becomes a nationalist voice, as one episode about conflict between local militia and a very British, upper-class commanding officer demonstrates on both the public and private levels. (Chapter 15 informally identified at the CBC as "Edgar and the Militia," 11/4/79.) When the Boer War comes back to haunt Edgar and his two close friends, who are still in the army, we see both the spectres of that particular colonial war and the experience of veterans in any war.

The show is that rarity, a series that appeals to many people on many levels. Although the CBC review tape is missing at present, I have vivid memories of when Edgar asks Sheila to go with him to a regimental party intending, as she is well aware, to ask her to marry him. In a brief scene that seems to last forever, these two normally voluble people chat in a desultory way. The pauses lengthen until, at last, silence falls between them. In a moment that is pure Chekhov, both wordlessly recognize that Edgar cannot make the move. Devastated but proud, Sheila walks out with another man, leaving Edgar sick with anger at himself and the whole family in confusion.

In the next episode I reviewed, "Edgar in Trouble," broadcast nine months later (11/1/78, dir. Herb Roland), she has married the other man, and Edgar, bedding down on the banks of the river in an old shack, takes to drink and women, bringing public disgrace on Clara and Lizzie. The episode shows his behaviour for the self-indulgent theatrics it is, yet also reveals the real agony of spirit underlying it. As the episode nears its end, Lizzie has fallen into a depression that is serious enough to send her to bed. James tries to comfort her, but it is Edgar she wants, and no one can find him. Clara expresses part of what the viewer is feeling when she says, decidedly, "I still think it is the most selfish thing I ever heard of. He's not the only one, he's not." Placating her in a voice tinged with envy and affection, John talks of how Edgar is likely stretched out on the side of the hill, soaking up the good spring weather. The camera cuts to a medium shot of Edgar on the riverbank, service revolver in hand, moves to a ground-level shot of his boots, and then up for an extreme closeup of his face. It is very clear to the viewer that Edgar may well use the gun.

A Hollywood touch would have had young Clement, who has been desperately hurt and upset by his uncle's inexplicable neglect, find Edgar just before he pulls the trigger in a "saved by a boy's faith and love" type of conclusion. Not in this series, however: it is seldom that predictable. We see the lad running along the riverbank with his treasured army bugle, the one Edgar had given him in the pilot. The camera searchs the bank, using Clement's point of view. Failing to spot his uncle, he tries unsuccessfully to blow the bugle. We hear the crack of a shot. Cut to Edgar on the opposite bank, with a dead rabbit.

Clement brings him home. In the kitchen, Edgar delays so he can mysteriously doctor the rabbit in some way before going up to see Lizzie. She looks at him; then turns away. Unperturbed, he starts to spin a tale about having to hunt down a rabbit with two heads for her. With a form of address he seldom uses, he concludes quietly, levelly: "So you see why I couldn't come home, Mother." Already understanding the subtext but unforgiving, she clings to the manifest absurdity of his story, and lo, he produces a two-headed rabbit. As always, he has made her laugh. "Fool!"—her favourite name for him: "Lizzie," he grins. He is indeed the "all-licensed fool," the unconventional yet honourable prankster, the restless warrior who has had his war, can't settle down and can't leave; who is completely without fear of the town's conventional morality but terrified of emotional commitment, an outsider for over half the series.

Needless to say, when, seven chapters later, Sheila's husband turns out to be a con man who dies in prison, freeing her to marry again, and Edgar has finally found a new occupation (hotel-keeper) to still his restlessness, their wedding (Chapter 13, "The Wedding," 20/12/78) gives faithful viewers much satisfaction. A strike intervenes at the factory; Clara becomes ill and has to go to a sanatorium for months; James runs successfully for mayor. Then, in an episode written by Pinsent that is a wonderful, uneasy mixture of comedy, irony and tragedy, Sheila returns to her Catholic faith in a village rife with Orangemen. Chapter 18, "The Catholics" (25/11/79, dir. Jim Swan), opens with a portrait of a small-town "Glorious Twelfth" at the turn of the century, a nostalgic evocation of a village celebration with banners, King Billy on his white horse, bands, young men drinking behind the hay bales, ice cream, and bunting. But to John's factory girls, his Irish work force, on this "odd day" as they call it and to Sheila (now pregnant), it means chants of "Mick on the end of a stick" and "Pope on the end of a rope."

Later, in a touchingly awkward scene, James Sturgess, who has never had a son, brings Sheila a baby bracelet which belonged to him, signifying her complete acceptance into the WASP Sturgess family. When he has gone, Sheila talks over her decision to return to her faith with Edgar: "Strong as you are, I have to have my own strong right arm." Edgar:"Is this going to make any noticeable difference to me?" "No [pause] but I'm going to have to be remarried." Edgar, wryly: "Then I suppose I'll have to, too." "For her health and peace of mind," Edgar confronts his staunchly Orange mother, Lizzie: "Sheila is of the Catholic faith, pure and simple." Lizzie reverts to form: "There's nothing pure and simple about that . . . she's a Catholic Irish servant girl." Because he is mayor, brother James also refuses to come to the wedding. Cut to a shot of Edgar in uniform; the shot opens out to show Sheila in a white head covering. Then, after an out of shot "Amen," the camera pulls back further to reveal James, unhappy but still loyal.

The episode's climax occurs when Sheila delivers her baby prematurely. Summoned by Edgar after her long estrangement, Lizzie finds the baby dying and Edgar going for a priest to baptize it. Lizzie to Edgar, in the hall, tightlipped: "If you must." Edgar (urgent warning, pleading): "She has the greatest respect for your strength. I want her to see it in your face. I want her to see that again. She's been through enough." Sheila, upstairs: "Who's there?" "It's only me, Mother Sturgess." "Mother Sturgess?" "Yes."

In a scene I found very moving, the two women talk, as young and old mothers do, about the baby and his high forehead. Lizzie: "A sign of curiosity—good in boys, not in girls." Sheila: "Girls already know enough." The baby is thus turned into a little person, whose death truly affects us. Though Sheila asks Lizzie to hold her hand, her fierce concentration on the baby never wavers for an instant. Realizing that he is slipping away before the priest can arrive, she is determined to baptize him herself. Lizzie is shocked: "It's not proper!" Sheila: "Please, it's proper. I can do it. Please, please, hold the bowl." There is a closeup of two old hands holding the bowl, and the camera moves to Sheila signing him with the water and the appropriate phrases. Lizzie, very simply: "Amen." Sheila, her voice drained: "He's gone." Lizzie, who sits down, a world of compassion and loss in her voice: "Oh, child." Hearing Edgar come in with the priest, Lizzie goes to issue a wordless warning. Edgar presses on; she respectfully stops the priest and tells him what has happened. The sequence concludes with a shot of Edgar and Sheila together, looking down at their dead son. Gently, he touches her.

Seldom have the vexed questions of the religious bigotry, based on ignorance and steeped historically in two kinds of Irish pride, which in a few places still bedevils this country been so fairly or sensitively explored. There are no villains in this piece. We know all the characters thoroughly, and none of the questions or answers are easy ones. Class conflict, racial slurs, small-town narrowness, and family personalities grating together are all exposed in the context of their times. The episode also marks a turning point for the Sturgess family.

No wonder Pinsent decided that writing episodes (including the ones quoted above) as well as starring in the series was getting to be a burden. After the great Tamarack fire of 1905 (Chapter 20, 16 December 1979), with which he moved to end the series, many things change. With the factory burned down and many homes destroyed, half the population of the little village leaves. In the last chapter, James nearly leaves as well. In "The Old and the New" (10/12/79), the broadcast returns full circle. It is once again close to Christmas. Edgar, feeling trapped and restless for much of the episode, gets drunk, explodes, and demolishes the storeroom of his hotel. Sheila intervenes, articulates his dilemma and nudges him towards a different decision. In the end, he accepts a post at the Royal Military College, and Sheila finds herself an army wife after all. The episode also focuses on Lizzie's sense of betrayal as her family begins to scatter, Clara's struggle to be reacquainted with the family she has not seen for the two or three years she has been in a sanatorium, and James's sturdy determination to rebuild the town.

The last scene is another Christmas-New Year, marking a full circle from the Christmas depicted in the pilot. During a speech full of sententious saws by James, the camera roves around the family circle, pairing John and Clara, catching a military salute from Edgar to Clement, glancing at Jane (now quite the young lady) kissing Lizzie, and Edgar shaking hands with James, and pausing over a toast to the future (which is also to the viewer, seventy years later). From Edgar, next to Sheila: "Tomorrow is not a bad place to be." As they all sing "Auld Lang Syne," there is a cross fade to a shot of all of them together, posed as in a family portrait, and a bright flash fading into a sepia photograph. The credits roll over other photographs of characters as they appeared in earlier episodes. It makes an open-ended but very satisfying ending to a remarkable series.

Aside from his enduring appeal as an actor with considerable range, intelligence, and magnetism, Gordon Pinsent is one of the best writers in CBC television drama. It is not easy to write series television, much less a series that is successful family entertainment, evocative of a time and place now vanished, and complex rather than simply sentimental in its emotional structure. *A Gift to Last* was not like its contemporaries *Little House on the Prairie* (1974-84) or *The Waltons* (1972-79). Family values are stressed in the series, but they evolve through time and changing circumstances. They offer no protection from the way the world's accidents happen but they do provide the backbone for human response. The distance between Paul St. Pierre's *Cariboo Country* in the 'sixties and small-town Ontario in 1905 is not so great as one might think.

Home Fires

Home Fires (1980-84, pr. Robert Sherrin) was the most recent series to date to try to recreate a period family—here, enduring the stresses of World War II—over more than one season. The series had a loyal and devoted following, though several hundred thousand is small in television terms. Dr. Lowe, his Polish-Jewish wife Hannah, daughter Terry, son Sydney, and assorted friends, enemies, and lovers were created by writer Jim Purdy. The series had many individually good moments; among them, when Terry, as a nurse's aide, learns to deal with the war-wounded; when Dr. Lowe endures the indignity of internment for his socialist beliefs; when Hannah tries to mother a young Jewish refugee scarred by the war; and when Sydney, a successful fighter pilot, implodes with guilt, relief, and longing at the war's end. As time goes on, the focus shifts from Gerard Parkes as the doctor to Terry who is first a war bride, then widow, then newlywed. In the last episode she decides to study to be a doctor, a final and rather improbable touch that ironically reminded this viewer of the thousands of other women we had seen on the job in this series but who in real life were exhorted to go home and procreate, leaving the factories, trades, airplanes, radar, and all the other responsibilities they had shouldered to their rightful male heirs.

Period detail in the sets, costumes, music, dialogue, the vignettes of service clubs and factory floors, service life and refugee work was meticulously researched and recreated, adding considerably to the interest of the show. The problem was that, somehow, the senior Lowes remained opaque characters. Hannah (Kim Yaroshevskaya) probably suffered more on the home front than the others because she was Jewish, but there was very little sense of her cultural or religious heritage in her house, speech, even her attitudes. The good doctor was just that: good, with few other distinguishing characteristics. Neither Yaroshevskaya nor Parkes was given much to work with. Wendy Crewson (Terry) and Peter Spence (Sydney) had more to do as they grew up and matured, and there were some vivid subsidiary characters. The stories had nostalgic appeal for older viewers, some historical appeal for younger viewers, and occasional moments of intensity or humour to lighten the way. *Home Fires* did have its fans, but I was not one of them: I found the plots too thin, the dialogue pedestrian, and the characters often too predictable. Nevertheless, as reasonably serious family entertainment, the show filled a niche now empty in the schedule. *Backstretch* may have been intended to replace it but, as we have seen, it went astray in its second season.

Hatch's Mill

A bridge between sitcom and family entertainment series like *Backstretch* and *A Gift to Last* is an earlier series which doesn't really belong in any classification more precise than "light entertainment." *Hatch's Mill* (1967) was a half-hour series devised specifically for the Centennial year. It featured a pioneer inn-keeping family. The *CBC Times* (21-27/10/67) called it "a rollicking comedy-adventure series in colour, revolving around the mythical Hatch family in an Upper Canadian backwoods hamlet in the 1830's . . . a ten-part Centennial film series." Intended as "a refreshing escape from the tensions of the modern rat-race," it starred Robert Christie and Cosette Lee, with Marc Strange and Sylvia Feigel. Executive producer Ron Weyman attempted to further define the series' flavour when he described it as the portrait of a "free-swinging individualistic pioneering group. Although the series is based on historical incidents, we've tried to find those that have a strong entertainment or production value, using them as a point of departure." Documentary or direct cinema shooting style and topical subjects were not to be the distinguishing marks of this series, which seems to have been conceived in almost conscious contrast to its contemporaries, *Wojeck* and *Quentin Durgens M.P.* George Salverson created it, and the scriptwriters included veterans Leslie MacFarlane, Salverson, Munroe Scott, and Donald Jack. The associate producer was David Peddie, who went on to oversee the well crafted anthology, *Canadian Short Stories/To See Ourselves* (1971-76). A voice-over acknowledges that the production team also had the "advice and assistance of Pioneer Village."

I looked at two programmes, "Temperance" and the hour-long "Prophet." The series is an interesting, if uneasy, mixture of 'sixties formula sitcom, full of one-liners and slapstick, with a buxom blonde daughter, Silence, a handsome, dark son, Saul, jolly but all-wise patriarch Noah, and Maggie, the no-nonsense mother. Yet all this takes place against authentic settings, anecdotes, issues, and characters drawn from the tense and explosive year for Upper and Lower Canada, 1837. George McGowan, who also directed challenging plays for *Festival*, produced and directed this show.

"Temperance" (19/12/67, wr. Donald Jacke; in the festive season) begins with a slow pan over Noah's Inn regulars snoring away or muttering, "Winter's never going to end," and "I've read both books." They then break into a song popular in the 'sixties which begins, "Landlord fill the flowing bowl until it doth run over." The plot centres on the newly arrived Peppermuellers, whose cabin-raising bee attracts settlers just as in every pioneer and Western film, until the pioneers find that no liquor is being served. This would not have been an acceptable subject for sitcom in the U.S. at that time, particularly since the programme points out that drinking was fun and our ancestors, both men and women, did a lot of it. The episode also suggests, however, that drinking to stupefaction was a by-roduct of loneliness, fatigue, and boredom. Moreover, there is a glimpse of an ugly domestic dispute fuelled by drunkenness: the episode, then, is not ridiculing the temperance movement. Yet its treatment of Indians and liquor in the subplot is a surprisingly racist stereotype, with embarrassing dialogue, considering the CBC's sensitivity in other contemporary series. However, in their drunken rage, the Indians get the tippling hypocrite of a Tory judge to sign the licence for the inn to sell liquor, bringing about a happy ending. In the main plot an exasperated Mr. Peppermueller tells Mrs. Peppermueller to sit down and shut up, then invites her to have a drink. The episode ends with predictable revelry.

The episode is very uneven in tone, with slapstick as sexist and overplayed as in *The Beverly Hillbillies*. Yet, writer Donald Jacke makes some sharp comments on the class conflict fanned by the Family Compact: "Snobbish old Tory . . . you're the reason for all these 'reformers.' You'll get no forelocks tugged here." The script even reminds us that "we got Bond Head by mistake, instead of his brilliant cousin." What makes this episode so problematic is that the potential for sharp social comment, vivid historical anecdote and good-humoured farce is too often smothered by stereotyped jokes, attitudes, and characters.

"The Prophet" (26/12/67, wr. Munroe Scott) seems to have been a Boxing Day special. Suspense builds when a man, his wife, and young daughter enter the village on foot. On his back, he carries what looks like a coffin but is in fact a clock. Man: "We live by faith. Our destination is in the mind of God. Names are a burden we have chosen to discard." The acting skill of Douglas Rain makes these lines credible and gives the man's character overtones of dignity, rather than simple fanaticism. When the wife reads the palm of one of the continuing characters, a black—an unusual, evocative and historically accurate touch—his past and future remain private to the two of them. She then explains to the others that her psychic gift is limited to "some people, some times. [It's] a rare gift not to be squandered." As the mystery of husband, wife and child lodging in the barn near Christmas develops, the people of the village get drawn into the their tense, waiting silence. Eventually, we learn that the world is to end that night, 15 December 1837. As most of the villagers panic, get drunk, sit out in the snow in rocking chairs wrapped in white sheets, or whirl in ecstasy in the smoky firelight, Noah Hatch, as the "fix-it" character of the formula, gets out his Bible to find another date for the Second Coming. Comedy and moments of touching revelation fill in the time as he struggles to extricate the community from its hysteria and the proud Scot and his family from their vision. Breaking formula, Noah's research into *Revelations* and *Daniel*, and his last-ditch appeal to the husband to consult with the Presbyterian moderator, fail to work.

An ingenious plot twist concludes the play. Saul has pursued the girl to the roof to stop her trying to fly to God. As he flings himself on her to prevent her from jumping, the hands of the huge clock reach twelve. Just then, the central beam of the barn collapses under the couple's weight smashing the clock. Everyone rushes in to find the Scot covered in dust, holding up the beam over Silence, who is stunned, but not hurt. We see a shot of Saul and the girl kissing in the straw. This act of God, stopping the clock seconds before midnight, convinces the husband that he has misread Scripture and misunderstood his vision. This episode had a clearer point of view, a more pervasive historical flavour, and a stronger script than the previous one. It also had numerous fresh touches and an intriguing, suspenseful plot; it worked very well.

Overall, *Hatch's Mill* was an entertaining series which showed signs of developing away from the clichés of slapstick and stereotype towards a particular sense of period and place, as well as humour derived from character interplay flavoured with some very funny sight gags. It also had the potential for unfamiliar subject matter and salty observations about politics and class that were as relevant in 1967 as they were in 1837. For reasons I have been unable to discover, if there were reasons, the series was not renewed, despite the fact that there was nothing like it on Canadian or American television.

From this necessarily cursory glance at miniseries and serials, no obvious pattern emerges. This is appropriate because it is in miniseries where we should find both depth and diversity. *Empire Inc.* and *A Gift to Last* can be identified as two obvious successes which offered both, but many others gave large audiences considerable pleasure. There were, inevitably, a few resounding clunkers. If you are wondering where *Jalna* is, dear reader, it appears under anthology—and takes up very little space. *Jalna*'s failure may be responsible for the CBC's determination to keep its historical miniseries to six to eight episodes.

Some miniseries had a remarkable impact on viewers: John Kennedy, Head of TV Drama, relates that Jeanine Locke's *The Other Kingdom*, a two-part miniseries dealing with mastectomy, generated floods of mail. Though the programme was a good one, I am not sure in this case, or that of *A Far Cry From Home* (wr. B. A. Cameron [Cam Hubert of *Dreamspeaker* and *Ada*], prod. Anne Frank, dir. Gordon Pinsent), that an aesthetic evaluation is relevant. In the latter case, the script was didactic, the characters one-dimensional, and the camera work inventive. But the central focus should probably be on the impact of the piece on public awareness, which was exceptional. *A Far Cry from Home* (2/1/81) helped to put the social issue of wife battering on the public agenda. In most cases, factors common to the marked successes are completely predictable: they start with superior scripts and have good production values, sensitive directors and art directors, creative camerawork and sound technicians, imaginative producers and writers, and rounded characters played by actors who find ways to display subtext, context, and nuance for the camera. It sounds so simple, doesn't it?

NOTES

1. *The Producer's Medium: Conversations with Creators of American TV*. Oxford University Press, New York: 1983, 74-81.
2. Ibid., 132-33.
3. "CASEYITIS: How the Epidemic Started and Why It Spread," *TV Guide*, 65.
4. Ibid. "The Eternal Conflict between Good and Evil," 60.
5. But cf. John Kennedy's view of what happened in chapter 11. Kennedy has been Head of CBC Drama since 1977.

III

THE CBC (PROUDLY) PRESENTS

6

ANTHOLOGY: TO 1968
"Window on the world as the world looks in"

Anthology drama has had a long if rather uneven history at the CBC. A more complete account of its earlier years than the sketch which follows in an article I wrote for *Theatre History in Canada*.[1] For over thirty-five years, the CBC has been a cornucopia of individual programming spilling out dozens of television dramas a year (although fewer now than in the 'fifties and early 'sixties and significantly fewer in the late 'eighties with the vagaries of television financing and CBC cutbacks). They range from the incoherent through the inadequate to the amiable timewaster, from the intelligent story, to the challenging, innovative, demanding play that works on several levels. There are also the true classics (two or three in most years) which yield more and more when viewed repeatedly. If the metaphor of a map of a virtually unknown and very large country which I introduced at the beginning of this book is an accurate analogy for the size of the subject, it also describes the task of selecting, from among the thousands of hours of anthology, a few that accurately convey the realm's topography, and draw attention to some of the ones that stand out in the general terrain.

In the end, certain choices were made for me. Some of what seem to have been the most interesting dramas, like *Slow Dance on the Killing Ground* (18/10/67, dir. Melwyn Breen), *The Puppet Caravan* (1/3/67, wr. Marie-Claire Blais, dir. Paul Almond), *God's Sparrow* (11/70, wr. Philip Child, dir. Peter Carter), and many others are mangled or else missing, probably gone for good. Of what is left, I can claim to have seen most of the "premier" prestige programmes. But I could not possibly see all the surviving dramas of *General Motors Presents* (1953-61), *The Unforeseen* (1958-60), *Playdate* (1961-65), *On Camera* (1954-58), *Royal Suite* (1976), and the other anthologies. Instead, using what information I could unearth, I looked at a selection of programmes chosen for the director or playwright or because the material was difficult to adapt for television. From those already winnowed by the criteria just noted, I have used a tiny fraction for analysis in this chapter. Even so, I will introduce you to a sample of theatrical plays adapted with skill, short stories or novels turned into drama, original scripts, fantasies and topical dramas. I will also provide short descriptions of the major anthologies.

The evolution of CBC television drama is divided into two periods: first, 1952-68, when live and live-to-tape television evolved, filmed drama was introduced, and morale was for the most part high. According to Hugh Gauntlett (in middle management for over twenty years), most producer-directors were on staff, not contract and therefore more committed to the CBC, though staff rather than contract status also foretold a degree of burnout at a time when sabbaticals were unknown. A consistent level of commitment to all kinds of television drama was also evident on screens at home which represented, in part, the vision of Doug Nixon, director of entertainment programmes, who pulled together the whole programme mix to form a schedule. During this period there was no "Head of TV Drama." In the earliest days CBLT programme director Stuart Griffiths and Mavor Moore prepared the ground and hired the first generation of producers from CBC radio or from the U.K. Very rapidly under Henry Kaplan a second generation of producers learned their trade and went on to do excellent work. Some worked primarily under Robert Allen doing more serious, classical material. Others did the more topical and commercial anthologies supervised by Sydney Newman. Most worked with both men but had preferences and an aptitude for one or the other kind of drama. Newman left for England in 1958. By the early 'sixties Ronald Weyman had under his wing much of the popular drama with his repeated efforts to get into film first with *The Serial* and then series and specials. Ed Moser also supervised some of the more popular anthologies and Robert Allen continued with *Folio*, then *Festival*, to oversee the prestigious and experimental television drama.

The period 1968-73 was transitional, and will be treated as such. Fletcher Markle presided over this period as TV drama's first "Head." The years 1974-86 saw a return to more consistently high quality, but 1975 also marked the beginning of an inexorable squeeze on resources, including broadcast hours for drama of any kind. John Hirsch was chosen from the theatre in 1973 to get TV drama back on track by introducing some innovation and developing popular series ideas. He inherited a department adrift. He did try new things, many successful, some not. He had a revitalizing effect on the programme mix yet paradoxically had a demoralizing effect on some of his most creative people. He left four years later to be succeeded by John

Kennedy in the 1977-78 season. Kennedy brought many things to the drama department, not the least of which included stability, encouragement, pragmatism and a populist sense of television's potential that did not exclude experiment or innovation in subject or form. Yet in this period, television anthology at the CBC was an endangered, though by no means extinct, species, primarily because of pressure to produce sure-fire series and thus ratings. Inevitably, the wide range of subjects and styles which had characterized the earlier years was also narrowed, to our considerable loss as viewers.

Canada 1952: Movies and Theatre—The Unborn Rivals to Television

In 1952, Canadian film was not really a rival to Canadian television. The handful of independent film producers in the country were reluctant to touch dramatic films, and the National Film Board had not significantly moved into drama. Nor did Canadian television face the entrenched opposition from big film corporations that forced early American television to create new material instead of rerunning old movies or restaging Broadway hits. On the other hand, American television did not have to contend with an audience that had already turned its antennae to foreign stations in Buffalo or Detroit. As in any ex-colonial culture, Canadian viewers, unlike American or (British) television audiences, were conditioned to expect the kinds of stories and conventions of storytelling already established by a different country with different attitudes, goals, and cultural traditions.

We must also remember that the pattern of culture for Canadian drama developed quite differently from those of Britain, the U.S., and Europe, all of which had fully developed theatrical traditions long before mass media arrived. What professional theatre we had was seriously reduced in the depression. Moreover, as Raymond Williams points out in *Drama in a Dramatized Society* (1975),[2] speaking of audiences who have had drama available at the turn of a switch for over fifty years: "what we have now is drama as a habitual experience. . . . The slice of life, once a projection of naturalist drama, is now a voluntary, habitual, internal rhythm; the flow of action and acting, or representation and performance, is raised to a new convention, that of basic need." Our television drama, as a collectively observed and widely known body of work, took shape after this fundamental change in the function of drama occurred. Thus television, following its elder sister radio, helped change our expectations about all forms of drama well before our professional theatre had matured. From the late 1950s on, in fact, television drama provided the pervasive context for the average person's experience of theatre.

A glance at the economics of Canadian theatre adds another detail to television's role as a developmental influence on live performance in Canada. Canada is still a relatively thinly populated country without many theatre centres, theatre patrons, or in many areas of the country, knowledgeable, sophisticated audiences. So far, it also lacks a body of dramatic literature that is familiar to the average citizen. Actors, writers, musicians, designers, and technicians find it difficult to make a living here. Since the 1930s, CBC radio and then the regional centres of television drama have kept such people working.

To take actors as perhaps the most familiar example, a cross-check of Stratford's cast lists for the early 1960s reveals that many members of the company—including Frances Hyland, Douglas Rain, Leo Ciceri, Hugh Webster, Eric House, Martha Henry, William Hutt, William Needles, Kate Reid and John Colicos—carried their ensemble playing from the Stratford stage to CBC television and back again. They managed to live in Canada year round because radio and television work was available between the short Stratford seasons of those early years.

Television also created an outlet for Canadian writers. It was a training ground for new playwrights as well as bread and butter (and a training ground in new techniques) for the older ones. Jack Winter, Bernard Slade and Patricia Joudry were among those who wrote extensively for television before their stage successes. In 1962, David French graduated from the position of CBC mailboy to CBC script writer through Montreal's *Shoestring Theatre* series (1960-67). Carol Bolt wrote for *The Collaborators* in the early 'seventies. George Ryga wrote scripts for *Shoestring Theatre*, *Q for Quest* (1961-64), and *Festival* (1960-69), as well as *The Newcomers* (1977, 1980); and Timothy Findley wrote scripts for *The National Dream*, *Jalna*, and *The Newcomers*. Judith Thompson, a new talent for the 'eighties (*Crackwalker*) has joined the list with scripts for the domestic comedy *Airwaves* (1986) and the John Kastner production *Turning to Stone* (1986).

In the 'fifties, television drama provided a different kind of challenge for designers, who found there were very few opportunities in Canadian theatre. Rudi Dorn and Nicolai Soloviov in particular formed creative partnerships with Harvey Hart, Paul Almond, Mario Prizek, and other directors to give many of the more challenging scripts a visual dimension that was often startling and imaginative. Even now, some designers move back and forth between television and theatre—again, one of the ways to remain in or return to Canada. Finally, television drama has been a proving ground for a flock of producers-directors whose careers took them from television into theatre or film and often back again: George Bloomfield, David Gardner, Eric Till, and Paul Almond are four among many. Technical people are less likely to work in both media, since sound, lighting, and special effects for television call for quite different skills from those required for the theatre. These television specialists are more likely to alternate with film.

Television gave its directors freedom to experiment with stage and cinematic conventions. It also demanded that they discover the dramatic conventions best suited to the new medium and its unique characteristics. As those characteristics were identified, and as writers and directors became more experienced with the medium, the forms of television drama began to influence the structure and conventions of stage plays: examples are the narrative conventions borrowed from situation comedy; the use of voice-over as developed in both radio and television (which borrowed asides and soliloquies from theatre); the familiar radio-television docudrama structure of sequenced vignettes and television's rapid pacing, enhanced by jump cuts and montage from film vocabulary which has become the norm against which we measure the pace of most dramatic forms in the 'eighties. Perhaps reflecting the emphasis on realism that characterized the development of television drama in Canada from 1975 to 1986, many realistic Canadian plays of the 'seventies moved easily from their alternate theatre forms to television—among others, Bolt's *One Night Stand* (8/3/78, dir. Allan King) and Fennario's *On The Job* (27/3/76, dirs. Martin Kinch/Allan King).

A good many Canadians think that we have too small a population to afford a television establishment, a theatre establishment, and a film establishment, a mindset that could eventually becomes a self-fulfilling prophecy. Yet, since our creative people often work in all three media, they bring from theatre to television a more balanced regard for the values of dialogue and subtext than is usually noted in programmes from the United States. From television to theatre, they bring the sense of the fluidity of time and place provided by film and tape and the sense of intimate address that characterizes the close-ups prevalent in television drama.

The CBC Presents

In the early days, television drama performed another important service for theatre by introducing a significant proportion of the best contemporary and classical world theatre, adapted for broadcast, to the scattered and rural corners of this very large country. In this respect, as with the development of writers and many actors, CBC radio had already established itself as what Howard Fink aptly calls, "Canada's National Theatre of the Air." In the 'forties and 'fifties Canadians had enjoyed some of the best radio drama in the world, thanks to producers Andrew Allan, Rupert Caplan, J. Frank Willis and Esse Ljungh.

Television had a different contribution to make. The opportunity to see classical and contemporary drama was particularly important to the potential playwrights, directors, and actors in inaccessible parts of the country, who, up to that point, had had no other experience of professionally staged theatre. From the beginning, television also brought an increasingly large audience two Canadian dimensions: the country's history, either in fictional form or as documentary drama, and controversial issues explored through topical dramas. It is now a critical commonplace that these two foci have also been central to the development of contemporary Canadian theatre.

Canadian television drama also introduced the country at large to plays which had originated in the Canadian theatre. In the 1952-53 season these included Morley Callaghan's *To Tell The Truth*, Ted Allan's *The Money Makers*, Robertson Davies' *Fortune My Foe*, and a version of some of the Coventry miracle plays by Dora Mavor Moore's New Play Society. Later, television plays became popular stage plays: for example, *The Black Bonspiel of Wullie McCrimmon* (9/10/55, dir.-pr. Robert Allen, which began in radio and had several repeat broadcasts on television), and *Back to Beulah* (1974, dir. Eric Till); the Norman Campbell-Don Harron musical, *Anne of Green Gables* (1956), Tom Hendry's *Fifteen Miles of Broken Glass* (22/2/67, dir. John Hirsch), and Gordon Pinsent's opening episode of *A Gift to Last* (1979) which has been both a stage play and a musical.

Finally, and most important of all for the maturing of the new medium, television produced a handful of brilliant, innovative dramas that were written for or are best performed on television, which stand up superbly to viewing fifteen or twenty years later and which form the core of a valuable heritage that is often ignored and neglected. It is an absurd, regrettable, and perhaps a quintessentially Canadian fact of life at the CBC that almost no one making television drama in the 1980s remembers with pride, reviews with interest, or ever reflects upon this remarkable record. In *Telling Our Story* (30/12/86) the CBC broadcast a part of its presentation to the 1986 CRTC hearings that made a very emotional, probably effective statement on CBC drama. Not one clip came from a drama predating 1982—even though the CBC has been telling our stories on television since 1952.

It should be a condition of employment that directors, designers, cinematographers, producers, and writers look at a dozen or so plays selected from the rich store available. Would a filmmaker confess ignorance of the work of Welles? Ford? Eisenstein? Bergman? Would a theatre director be unaware of Stanislavsky? Meyerhold? Brook? Not knowing a national history and an aesthetic context makes for shallow cultural roots. Unhappily, this CBC obsession with the present and the immediate future is universal.

A television anthology is a collection of plays, different each week, shown in the same time slot under one title. Some anthologies have varied considerably in content. Over the ten years of *General Motors Presents*, the viewers saw new Canadian scripts that included comedies, tragedies, thrillers, domestic dramas, a Western or two, topical dramas, and short-story adaptations. *The Unforeseen* specialized in ironic twists and psychological fables. *Q for Quest* was an experimental anthology of half-hour dramas, films, concerts, and readings. *Festival* was the proudest ship of the fleet. Here, the viewer would find the classics, modern drama, contemporary experiments, and some Canadian plays, all free of the 30, 60 or 90-minute strictures or the commercials imposed on other anthologies. *Festival* also presented concerts, opera, and ballet. By the mid-'eighties the Arts Music and Science department had virtually no money to make new programmes, or even develop new projects. Distinguished series like Vincent Tovell's *Hand and Eye* (1982), biographies of artists and scientists, concerts, Norman Campbell's ballets, "ideas" specials have slowly faded and vanished and with it the cultural context of drama as an art form.

Out of the first fifteen years of television drama, 1952-67, came such television playwrights as Paul St. Pierre, Mac Shoub, Charles Israel, M. Charles Cohen, Munroe Scott, Philip Hersch, Joseph Schull, Lister Sinclair, Len Peterson, Mavor Moore, and George Salverson. Some— Tommy Tweed, Shoub, Schull, Petersen, Sinclair, Mavor Moore—had made their mark in radio drama for many years before television came on the scene. Others found television more compatible to their talents and wrote chiefly for it. A few continued to write extensively for both.

For most people now under forty, television provided the first formative, direct, and continuous experience of drama. Whether or not this was the case in any particular instance, Canadian television drama still played a crucial role for all of us as a window through which we saw glimpses of ourselves, acted for and by ourselves. Next door on the dial to *Milton Berle* (1948-57) or *Maverick* (1957-62) were plays written, produced, acted, directed and watched by Canadians; hundreds of plays, performed for millions over the years. All of this was taking place when, for most parts of this country, "Canadian drama" meant a couple of plays in the local tryouts for the Dominion Drama Festival and "theatre" meant a trip into a bigger city for an American touring company version of an old or an untried musical.

Consider the context of international drama in the 'fifties as television drama in Canada came to life. Most plays fell into the categories of realistic, well-made play or drawing-room comedy. Tennessee Williams, Thornton Wilder, William Saroyan, Arthur Miller and Noel Coward were seen only by the handful of Canadians attending theatres in Vancouver, Toronto, or Montreal; Pinter, Beckett, and Albee had barely begun to write. Yet in its first fifteen years, CBC television introduced all of these playwrights to the growing television audience, as well as Brecht, Anouilh, Pirandello and

Lorca, together with Chekhov, Shaw, Molière, Wilde, and Ibsen.

As Graham Murdock notes in "Radical Drama, Radical Theatre,"[3] although many years younger than BBC television drama, our early television drama out-performed the BBC in both originality of concept and execution. In 1956, the BBC broadcast ninety-nine full-length plays. Sixty-two of them were derived from a stage dominated at that time by a set of formulae revolving around the conversation in upper-or middle-class drawing rooms. Murdock quotes Kenneth Tynan's characterization of that form of theatre as a "glibly codified fairy world." By contrast, for the first two seasons, Canada was producing one new ninety-minute play a week on the *CBC Television Theatre* (1953-57) alone, plus numerous shorter plays in other anthologies. Murdock is accurate when he qualifies Canadian television drama as "adventurous." That the BBC also thought so was also apparent in its purchase of CBC kinescopes, sight unseen, mostly from *On Camera* including work by Elsie Park Gowan and Patricia Joudry. Twelve million viewers and critics enjoyed them so much that the BBC bought 26 more not yet produced for an anthology called *Canadian Television Theatre* (November 1957). Soon afterwards, the most innovative of British commercial television networks, Granada, imported CBC drama supervisor Sidney Newman to give it a jolt. Sean Sutton, later head of the drama group at BBC, clearly admired his boss: "Redoubtable, rambunctious, full-blooded Sydney Newman . . . [then] burst into BBC drama at its moment of expansion, seized the opportunity and set a match to a dramatic bonfire that has warmed us all since."[4]

A quick skim through Halliwell's *Television Companion* reveals that kinescopes, then tapes or films of most of our series television, along with many plays from anthologies and drama specials, have appeared on the BBC or ITV. The first wave of CBC television producers: David Greene, Charles Jarrott and others were English. During the 'fifties at the CBC supervising producers Newman, Henry Kaplan, Franz Kramer, and Robert Allen brought along a new group of very talented Canadian producer-directors: Harvey Hart, Paul Almond, Mario Prizek, Daryl Duke, Silvio Narizzanno, Eric Till, David Gardner, Melwyn Breen, and Leo Orenstein. Together, they commissioned, adapted, wrote, or coaxed into being much of our early television drama.

Legendary Drama From the Golden Age of American TV Drama

On the whole, the American television dramas which were influential in Canada and Britain were mid-'fifties works like Paddy Chayevsky's *Marty* (24/5/53, dir. Delbert Mann), Reginald Rose's *Twelve Angry Men* (20/9/54), Rod Serling's *Patterns* (12/1/55), and J. P. Miller's *Days of Wine and Roses*

(2/10/58). Though often turned into films that replaced the rough images and sharp edges in performance of the originals, the television plays are still fondly remembered by many, and were rebroadcast on PBS in the early 'eighties. They formed the immediate viewing context for Canadians within reach of the U.S. border stations. The emphasis in most of them was on psychological realism and a naturalistic or representational style. Regrettably, as the decade progressed, American television drama retreated from engagement with social or political issues under pressure from sponsors and the infamous blacklist generated by McCarthy sympathizers during the American anti-Communist hysteria of mid-decade.

In Canada, meanwhile, television playwrights and directors were trying their hand at a wider variety of styles: fantasy and surrealism—as in W. O. Mitchell's *Honey and Hoppers* (7/11/57), or Paul Almond's version of Clive Exton's complex satire on television, *The Close Prisoner* (11/11/64); dramatized documentary, such as Charles Israel's *The Odds and the Gods* (1/16/55); drama restricted to a single point of view, like Frank Freedman's *Night Admission* (19/11/63); or collage of voice, sound, music and expressionist drama techniques, as used in Hugh Webster's *Kim* (27/1/63) and absurdist drama like Jacques Languirand's eerie drama *Grand Exits* (16/4/62). Most of these plays are discussed below. Till, Almond and Mario Prizek all had highly surrealistic imaginations, ably assisted by designers like Dorn and Soloviov. More important, when the golden age of American anthology withered away, the medium's potential for individualistic expression was extended in Canada long after the experiments in style and the variety of content afforded by anthology had disappeared in the United States.

Looking back from a vantage-point of well over thirty years, we can see the reasons for this early success. From the beginning, the range of dramatic material available every week on the CBC was a contrast with the fare offered in Britain and the United States. Mavor Moore was the executive producer of drama in the first season (1952-53), supervising, among other things, *CBC Television Theatre*. Of thirty-four plays presented in this anthology, seven were Canadian, four of them adapted from the stage: Morley Callaghan's *To Tell the Truth*, according to the press release the first Canadian play to be staged at Toronto's Royal Alexandra Theatre; Ted Allen's *The Money Makers*; Lister Sinclair's *One John Smith*, and Robertson Davies's *Fortune My Foe*. Alex Dyer's *The Case of Prince Charming*, Joseph Schull's *The Bridge*, and Sinclair's *Hilda Morgan* had been adapted from radio, the source of many early scripts. That season also included North America's first full-length television production of *Othello* and a two-hour production of the opera *Don Giovanni*. In the same season, Don Harron, Rita Greer, Stanley Mann, Leslie MacFarlane, and others made very faithful half-hour adaptations of the stories in Leacock's *Sunshine Sketches*. This first season gave viewers a remarkable mix of television drama to enjoy.

Hilda Morgan: If You Change the Medium, Do You Change the Message?

The heated controversy that arose over the television broadcast of *Hilda Morgan* (about an unwed mother-to-be), despite its careful presentation complete with a panel of experts to discuss the issue following the show, pointed to the fact that standards of acceptability for television audiences differed from those applied by radio listeners (who had heard the play on 22 May 1949, rebroadcast 11 February 1950). This may have been because sponsor pressure was already making American television drama generally more and more bland, thus conditioning Canadian viewer expectations. It was a sensation in the early 'fifties when Lucy Arnaz was allowed to be pregnant on television, even though she was married, both on and offscreen, to the father. The word "pregnant" was never used on the programme. Perhaps too, television viewers were less sophisticated than the well-versed audiences of CBC radio anthology. Perhaps the dust-up arose simply because pictures have a different impact from sound. Radio's great strength is that a listener can personally and precisely decide how much he/she chooses to imagine and in what kind of detail. From hotline shows to complex dramas, it is a participatory medium. Television, however, shows its wares. It is easier to be conscious of heads turning away from a picture on a television set than to be conscious of the thresholds over which listeners' imaginations will not venture. Whatever the reason, the difference in *Hilda Morgan*'s reception by radio and television audiences indicated the arrival of a new medium with a new audience and new expectations.

EARLY DAYS

During its second season (1953-54), CBC television announced that it would present "an original ninety-minute play every Tuesday night . . . broadcast from 9 o'clock on so that . . . the play might not be hampered by the rigid sixty-minute structure adapted by the majority of television dramas in the United States."[5] For the most part, *CBC Television Theatre*, as the series was called, continued on a sustaining basis—that is, without commercials. On 20 April 1954, however, part of the run of the anthology was retitled *General Motors Theatre*. In these plays, "distinguished American and British personalities appear as hosts and sometimes as lead actors." The link between the sponsor and guests "from abroad" was probably not accidental. Just like film and theatre, television was mesmerized by the star system and made imitative by lack of confidence. Even so, it was Gratien Gélinas's, "Montreal's Beloved Fridolin," who introduced *The Blood is Strong* (1954, wr. Lister Sinclair).

Using the CBC's own classifications, we find twelve "human interest plays," ten "comedies," five "tragedies," and six "melodramas" in this season. The anthology was described as maintaining "a high emphasis on the realist plays of ordinary life and light comedies." Canadian originals written specifically for television that year included Rod Coneybeare's *The Grown Ones* (26/1/54) and *The Man Who Ran Away* (6/4/54), Stanley Mann's *A Business of His Own* (17/11/53) and *The Haven* (13/4/54), and Edward Rollins's *Feast of Stephen* (19/1/54). Patricia Joudry adapted her radio play, *Teach Me How to Cry* (15/10/1953) for the anthology, a significant step on its way to production in New York and London. Meanwhile, script editor Nathan Cohen and producer Sidney Newman were out in universities and journalism schools prodding people into submitting scripts. R.L. Jackson quotes Sidney Newman as saying that "the primary fact [was] that the audiences were Canadian . . . Canadians seeing themselves in dramatic situations always seem to be to me the best way to get them to watch my programs."[6] The anthology's success seemed to vindicate his point of view.

In 1954-55, the CBC experimented with a new anthology, called *Scope*, which was described by its director of programmes, Charles Jennings, as "somewhat similar to CBC's *Wednesday Night* on radio . . . a series intended to offer something for everyone. . . . It will not adhere to any one theme but will present a wide range of subject matter in various ways, some of which will be experimental."[7] The first programme was to be a ninety-minute musical—originally presented on CBC radio's *Wednesday Night* in March 1954 as *The Hero of Mariposa* —called *Sunshine Town* (19/12/54) and written by Mavor Moore, orchestrated by Howard Cable, produced by Norman Campbell, and introduced by Tommy Tweed. Thus, Leacock continued to nourish the infant Canadian medium with a shot of Canadian wry. *O Canada* (wr. Eric Nicol, prod. Norman Campbell) was "an uncensored history with music," with Indian and francophone stereotypes that would trouble us now. Nevertheless, it was a lively sendup. To quote from the kinescope, Canadian history "usually begins with a foot or more of snow. The result is a saga which has to be shovelled out before it gets moving." *Folio* succeeded *Scope* in the winter of 1955, and lasted until 1960, in its five years offering seventy programmes. Of these, twenty-four were original Canadian dramas and several others were Canadian adaptations of short stories and other material. There were also fourteen musical comedies and revues, all of Canadian origin and most written especially for television.[8]

First Performance

Another anthology of very special television dramas in the 'fifties, called *First Performance*, was sponsored by the Bank of Canada during its Canada Savings Bond drives. *First Performance* had four seasons of four programmes each, run during October. Sydney Newman's presence as the supervising producer may have accounted for the increasingly contemporary slant to plays and the emphasis on Canadian scripts. *The Colonel and the Lady* (wr. Ronald Hambleton, prod. Norman Campbell), starring Douglas Campbell and Katherine Blake, opened the series on 6 October 1955. The "Colonel" was Tiger Dunlop, the "Lady," Anna Jameson, and the script a celebration of the witty encounter between the legendary pioneer and the determined English gentlewoman. The pace is a little slow and the blocking clumsy to allow the cumbersome cameras to change position between shots but it is an entertaining piece nonetheless.

An introduction by Robert Christie to all four linked the anthology rather vaguely to the Canadian Players. However, Lillian Hellman's *Monserrat* (20/10/55, ad. Mac Shoub, dir. Rupert Caplan) had some of the same cast as that of the Théâtre du Nouveau Monde production. Ibsen's *The Doll's House* followed (27/10/55). In the fall of 1956, all four plays were by Canadians: Arthur Hailey's *Time-Lock* and Joseph Schull's *O'Brien*, both now lost, Leslie MacFarlane's unintentionally racist and unfunny version of the Riel rebellion, *Black of the Moon* (17/10/56, prod. David Greene), which was balanced by a CBC presentation of John Coulter's *Riel* a few years later; and *The Discoverers* by Max Rosenfeld and George Salverson, a fine docudrama about the discovery of insulin. In 1957, the plays included Arthur Hailey's *The Seeds of Power* (3/10/57, prod. Paul Almond), an effective thriller about the political sabotage of a Canadian power plant in India; *Ice on Fire* by Len Petersen, a searching and anti-heroic look at professional hockey; a light comedy (also lost) adapted by Leslie MacFarlane from Stuart Trueman's funny novel *Cousin Elva*; and *Janey Canuck* (24/10/57, prod. Leo Orenstein), Lister Sinclair's adaptation of a book by Byrne Hope Saunders that gave an amusing account of Emily Murphy, a feminist who led the fight to have Canadian women declared "persons" by the Privy Council and thus made eligible for all offices in the country. All three plays to survive in kinescope from the 1958 season were topical: Lester Powell's *Panic at Parth Bay* (7/10/58, prod. Harvey Hart) dealing with the still topical issue of nuclear contamination; Mavor Moore's *The Man Who Caught Bullets* (4/11/58, prod. Mario Prizek), about the Korean War's aftermath for a veteran of that so-called U.N. police action, and Ivor Barry's translation of Marcel Dubé's *Man in the House* (31/10/58), set in working class Montreal.

Ice on Fire (10/10/57, wr. Len Petersen, prod. Ted Kotcheff) shows a contemporary viewer how little has changed in some areas of Canadian life. It is a well-written piece about an aging hockey player (Bill Walker) with an expensive wife (Toby Robins), who is trying to hold his place on an NHL team. Film clips of players practising in an arena are sparingly used. The live drama depended on actors on rollerskates, cleverly concealed for the most part by fluid camera work. When it was broadcast on the BBC the subject of an aging athlete would not have been unfamiliar, but the game of ice hockey must have looked very exotic to British viewers. Ivor Brown in *The Listener* (4/12/58) enjoyed "plenty of startling shots of a team in action as well as a boardroom in acrimony. The pace was suitably frantic and tempers high. The ice was indeed ablaze. There was a fine, hard-bitten performance of an angry old man by Mavor Moore."

In the 1956-57 season, the half-hour *On Camera* anthology drew large audiences with plays like Patricia Joudry's *A Woman's Point of View* and Elsie Park Gowan's *Stagecoach Bride*. The anthology revived the documentary drama and Living Newspaper forms from the theatre of the 'thirties by adapting similar material to modern circumstances in a fresh medium. Writer Mac Shoub, for example, took the news headline, "Why Big League Goalies Are Cracking Up," and developed the play, *Big League Goalie*. All kinds of subjects were covered: juvenile delinquency, a current murder case, Italian peasants arguing whether or not to plant crops on the grave of a Canadian soldier, and the tension between a deeply religious Scot and other citizens of an isolated British Columbia hamlet. Yet *On Camera* also did plays like *The Bottle Imp*, adapted by R. Denis from a story by Robert Louis Stevenson and produced by Arthur Hiller.

Moreover, "drama" at this time was not confined to one CBC department: arts, science, and current affairs also used it as a way of getting across their concerns. *Explorations* (1957-63) presented a programme on masks and their uses, one with Bruno Gerussi using *Hamlet* to show acting technique and two half-hours on the Canadian theatre of the nineteenth and early twentieth centuries called, respectively, *The Golden Age* and *The Canadian Mirror*.

Topical drama continued a major focus in the anthologies of the 'sixties, particularly *G. M. Presents*. There was also an emphasis on topical material in series like *Cariboo Country*, from which came two *Festival* programmes,

The Education of Phyllistine and the specially commissioned *How to Break a Quarterhorse*, and in *Wojeck*, *Quentin Durgens* M.P., and *The Manipulators*. Dramas on such themes as being laid off and race relations also turned up more often on *Festival* in its later years than in the earlier period which tended to emphasize adaptations of classic and contemporary theatre.

Francophone Drama in English

Festival (1959-70), *Folio*'s successor continued to introduce large audiences in English Canada to plays from contemporary French Canadian theatre, regularly featuring French Canadian actors: *The Endless Echo* (11/3/63) by Jean-Robert Remillard; *Bousille and the Just* (26/2/62) and *Yesterday the Children Were Dancing* (3/5/68) by Gratien Gélinas; *Two Terrible Women* (24/3/65) by André Laurendeau, and two very inventive Theatre absurdist plays by Jacques Languirand, *The Offbeats* (5/6/61) and *Grand Exits* (16/4/62). Some of Quebec's most popular television playwrights, however, did not find their way to CBC English television. The York archives have no scripts by Guy Dufresne and Yves Thériault, even though the latter is listed as having written over a thousand TV and radio scripts by the *Oxford Companion to Canadian Literature*.[9]

Marcel Dubé has been one of Quebec's most prolific and successful playwrights for stage and television. The CBC English network produced at least three of his plays, two scripts of which are in the York archives—*Zone*, aired on *G. M. Presents*, and *A Man in the House*. *Zone* was one of the most readily available in English, "best play" in the 1953 Dominion Drama Festival, and broadcast by the CBC in May 1956. *A Man in the House* was produced for the prestigious Canada Saving Bonds anthology, *First Performance* (31/10/58). In 1963, *Playdate* produced *Not for Every Eye* (14/10/63), an absorbing study of the effects of church censorship on the lives of people in a small town. Jack Creley played a bookstore clerk from the big city with a secret; he allowed selected people of a small town access to books from the Vatican's Index on Forbidden Literature, thereby losing his tenuous hold on his place in the community. It is a small-scale tragedy with a bitter flavour.

In 1957, Jacques Languirand wrote *Les Grands Départs*, a Radio-Canada television play afterwards put on stage in 1958. In 1962, the television version, translated by Ivor Barry, was carried on *Festival*. Highlighting this mixture of wry comedy and pathos, producer Mario Prizek got some extraordinary performances, full of vulnerability and subtle nuance, from Frances Hyland, John Drainie, Norma Renault, Robert Christie, Toby Tarnow, and Alex McKee. In each sequence, the random objects of the torn-up room framed, even shaped, the action. It was a highly visual, fluid production, full of the startling beauty and pathos of displaced or half-forgotten objects.

My edition of André Laurendeau's *Théâtre*, containing *Deux Femmes Terribles*, *Maria-Emma*, and *Les Vertu des Chattes*[10] has a brief introduction by Réal Benoit which mentions rather enigmatically that "peine perdue, l'échec que connût *Deux Femmes Terribles* au Théâtre du Nouveau Monde (1961) l'avait blessé profondement." The CBC version was an adaptation by Gwethlyn Graham for Mario Prizek on *Festival* (24/3/65).

CANADIAN, AMERICAN, AND BRITISH ANTHOLOGY:
CONSTRAINTS AND COMPARISONS

In 1961 anthology drama, previously one of the most popular television forms in the United States, all but disappeared. Yet in Canada, anthologies, though they thinned out considerably, continued for another twenty years. It is only since the late 'seventies that anthology drama has been a rare bird in our television aviary. In Britain, on BBC-2 and, to some degree, on ITV and Channel 4, the form still flourishes.

The reasons for this highlight the differences in broadcasting systems. Originally, sponsors liked being identified with a prestige "product." In the 'fifties, for example, U.S. Steel was selling, not hunks of metal, but an image. Since the essence of anthology is a "different story every week," audiences are not able to build identification with a particular character week after week. In fact, they may well turn off the set in the middle of a story they do not like, and thus cannot be delivered to sponsors in the predictable numbers which became a paramount consideration when television succeeded radio as the chief advertising medium. Moreover, it is not as easy to syndicate anthology drama as it is a sitcom or copshows on the late afternoon or 11:30 p.m. slots where reruns predominate.[11]

Most of the seventeen to nineteen hours a week of drama on American anthologies were based in New York and included many theatre adaptations. In the 'forties and early 'fifties the youngest and brightest wanted to go with radio where the action was, but ended up in TV where work was available—a common situation in Britain and Canada as well. Since there were no reruns because everything was presented live, television in those days had an experimental, "anything-goes" atmosphere. Good, bad, or ridiculous, something had to go out every week. Every script had lines that were expendable and lines that were expandable, depending on small variables in timing which error, accident, or nerves could make. When the production staffs took a holiday, the programmes went off the air. A good example of the freedom enjoyed by writers and producers on American networks during this brief period is this anecdote about the genesis of *Marty*.

According to Rod Steiger, producer Coe was shooting another Chayevsky script in a ballroom because studio space was in short supply. Chayevsky happened to mention a story idea involving the girls who lined the walls in these places, waiting to be asked to dance. His idea had been prompted by a sign requesting girls not to turn down any invitation. Coe said, "Yes, yes, write it!" and got on with the show at hand. Within two or three weeks, the first two acts of the new play were ready just as the cast went into rehearsal; the third act barely arrived in time to be rehearsed for its live performance. Yet *Marty* is the script most likely to be remembered from that period, and it is often cited as influencing writers, producers, and critics in the U.S., Britain, and Canada. *Marty's* realism, the small-scale domestic situation, its regional, working-class speech, small number of sets, contemporary social observations, and emphasis on psychology, the closeup and the human face set the style for much of American anthology.

In the early 'fifties, audiences of three or four thousand people, incapable of measurement by today's network standards, were common. Television sets were first owned by the more affluent and educated upper middle classes who could afford them. With its emphasis on internal psychology, American anthology drama rarely saw a violent resolution. In fact, several of the most famous, including *Requiem for a Heavyweight*, the tragedy of a used-up boxer, had downbeat endings—rarities by the mid-'sixties. Performances of actors like Rod Steiger, Jack Palance, Jack Lemmon, Paul Newman, Joanne Woodward, and Julie Harris matched the quality of writers like Reginald Rose, J.P. Miller, Tad Mosel, Horton Foote, and Rod Serling and directors like Sidney Lumet, Arthur Penn, and Delbert Mann. Paddy Chayevsky: "[We] were stuck with every possible technical difficulty, [and] none of the Hollywood efficiency. We were doing those shows out of old radio stations that had been rebuilt with no place to put the lights, nothing elaborate in the way of scenery. But when you were stuck with those restrictions, the writing had to be better. . . . Go to Hollywood and they hand you everything in the world—except the need to do imaginative work."[12] The strengths and limitations of those days are summed up in his comment that "You create character and that becomes your plot."

American technology and Canadian television techniques leapfrogged one another, each learning from the other through the exchanges of personnel and kinescopes. In 1956, videotape came to the U.S., and soon after Canada went live to tape. In 1966, the CBC had acquired videotape that could be more easily edited. As our producers moved out into the U.K. and the U.S. looking for new challenges, their kinescopes and tapes were their audition pieces. Since no one at the CBC kept track of this material in those days, copies of programmes that appear to be lost may still survive in private hands.

The breakthrough in British television drama came well after the American and Canadian television drama had reached its first peak. What has been called an "annus mirabilis" season (1960) was directly attributable to expatriate Sydney Newman, the Head of TV Drama at Granada, who looked for new populist scripts about topical subjects and real lives and was willing to see a range of styles from realism to the absurd. Plays by voices new to television, like Alun Owen, Harold Pinter and Clive Exton (*Contrast*, Autumn 1965) appeared. By then, British critics and playwrights were articulating what they considered television drama to be: "plays in which the camera is an interpreter rather than an onlooker . . . the small screen, the small audience, the semi-darkness, all encourage a drama of high emotional and poetic intensity," as director Don Taylor wrote in "The Gorboduc Stage" (*Contrast*, Spring 1964). In Britain, the writer has always enjoyed special status. Writers in the United States lost this kind of influence over their scripts in production when series television arrived.

In Canada too, few writers are given professional respect by either producers or the corporation. Yet the basic fact is that no good "single" or series drama can succeed without a good script. In Britain, television scripts are also published[13] as opposed to "novelized"—a rare occurrence in the U.S. or Canada. In 1981 Shaun Sutton summed up the debt owed to the people who wrote television scripts this way: "If television drama means anything today, and it does, it acquired that meaning from its early writers."[14]

In the years 1955-60, as Canadian television was just getting under way, several factors changed the structure of network programming in the United States. Westerns, crime and punishment programmes, and action-adventure stories—many of them taking advantage of locations in sunny Hollywood—began emerging as series which were built around the same sets of characters. *Dragnet* had shown the way in 1951, with its details about police work and a theme song that kids thumped out in playgrounds all over North America; Joe Friday's "Just the facts, Ma'am," became a catch phrase. The police jargon and the emphasis on authenticity set the mould for the genre, but a more important convention may have been the reassuring epilogue that told the audience exactly how and when the criminal was punished. Film, pioneered by *I Love Lucy*, made reruns and syndication to new markets possible. Eventually, the big film studios got interested in their powerful new rival and started making series.

Canada offered no perpetually sunny climate for year-round location shooting and no movie empires with huge sound stages. Studio facilities needed to be built, but they were not. Instead, they were "adapted" from existing space. To the disgrace of the government and CBC policy makers, there are still no properly equipped drama studios in CBC Toronto, where most English CBC drama is made. Over the years, the drama department has actually lost drama studios until, in 1987, only one is left.

What the CBC did finally gain was the right to make drama films as Americans had been doing since the 'fifties. From 1952 to 1962, it had been forbidden to make long drama films, supposed to be the National Film Board's prerogative. Why good full-length films from both institutions were not in the national interest is a mystery to someone not in the civil service. The NFB did not, however, control the country's movie distribution. Unlike most other nations, Canada did not (and does not) have a quota system; when we do make excellent full-length films, few people see them. The CBC, meanwhile, could distribute films into millions of homes but could not make dramas that called for the freedom and realism of film. Its directors worked under quite unnecessarily difficult conditions to make their programmes in the 'fifties and early 'sixties; this was a situation that Ronald Weyman and others fought successfully to change, first with *The Serial* in 1962-66. Canadian novels like Thomas Costain's *Son of a Hundred Kings* were adapted into serial format and original series like *Cariboo Country* and *Quentin Durgens M.P.* first appeared on *The Serial*. Weyman then went on to film series like *Wojeck* and *Corwin*.

By the mid-'seventies, producers were having problems with studio space and reliable electronic equipment. As a result, by the 'eighties, most anthology drama we see was made using the expensive, and in some ways, limited medium of film. However, tape made such advances in the 'eighties that film lighting, single camera shooting techniques and relatively portable technology could be used on location. Costs dropped again. Once more forms of television drama continued to adapt to the new technology. In Britain and on the Continent, film and studio drama flourish side by side in both anthologies and specials or what the British call "single" plays.

By the time series drama on American television was largely being put on film and far larger audiences were buying sets, television drama was recognized as an industry. The emphasis shifted towards making money rather than simply meeting expenses. Good programmes died and still do, not because no one is watching or the show is losing money, but because the profits are not big enough. In the mid 'sixties and 'seventies, with the growth of the science of demographics, if the wrong people were watching—that is, not the 25- to 40-year-old urban $25-45,000 earners who were the market for a particular sponsor's product—a show with a respectable audience share would still disappear. By the late 'sixties such attitudes had infected CBC thinking—which unlike the BBC depended both on Parliment and on commercials for its revenue. This explanation was offered by the CBC when it killed the immensely popular *Don Messer and His Islanders* in 1969. It was not "youth-oriented" like the hootenannies then in vogue. As John Gray's popular musical about Don Messer (1983, toured nationally in 1984) points out so vividly, one more link with our own regional tradition was displaced by a transient imitation of foreign popular culture. A genuine alternative was eliminated by sponsors, ratings, demographics, and the failure to value the indigenous and distinctive voice.

As attitudes to television changed with film, the arrival of series, and increasing pressure from sponsors, programmes on both sides of the border became "inventory." The programme was now simply "cheese for the mousetrap." The reasoning went this way: since the product being sold solves a viewer's day-to-day problems and often sells fantasy, or what the 'seventies learned to call lifestyle, the programme next to the commercial should not address the same mundane problems the commercial promises to solve unless there is a witch, a genie, a Martian, a wash of nostalgia, a cruise ship, or a fantasy island to soften the impact. In the case of *Loveboat*, the "product" (programme) increased sales of cruise trips to Acapulco. Most of children's Saturday morning American cartoons in the 'eighties are simply half-hour toy commercials. The toy precedes the television programme, which is devised to sell the toy.

The economics of privately owned networks in the United States dictated that television anthology rule the airwaves for only a dozen years. At the same time, production moved from the theatrical heart of America, New York, to Hollywood. Actors could no longer comfortably appear in both television drama and the theatre and had to choose between the two, a dilemma never faced by London or Toronto based actors.

Speculations in 1960 on the Future Shape of TV Drama

Back in 1960, some CBC writers and directors took the risk of predicting where TV drama would head in the following decade. In an article called "Where Will It Go From Here?" (*CBC Times*, 10-16/12/60), Brian Stewart tried to find out just "what are the ingredients of a good hour-long television drama." Writer Charles Israel said that the longer, sixty- and ninety-minute form needed conflict; the implication seemed to be that a lyrical mood piece could fill twenty-six minutes but not a whole hour. Melwyn Breen, who directed for both *G. M. Presents* and later *Festival*, commented that the need was for a strong story line. Thus, both men gave a traditional response which would still hold true for most theatre, film, radio drama, and television. Writer-actor Mavor Moore said with more imagination: "There is practically nothing you can say that applies to all dramatic works except that they must be audible, visible and they must relate . . . no, that's too strong a word . . . they must convey a story." What he thought television could not do well was ritualistic drama and drama heavily charged with symbolism and poetic language. This could be said to be true of such later productions as the adaptation of Lorca's *Yerma* (8/2/67) and George Ryga's *Man Alive* (30/3/66), but not of *Point of Departure* (1960), the adapation of Michel Tremblay's *Les Belles Sœurs* (1970s) or contemporary BBC versions of plays like *Titus Andronicus* (1982)—not to mention Ingmar Bergman's *Fanny and Alexander* (late 1970s), or several pieces of German, Swedish, Israeli, Jewish and Spanish television made in 1985 and featured in the 1986 Banff Festival competition. Still, Moore was right on target when he commented that in the context of the rather flaccid state of commercial theatre in 1960, its "tepid drama" made even sport seem more "theatrical" than the real thing.

Herbert Whittaker, for years the theatre critic of the *Globe and Mail*, thought that television had to discover its own technique, "perhaps a fragmentary and enquiring technique because its strength is in the you-were-there, reportorial style." Anticipating the breakthrough in format and content of *The Open Grave* (1963), his chief objection to adaptations, still valid for much television drama, was that "TV suffers from the apparent need to tie up plots. You see this when TV tries adapting a novel. All you get is the narrative, which is often the weakest part of a novel." Again, Eric Till and others had occasionally proven Whittaker wrong in dramas like *Ward Number Six* (3/2/59), the adaptation of Chekhov's moody short story, a piece about a Russian insane asylum. With *Pale Horse, Pale Rider* (23/10/63), Till went on to demonstrate how atmosphere and ambiguity could be superbly evoked within the limitations of black and white, live-to-tape-television.

Ed Moser, executive producer for the more ratings-oriented, commercially sponsored and more popular *G. M. Presents*, pointed out that three acts of roughly sixteen to seventeen minutes, with a climax in each, create a different structure from fifty unbroken minutes. He also emphasized that television audiences, unlike theatre audiences, were fragmented, apparently meaning by this that they did not participate in collective interactive experience. All of us can remember, however, the sense of community involved in watching crises (assassinations, kidnappings), elections, hockey games, moon walks, the ends of favourites like *M*A*S*H*, or mass-market miniseries like *Holocaust* or *The Winds of War*. Even in the 'fifties, from the McCarthy hearings to the Diefenbaker era, most people were watching and talking the next day about widely advertised drama repeats that were fondly remembered years later. Mention "I told him, Julie, don't go," or Trudeau on the steps of Parliament saying, "Watch me," and the country remembers these moments.

Melwyn Breen complained that there was a dearth of experimental TV scripts. Israel, although picking up on the general feeling that the first wave had passed by 1960 (and many creative people like Newman, Silvio Narrizanno, Ted Kotcheff, Arthur Hiller and David Greene had left for the U.K. or the U.S.) disagreed: "CBC is a good place to try experiment; not as good as it once was perhaps, but still very good." The producers of Montreal's *Shoestring Theatre* or, later, the nationally broadcast *Q for Quest* would not have agreed with Breen, even though, like other television producers and story editors, they conscientiously read their way through mounds of imitative scripts which echoed the film, theatre, and radio hits of the day, the trend of the year, or, with few alterations, reproduced current headlines. They knew that experimental television drama was still flourishing because they were doing it.

George McGowan, commuting at that time between Stratford and the CBC, summed up the picture as most directors saw it. Like the others, he knew the wonderful days of television anthology in the United States were gone: "In the best period of TV drama, producers like Fred Coe gathered around themselves writers like Chayevsky and Serling and let them work out ideas without too much pressure. Producers [then] had some of the same functions as the great editors of the publishing houses." This was a perceptive comment, despite the many examples of fine television drama where images and performances masked weak narrative and wooden dialogue.

Interestingly enough, all these creative people voiced the fear that television's new film and tape technologies would lock the medium into the strait-jacket of naturalism; Moser called it "the Ibsen hangover." In this respect, they were prophetic of the 'seventies, not the 'sixties. Surface realism did begin to prevail as film and tape replaced live studio drama, but the personal documentary filmmakers of the 'sixties, including Ron Kelly and Allan King, tried to adapt documentary's direct-camera techniques to television fiction. They discovered that when hand-held cameras "found" lighting, and unstaged and unrehearsed conflicts were used to approximate our fragmented sense of everyday reality, the drama becomes fragmented as well. Paradoxically, the result is perceived as highly stylized by the viewer. The best example is probably Eric Till's *Freedom of the City* (1975), discussed below, which uses most of the old and new techniques and then some.

The survival of anthology as a form differentiated Canadian television from the American model, providing genuine alternatives for viewers and significantly inflecting the series derived from American and British genres. It is true that Canadians watch a lot of situation comedies, soaps and copshows, as well as T.V. movies, for relaxation. Unlike Americans, however, we also welcomed serials that demanded viewer commitments of time and concentration as early as 1964. Even in series, we have accepted episodes lacking the mandatory, poetically just endings. We tune in to experiments, mixed genres like the satirical, comedy-thriller series *Seeing Things* and successful period dramas. Perhaps most important of all, we still produce and watch anthology drama that fits no week-to-week formula, portrays no set point of view, relies on no familiar characters, and, in fact, is much more like live theatre in range and potential. In the increasingly intense debate over the future of broadcasting in the mid-'eighties, the basic premise no one disputes is that we have to be different and better in order to hold our own. Yet to my knowledge, no one in this debate has clearly and confidently stated that we have been making programmes that were different and often better than most of our competitors for over thirty-five years.

Contemporary Assessments of CBC Anthology

One major reason why anthology drama has survived in Canada this long is that it got off to a cracking good start. For the first seventeen years of CBC television's history, regular weekly slots were always set aside for the form on *CBC Television Theatre, Scope, Folio*, and *Festival*. In the early 'sixties, *Festival* was joined by the experimental half-hour anthology *Q for Quest*, a division of labour analogous to the traditional spread of production in most of our regional theatres, with *Festival* functioning as the main and *Q for Quest* as the second stage. Given this commitment, theatre critics of the 1960s at least seemed to recognize the contribution of CBC radio and television drama as our national theatre of the air. *Globe and Mail* drama critic Herbert Whittaker, writing in 1961 about the whole spectrum of radio and television drama on the CBC, on the twenty-fifth anniversary of the corporation pointed out:

> For the past twenty-five years, the CBC has supplied most of the dramatic intake of Canada. No other country has had to rely so heavily on one single source for theatrical knowledge, experience and expression. It is undeniable that the CBC has been the major employer of playwrights in this country. For the past twenty-five years how many people would have been able to earn a living acting except through the CBC? In short, the CBC has subsidized a whole theatre for us for a quarter of a century.
>
> How good has it been? In the field of drama, its taste has been high, its approach both serious and creative. . . . The Canadian theatre can celebrate the Twenty-fifth Anniversary of its greatest single benefactor this week. Let it be done wholeheartedly. For an organization that is

part of a civil service, it has done better than any other untrammeled body you can name. In the field of drama, battling a quarter-century of complaint and criticism, the Canadian Broadcasting Corporation has done a gallant job for the theatre in Canada.[15]

As one might expect, the perspective of the *New York Times* (19 January 1964) was rather different; however, it was equally complimentary. Three years after the demise of most American anthology drama, the three hours of more challenging television drama available each week on the CBC were cited with envy. Kenneth Brown's *The Brig* (21/1/64), Fletcher Markle's adaptation of Katherine Anne Porter's *Pale Horse, Pale Rider*, Wesker's *Roots* (4/12/63), Anouilh's *Antigone* (9/10/63), Albee's *The American Dream* (13/5/63), and the musical version of Molière's *Le Misanthrope* (called *The Slave of Love* (27/11/63), were listed as recent instances of "very sophisticated programming." Noting "the opportunities [for actors] to expand their talents," the New York critic admired the CBC's unwillingness to build stars and lock actors into roles. Supervising producers for programmes such as *Q for Quest* and *Festival* treated their audience not as a mass, but as groups of people with different tastes and different views. If a viewer failed to find something to enjoy one week, he or she would likely find something the next week ranging from comedy to high tragedy, folk to opera.

As one might also expect, respected theatre critic, host of *Fighting Words* and ex-story editor for Sydney Newman, Nathan Cohen did not agree with Whittaker or the *New York Times*, and he would probably not have agreed with many of the arguments I advance in this book. In 1966, he wrote: "There was no golden age of TV drama for us. We never produced a group of recognized authors identified with a specific programme."[16] This is true enough, at least as far as the public was concerned, and is a strong contrast to the British emphasis on the writer from the mid-'fifties on. Even in the United States, as long as anthology survived, the playwright's reputation was often what drew audiences. Probably few readers could name a current CBC playwright, not because our playwrights are not doing fine work, but because we have followed the American shift in emphasis, in both publicity and criticism, from writer to actor or, occasionally, to the producer or director. Since status is important in human affairs, this tendency to overlook the most fundamental contribution to good television drama, a good script, has neither served our playwrights justly nor our television drama well. Treat every writer as a journeyman in terms of public recognition and the writer may well settle for journeyman standards rather than try for mastery of his/her art.

In Cohen's view, CBC-TV drama production never achieved "the glossy consistency of such U. S. anthologies as *Kraft Theatre* (1947-55) and *Studio One* (1948-58) and *Playhouse 90* (1956-60).'' Frankly, he was just wrong, as a comparison of surviving kinescopes from *Festival* or *Playdate* with the rebroadcasts of American classics already cited shows. He did admit that, as the network began to grow and production passed its infancy, TV drama quickly became the closest thing to a national theatre we had in Canada, offering plays that could be seen simultaneously across the country. Cohen went on to mention specifically the excitement, tension, and feedback from Arthur Hailey's very competent thriller, *Flight Into Danger* (3/4/56); as a story editor, he had discovered Hailey's piece for Sydney Newman, but thought it unproduceable because of the multiple film cues required. Newman insisted that the technical difficulties daunting Cohen could be overcome.

Cohen's own observations on what constitute the inherent rules of dramatic form for television are also open to question. I must disagree with his generalizations that television drama was never first class and seldom indigenous in this country. On the latter point, the bulk of the evidence indicates otherwise. I do agree with him wholeheartedly, however, when he says that much of Canadian television anthology held an ''irresistible appeal for those of us who like a good story well told.'' From the week in, week out offerings, one could not reasonably expect more; and out of such plenty can come the extraordinary.

Festival

As the 'fifties shaded into the 'sixties, television drama became the first line of defence along our increasingly tenuous electronic border, as well as a nursery for some distinctively Canadian drama. *Q for Quest*, for example, commissioned George Ryga's *Indian* (25/11/62) and *Two Soldiers* (8/10/63) and provided the first professional production of James Reaney's *One Man Masque*, called *An Evening Without James Reaney*. *Festival* introduced Reaney's *The Killdeer* (12/6/61) to a larger audience. Both anthologies commissioned successful adaptations of minor literary works that turned into major television plays: Fletcher Markle's version of Katharine Anne Porter's *Pale Horse, Pale Rider*; Mac Shoub's adaptation of Chekhov's short story *Ward Number Six*, Alvin Goldman's treatment of an obscure Scottish theological thriller, *The Private Confessions of a Justified Sinner* (13/5/64), and *Behind God's Back* (26/1/69), adapted by Jan Carew from a short story by Austin Clarke.

Festival had the longest run as the corporation's flagship anthology series (1960-69). The title reappears sporadically in the 'seventies, but bears no relationship to the original prestige anthology. By 1960, the CBC recognized that television was the preeminent mass medium on the continent, the one most people turned to for their drama. The corporation's response to these circumstances was a regular programme of ballet, opera, concerts and plays offered weekly, nine months of the year. *Festival*'s emphasis on the classics, important modern plays, and contemporary drama, which was criticized at the time by nationalists, served useful purposes. It helped legitimize television as a medium. It educated large audiences to a wide variety of dramatic conventions: unit sets, area lighting, the cross-cutting of scenes, the counterpointing of image and word, direct address, surrealistic design. As well, *Festival* satisfied knowledgeable audiences by keeping them abreast of developments in modern theatre. Perhaps most significantly, it set a standard for Canadian playwrights and put our own plays in a historical, geographical, and geopolitical perspective. In their three-month break, many CBC producers went to the United Kingdom to work on different materials in different conditions, an exchange which was mutually beneficial and virtually unknown by the 1970s.

The 1968 Mandate

At the end of the 1960s, the CBC was given a revised mandate in the form of a new broadcasting act. To paraphrase the legislation, CBC television was expected to contribute to national unity, strengthen our sense of identity, provide a regional balance, show a cross-section of Canadian culture, present controversial issues in a comprehensive and balanced way, strengthen our cultural fabric, and serve as a patron to the arts—a set of criteria we would never dream of applying to a national theatre or a national cinema, if we had such a thing. Expanding our vision to help effect social change, battling censorship and experimenting with form were not specifically part of the mandate. Nevertheless, there were times when the CBC broke new ground in these areas for Canadian theatre and Canadian film.

I take it as a given that when there is a modicum of contention, controversy and risk-taking Canadian television drama is healthy, alive and doing its job. Despite their dependence on government subsidy, Canadian theatre and even Canadian film have been remarkably free, although not wholly free from interference. However, at a time when few Canadian stage playwrights were tackling challenging material or experimenting with new forms and few people saw what little experimental film was around, CBC television was developing and broadcasting its own experimental scripts. For theatre and film, the CBC's *Festival*, *Quest*, and occasional drama specials pushed back the barriers of censorship. In the 1962-63 season, for example, the Cold War had reached its deepest freeze, and sponsors were censoring the content of United States television without mercy.[17] During that same period the CBC presented three plays by Canadian playwrights: Rudi Dorn's introspective and ironic *The Neutron and the Olive*, Jack Cooper's surrealistic anti-war fable *The Wounded Soldier*, and George Ryga's *Two Soldiers*, a superbly realistic worm's eye view of the possibiiity of war. In the early 'sixties, the CBC also produced Marghanita Laski's post-World War III parable *The Offshore Island*, Bernard Kops's anti-war play *The Dream of Peter Mann*, and *Sergeant Musgrave's Dance*, John Arden's examination of the causes of war, as well as Alan King's adaptation of the transcripts of the U.S. hearing on the loyalty of atomic physicist J. Robert Oppenheimer. The drama department also presented a remarkable *Galileo*, with CND posters, ironic captions, and other updated references to the Cold War.

Other examples of subject matter challenging the accepted North American television norms include *The Open Grave*, which treated the Resurrection as a fast-breaking new story, using *cinema verité* techniques; much later, *Reddick*, a topical drama about a United Church minister's crisis of conscience when confronted with hippies and bikers; and British writer James Saunders's racially mixed *Neighbours* (15/1/69), a play whose subject had been anticipated ten years earlier by Canadian Joseph Schull's *The Concert* (6/2/58). When it came to controversy, the CBC adopted the same strategies as regional theatres do to this day. If Robert Allen anticipated that there might be questions in Parliament, letters to the editor, or switchboards flooded with complaints, he admits that he would sometimes temper the wind by scheduling right after a controversial play, scripts which were clearly safe, and justifiably popular by authors like W. O. Mitchell, Arthur Hailey, or Bernard Slade.

In this respect, on the less frequent occasions when it takes a risk, CBC strategy has not changed. Regrettably, the Head of TV Drama no longer has decisive input into the scheduling. CBC public relations tends to stress less controversial drama like *Chautauqua Girl* and *Anne of Green Gables*. Nevertheless, it still makes programmes like *Oakmount High* (1985), which was banned from broadcast by the courts in Alberta because it tackled the issue of anti-semitism being taught in the schools while the James Keegstra case was still *sub judice*.

Performance Treasures on the Record

From the theatre historian's point of view, one of the most important contributions television drama made was to preserve, on fragile, scratched, often distorted kinescopes, the performances of some of our finest stage actors in some of their best roles—to name only a few, Martha Henry as Viola in *Twelfth Night* (8/4/64) and Hilde in *The Master Builder* (18/11/64); William Hutt as Uncle Vanya and Ivanoff; Frances Hyland as the Duchess of Malfi and Olga in *The Three Sisters* (13/2/61); Douglas Rain as Bosola in *The Duchess of Malfi* and as the Narrator in *Under Milkwood* (17/2/59); Leo Ciceri as a variety of Wildean heroes and the Cardinal in *The Prisoner* (1962); Kate Reid as Masha in *The Three Sisters* (13/2/61) and Mother Courage (20/1/65); Barry Morse as Vladimir in *Waiting for Godot* (2/12/64) and Don Harron as Tigius in *A Phoenix Too Frequent* (22/5/58). Among other treasures we find Lorne Greene, Kate Reid, Lloyd Bochner, and Jack Creley in Sartre's *The Unburied Dead* (24/4/57), Bruno Gerussi as Peer Gynt, Eric House as Sammy, Joseph Shaw as Voltore, Leslie Nielsen as John Proctor in *The Crucible* (10/10/59), and Lloyd Bochner as Duke Ferdinand in *The Duchess of Malfi*. John Colicos, Douglas Campbell, and Barbara Chilcott are also captured on kinescope. In the early 1970s, *Program X* continued to present well-known performers on tape in programmes like *An Evening with Kate Reid*, who played everything from Lady Bracknell to Saint Joan, Evenings with Barbara Hamilton and Gordon Pinsent, Mia Anderson's stage hit *Ten Women, Two Men and a Moose*, and a Dinah Christie/Tom Kneebone/Noel Coward Revue from Toronto's Theatre in the Dell.

The next generation of fine stage actors is also well represented in the anthology drama of the 'seventies and 'eighties though not in adaptations of modern or classical theatre: for example, R.H. Thompson as *Tyler* (1978) and Brent Carver in *Crossbar* (1979), Janet Amos and Anne Anglin in *Ada* (1977), Eric Peterson in *1837* (1975), *The King of Friday Night* (1984), *Charlie Grant's War* (1985), and *Grierson and Gouzenko* (1986); Tom Butler as Bernier in *Cementhead* (1979), Janet Laine-Green in *Chautauqua Girl* (1984), and Diane Belshaw in *Ready for Slaughter* (1983).

As the 'sixties progressed, *Festival* tried to balance the priorities of showing ourselves to ourselves and showing our best to the world in an international context, a bill of fare not unlike that of most regional theatres and attended by similarly mixed results. In response to changing tastes, however, many other CBC anthologies disappeared. As we have already seen, the CBC often inflected the forms made familiar by American genres—or simply invented their own. In fact, many of the scripts of the series are beautifully crafted one-act plays that could stand on their own beyond the context of the series.

The long-running anthology *General Motors Presents* included many Canadian scripts, of which I sampled only a few. As one would expect, they ranged from the inadequate to the quite good. *Flight into Danger* (3/7/56), produced by David Greene, with its skilfully crafted suspense and fine sense of entrapment, was a major hit in Canada, Britain, and the U.S. where a new production dropped a crucial line and thus lowered the suspense. Critic Philip Hope Wallace told readers of *The Listener* (4/10/56) that the CBC kinescope had "scared the daylights out of us." He praised James Doohan as the rusty fighter pilot who has to get the plane back down on the ground, Cec Linder as the air controller and director David Greene: "Here was something to sell sets by the millions . . . a short, pithy, and gruesomely exciting little piece." Wallace did not mention that everything depended on the technicians and director who had to time exactly the cues showing the plane's takeoff, flight, and safe arrival. Veteran actor Ed McNamara remembered that when the "safe arrival" film inserts came up on cue all present had to suppress their cheers (*Morningside*, 27/12/84). This first script by Arthur Hailey became the prototype for dozens of television plays and movies. As Halliwell says, "It was one of the first dramatic entertainments to prove that what's on at home can keep the cinemas empty." Back in Canada, it had been the first widely discussed popular television play: a real breakthrough.

In the 1950s, many other *G. M. Presents* were topical Canadian scripts. This anthology did, however, air adaptations of material like an adaptation of *Billy Budd*, *The Blood is Strong*, and a two-part special of Coulter's *Riel* historical drama (23 and 30/4/61, prod. George McGowan, with Bruno Gerussi) which captured Riel's charm and fanaticism, as well as clearly presenting both sides of the issue. Canadian playwrights produced on *G. M. Presents* included Hugh Kemp, Fred Edge, Stanley Mann, Bernard Slade, Arthur Hailey, Joseph Schull, Leslie MacFarlane, Patricia Joudry, Ted Allan, George Salverson, Len Petersen, Mavor Moore, Mordecai Richler and Jack Keyser.

Executive producer Ed Moser's *Playdate* also included Canadian scripts on largely contemporary themes. Here too there were some surprises, including a very funny production of Sheridan's *The Critic* (1963), produced by Norman Campbell, which opened with a close-up of a toy theatre where two cut-outs of ships fire at one another until one sinks. *The Thirteenth Laird* (2/5/62, by Munroe Scott, prod. George McGowan), was a historical drama about a near feudal settlement in nineteenth-century Ontario, with the general flavour of light comedy and domestic melodrama. There were also a few now very dated dramas like Slade's *Men Don't Make Passes* (1963) about a woman of "a certain age" who, by definition, had to feel insecure and incomplete without a man in her life. For some reason, this piece was produced twice by the CBC.

As a critic, what puzzles me is how the programmers thought that adaptations of Henry James, Marguerite Duras, Alun Owen, or August Strindberg, together with a whole variety of Canadian plays could mix every fourth week with an hour of singer Jo Stafford (1962-65) or comedian Red Skelton (1963-64). Both *Playdate* and *G. M. Presents* provided solid, often relevant entertainment. Even here, however, censorship could present

1. Ford Startime's presentation of *The Crucible* with Leslie Nielsen as John Proctor, Diana Maddox as Elizabeth Proctor. 2. In the 1954-55 season *Hamlet* appeared in the series *Scope* with Lloyd Bochner as Hamlet "live." 3. *Folio* appeared from 1955-60. In *Ward Number Six*, Jack Klugman played as Gramov, Mavor Moore as Dr. Ragin. Plates 4 to 11 are all from *Festival*, which ran from 1959 to 1969. 4. *Mother Courage*: Kate Reid as Courage, Jonathan White as Swiss Cheese, Len Birman as Eilif, Jackie Burroughs as Katrin, and a soldier. 5. *The Duchess of Malfi*: Frances Hyland as the Duchess, Michael Learned as Cariola.

6. *The Paper People*: Marc Strange as Jamie surrounded by "paper people." 7. *Grand Exits*: John Drainie, Frances Hyland, Norma Renault, and Toby Tarnow. 8. One element of the drama sequence in *Pale Horse, Pale Rider*. 9. *Galileo*: Hugh Webster as the narrator aka Brecht. 10. *The Offshore Island*: "outdoors" created in a studio with Irene Worth as Rachel and Tony Van Bridge as Martin.

11. *Reddick*: Don Harron as Reddick, Gary Reineck as Mark (centre above Harron), Don Barsenko as Gower (helping him). 12. *Cariboo Country* ran from 1960 to 1967. This scene from "All Indian" shows the Spirit Dancers. 13. *Q for Quest* ran from 1961 to 1964. This is a scene from *The Brig*. 14. Len Birman as the Indian, and Sean Sullivan as the agent in *Q for Quest*'s *Indian*.

15. In "The Last Man in the World," Sabina von Fircks as the prostitute. 16. *Wojeck* on location. 17. "Listen, an Old Man Is Speaking," a *Wojeck* episode. 18. *Vicky* played in *Sunday at Nine*'s 1971-72 season. Jackie Burroughs as Vicky, Sean Sullivan as her father, Kay Hawtree as her stepmother.

19. *The Beachcombers*: the original "family": Rae Brown as Molly, Robert Clothier as Relic.
20. *The Beachcombers*: Bruno Gerussi as Nick, Pat John as Jesse. 21. *King of Kensington*: Peter Boretski as Jack, Helene Winston as Gladys, Al Waxman as King, Jayne Eastwood as Gwen.
22. *Freedom of the City* (aftermath) appeared in the 1974-75 *Opening Night* series. 23. *Ice On Fire*: Bill Walker as Nick Philips, from the *First Performance* series.

24. *A Gift to Last*: Gordon Pinsent as Sergeant Edgar Sturgess, Mark Polley as Clement.
25. *A Gift to Last*: final chapter—the family portrait: Mark Polley as Clement, Janet Amos as Clara, Ruth Springford as Lizzy, Kate Parr as Jane, the new baby, Gerard Parkes as James, Dixie Seatle as Sheila, Gordon Pinsent as Edgar. Plates 26 to 34 are all from CBC specials.
26. *Peer Gynt*: Bruno Gerussi as Peer Gynt. Note the lighting and unit set. 27. *Point of Departure*: William Shatner as Orpheus, Lloyd Bochner as Death.

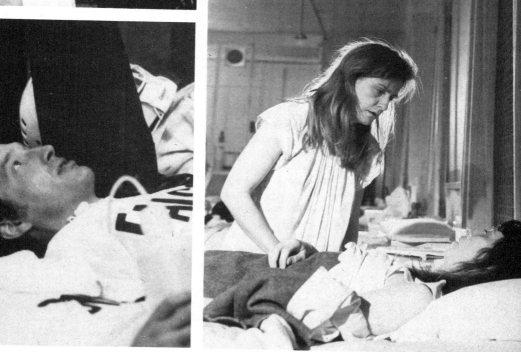

28. *Gentle Sinners*: Christopher Earle as Eric, Charlene Simiuk as Melissa. 29. *Red Emma*: Chapelle Jaffe as Emma 30. *Cementhead*: Tom Butler as "Bear" Bernier. 31. *Ada*: Janet Amos as Ada, Anne Anglin as Jenny.: Chapel.

32. *Turning to Stone*: Nicki Guadigni as Allison. 33. *Bethune*: Donald Sutherland as Bethune with Chinese soldier. 34. *Ready for Slaughter*: Gordon Pinsent as Will, Diana Belshaw as his wife, Boothe Savage as a farm "survivalist." 35. *Chautauqua Girl*: Janet Laine Green as Sally and her supporters.

problems. Newman consistently edged into, but usually avoided, conflicts with the sponsors: nevertheless, a play by Ted Allan about basket weavers in Mexico was scrapped on *G. M. Presents* because it was taken to suggest support for trade unions. Under the titles *CBC Theatre* (1953-57), *CBC Playbill* (1953-54), *Playdate* (1961-65), *First Person* (1960-66), *The Unforeseen*, *On Camera*, and their successor, *The Serial*, sponsored anthology dominated the CBC from 1952 until the late 1960s, and the bulk of the scripts were Canadian.

Television's role in the development of Canadian theatre and its achievements in the development of Canadian dramatic literature cannot be denied. Nevertheless, except for a brief fling under John Hirsch, it must also be admitted that television failed to do justice to Canadian theatre as it matured. In the first two decades, formal links with theatre were few. True, CBC television did cover the Dominion Drama Festival finals in documentaries and news features, and script editors scouted them for writing and acting talent. Very occasionally successful stage productions were adapted for television: for example, *Folio* broadcast a shortened version of the Crest Theatre production of John Osborne's *The Entertainer*. Only twice, however, did the CBC televise a Stratford Festival production. First, a flyer in my possession announced that *Peer Gynt* (29/12/57) would be "produced by the Stratford Festival Foundation in association with the CBC for this single presentation and sponsored by the International Nickel Company. The play will be directed by Michael Langham in association with Douglas Campbell." Twelve years later the CBC telecast *The Three Musketeers* (19/3/69, prod. David Gardner). The CBC's *Twelfth Night* (starring Martha Henry) had many of the same actors as those in the Stratford production, but it was not the same play. There was an odd indifference in the corporation to the task of recording our theatre history and bringing our best classical theatre to the country at large. At last, from 1981 on, one or two Stratford productions a year were reaching the screen. Unresolved problems with ACTRA, obtrusive technology, and greatly reduced budgets explain but do not excuse this.

Looking over *Festival*'s lists of plays and playwrights from the first fifteen years, one finds few Canadian works commissioned from playwrights like Robertson Davies, Merrill Denison or, later, Mavor Moore and Lister Sinclair, much less from internationally successful playwrights like John Herbert of *Fortune and Men's Eyes*. Even the Centennial year produced only a handful of Canadian plays in the premiere time slot of *Festival*. Instead, there was one repeat from *Cariboo Country*—St. Pierre's *The Education of Phyllistine*—and together with *How to Break a Quarterhorse*; Marie Claire Blais' *The Puppet Caravan* (1/3/67); *The Painted Door* (17/1/68), adapted from a short story by Sinclair Ross; *The True Bleeding*

Heart of Martin B (26/4/67, justly notorious around the CBC as a
pretentious failure, full of visual and verbal clichés); three collaborations
with the National Film Board, and one superbly original piece, Timothy
Findley's controversial *Paper People* (13/12/67). Yet *Festival* had the money
and freedom from a sponsor's pressure and rigid time slots which could have
encouraged Canadian playwrights to explore the medium in new ways for the
1960s.

By the end of the decade, *Festival* had run its course. In 1969, Bob
Blackburn, not a fan of CBC television drama, wrote a lengthy and in many
ways very moving obituary of the programme (Toronto *Telegram*, 5/3/69),
reminding his readers that it had served the purpose of "creating and
sustaining a new Canadian tradition of drama for stage and film as well as
television. . . . So it may be regarded as a past triumph rather than as a
present loss." As the early 'seventies proved, it was both.

A Critic's Album of Favourites

In the days when Canada was being mapped by bush plane and canoe, an
explorer would come home with a packet of snapshots. If they were in focus,
well composed, and full of memories, they would be arranged carefully in an
album to recall the small, personal moments that often stood out from the
official overview. Sometimes, ones that were just a little out of focus would
be included anyway because they still caught a special moment.

Now that satellites can read road signs from space, the ideal "album" of
television's landscape would be a montage of shots on videotape. Doing
justice to the achievements of Canadian television drama will call for many
different types of analysis by as many skilled critics over decades. In this
review of the first sixteen years of anthology, however, as in my review of the
other half of the record, I would like to search a tape of memorable
moments for my reader, in roughly chronological order, through a montage
of memorable moments.

Folio and *Festival* flourished under Robert Allen's supervision from the
mid-'fifties to the late-'sixties. He created the opportunities, recruited the
talent, fought and occasionally lost battles with management, and tried to
balance the mix of plays while giving producers creative freedom to do "all
the Chekhov" (Prizek) or lyrical mood pieces (Till).

Let us fast forward to the superb performances of Jack Klugman and
Mavor Moore, one a patient and the other a doctor in a mental ward,
backed by a strong supporting cast right down to the non-speaking roles in
Ward Number Six, (*Folio*, 3/2/59). The skilful hands of producer-director
Harvey Hart brought to life Mac Shoub's adaptation of Chekhov's short

story. Rudi Dorn's set allowed for long tracking shots, odd angles, glimpses of characters in the distance, in shadow, under beds or behind doors, creating an interplay of light and shadow that seemed to intensify the blurring of the boundaries between madness and sanity. The production was full of unexpected discoveries, just out of the frame or slightly out of focus.

Then there was John Drainie as Jake with Douglas Rain as Matthew the Hermit in *Honey and Hoppers* (*Folio*, 7/11/57) "for the first time live on Canadian television, an episode of *Jake and the Kid*," as the announcer said. It was a moving portrait of a Prairie eccentric. Jake: "He ain't queer. He's just decomposed mental and the prairie's took him back for its own." When Matthew, a combination of Moses and Noah and predestined protector of wild horses, launches into his own idiosyncratic version of the first chapter of Genesis, the viewer is not only moved but a little awed at the intensity of his vision.

In a very different mood, Rain also played the ironic, affectionate, passionately word-drunk narrator in a fine production of *Under Milkwood* (*Folio*, 17/2/59) which almost persuaded me that this quintessentially radio experience could be televised. The pictures were often as fragmented, as allusive, as the text itself. "Mary Ann Sailors" was simply a hand and a shadowed profile; "Gossamer Banyon" was all legs. The set was simple, beautifully textured, lit with evocative shadows. The unobtrusive sound effects supported the many voices heard in montage over dreaming faces. Above all, producer Paul Almond allowed the strong verbal magic of Dylan Thomas's script to dominate his slow, rhythmic camera work.

The Concert (wr. Joseph Schull, prod. Robert Allen, *Folio*, 2/6/58) was originally written for radio. In the play, a black novelist meets a nurse blinded in the war who falls in love with him, not knowing he is black—highly charged material for the 'fifties. On radio (7/3/48, rebroadcast 15/1/50 and 14/5/61), the listener does not know that he is black until the climax. In both the radio and television versions, the nurse never discovers why he leaves her. Allen changed the thrust of the play for television, making it a much more subtle and complicated study in irony by keeping her blindness a surprise for a short while and then by showing the viewer just who her new friend is. The script is clear and unsentimental, yet eloquent. Schull weaves his themes with considerable skill: how daily encounters with racism have shaped Jennings (William Marshall) and blindness has shaped Anne (Kate Reid), how their respective "handicaps" make their friendship possible, and how, in the circumstances, Jennings's love compels him to walk out without an explanation.

The impact a particular production can have on a good script may account for the following comment in Halliwell (though it may also be a product of critical fallibility): "Tricksy little play . . . the first television drama to enjoy international renown . . . repeatedly done in the United States and Britain as well as [Schull's] own country during the late forties and early fifties. . . . It depended on the denouement." Halliwell refers to her as a musician—which she may well have been in one of the British versions he saw and recalls the last British production as an early use of the subjective camera, though "the script was so rigged to score cheap points that it was hardly worth the trouble."

Harvey Hart's use of Helwig's drawings as a gentle commentary by the victim on the adult world of Ibsen's *The Wild Duck* (*Festival*, 25/2/63) was a very televisual device that more than counterbalanced the too literal treatment of the magic attic-forest where the duck is kept. Ibsen's later plays are not the most obvious choice for television, but an early play like *A Doll's House* is perfect, as British and American productions had demonstrated in the 'fifties. Geneviève Bujold was a memorable Nora for *Festival* (4/5/66): she gave her character vulnerability, sensuality, and a clearly developing determination to break free of Peter Donat's surprisingly lively, spoiled, wilful and equally sensual Torvald. In the last act, her sense of loss and disillusionment was balanced by deep anger, and we could sense the taste of freedom this anger lent her. It was a warm, conversational reading with many nuances, such as a key shot of Nora half undressed, which would suggest to first-time viewers that she had capitulated to Torvald's demands, and was going to bed: of course, she is changing to leave the house forever. Although we heard the famous slam of the door, producer Paul Almond did not follow Nora. Just as Ibsen does, he stayed with Torvald's reaction as the final image of the play.

The Crucible (*Ford Startime*, 10/10/59)[18] brought viewers a galaxy of the best actors of the period: Douglas Rain, William Needles, John Drainie, Leslie Nielsen as a tormented John Proctor, and Diana Maddox as his repressed wife. Producer Harvey Hart used an immense wooden set, solid with yeoman prosperity yet filled with corners, cobwebs, hidden layers, narrow passages, and steep angles. Slight shifts of camera angle discovered new objects or characters looking askance, aslant, the other way. Two of his most memorable shots are not in Arthur Miller's play at all: one is a very slow pan up from the braced feet to the set face of Proctor, wrists chained to a wall in the dungeon; the other takes the risk of following Proctor up to the gallows. Hart's last image is not that of Arthur Miller, whose's stage directions make sure that the audience focuses on his wife Elizabeth's face, streaming tears yet exalted, as the drums roll. Hart chose instead to use a closeup of Proctor, feet settled and hands bound. Yet neither shot was melodramatic: both made the viewer engage intimately with the emotions of the piece, contrasting Proctor's suffering and self-hatred in the first shot with his determination and hard-won peace in the other; both took advantage of the camera's ability to frame, tilt, and slowly pan over or close in on the actor.

In considerable contrast to Arthur Miller's apparent naturalism, two of Brecht's first presentational dramas also provided memorable moments for *Festival* viewers. Mario Prizek adapted to television a theatrical form in which episodic narratives are explained by placards and songs, film clips, or even direct address by on-stage narrators. Brecht's estrangement effect includes lights that are visible to the audience, unit sets moved about by actors, and absolutely realistic props which often contrast with costumes drawn from all periods. His songs and ballads are often intended to undercut the emotions and counterpoint or expand the themes. Brecht thought such devices would help the audience become aware of the process of play-making and thus encourage them to think about the play's issues rather than simply feel the characters' emotions. Producer-directors Mario Prizek (*Galileo*, 25/3/63) and George McGowan (*Mother Courage*, 20/1/65) approached the problem of translating these "new" theatrical conventions to television.

With the help of designer Rudi Dorn, Prizek found many ways to recreate Brecht's effects. His actors sometimes freeze into tableaux as the light cues fade; he uses balletic blocking; he intercuts slides of modern astronomical discoveries on a cyclorama at the back of the set with period woodcuts of the city as Galileo looks at the stars and starts to speculate. Realistic props are juxtaposed with a highly stylized garden or simplified table and chairs. Everyone is in vaguely period costume. However, the narrator (Hugh Webster) has been costumed and made up to remind us of Brecht, if we happen to know what the playwright looked like. The ironic allusion is never made explicit. We are shown closeups of flats representing modern hoardings covered with torn travel posters, a civil-defence notice and, in a last closeup, a shot of a battered CUCND disarmament poster. The play itself ends with a question from Galileo about whether or not the night sky is clear. Cut to a clip of the sun coming up, then seen to be a nuclear fireball rising above the horizon.

None of this would have worked had it not been for the strong cast: John Colicos as a cunning, voraciously sensual, brilliant Galileo; Bruno Gerussi as the anxious, excited, yet dignified Little Monk; Leo Ciceri as an attractive, menacing Barberini, and Gillie Fenwick as a purse-mouthed, oddly grandfatherly Cardinal Inquisitor. Sharon Acker gave the thankless role of Virginia touching naivety, real piety, and the sense that she eventually gets her own back with her father under house arrest (and her tutelage) for the remainder of his life. When Prizek sent Brecht's widow, Helena Weigel, a kinescope of the production, she asked for a copy to deposit in the archives of the Berliner Ensemble.

McGowan's treatment of *Mother Courage* was less complex in its detail. The orchestration of the songs was very contemporary and the sets expressionistic by contrast with the highly realistic wagon and props, but there were no startling interpretations. Instead, the director stressed the intellectual and moral issues. He also avoided the temptation to sentimentalize either Mother Courage's reaction to the death of her son, Swiss Cheese, or the self-sacrifice of her mute daughter, Katrin. Kate Reid was outstanding as the earthy, disillusioned survivor who saves her livelihood and loses her children to an endless war.

Eric Till's struggles to get Marghanita Laski's *Offshore Island* on air (*Festival*, 12/3/62, later rebroadcast) were many. According to Till (interview with MJM) CBC executives feared that it might be perceived as "anti-American," and hovered anxiously, making suggestions. Yet somehow Till got the play on the air as Laski had written it. It was not particularly innovative as a piece of television, but simply and memorably a well-written, beautifully performed, and subtly chilling anatomy of a small family that has survived, after a fashion, World War III. Irene Worth, Tony Van Bridge, and William Hutt gave ensemble performances that lent the unthinkable subject a particularized, very human colouring. In marked contrast to the highly publicized 1980s treatments in *The Day After* (1985, U.S.) and *Threads* (1984, U.K.), it did not emphasize visual horrors, relying instead on its quiet realism and mordant wit: "You never went and killed two pigs at once." "Of course not, the pig had two heads." There is even muted recognition that incest may be inevitable if the race is to survive.

The understated deprivations, tormenting memories, and the irony of a rescue attempt by those responsible for the holocaust are the precise elements that make this play work so well. The CBC was nervous about it right up to the moment of broadcast, with Lamont Tilden's opening announcement reassuring viewers that the play is European, that the BBC has broadcast it (which made it respectable, one assumes), that it is for adults, controversial, and so forth. Still a play of this kind could find a home on the CBC in 1962, amid the spate of safe TV Westerns, doctor and cop-shows and sitcoms on American networks, because a producer, Eric Till, and his supervising producer, Robert Allen, were determined to broadcast it.

Till had no such trouble with his well-known and much honoured success, *Pale Horse Pale Rider (Festival*, 17/3/64). He had this to say about the production (*CBC Times*, 14-20/3/64): "It is adapted from an impressionistic novel by Katherine Anne Porter [about World War I] and I felt that the production should lean towards impressionism. Miss Porter writes into her stories a special kind of mood vital to the story, without ever putting it obviously into the plot and this is difficult to create in pictures. The play has a most curious pacing—very slow. We did a lot of camera work in flashbacks and would use immobile cameras for long periods with none of the usual inter-cuts. I strove for a bareness of sound to isolate a particular moral or dramatic point.'' At the climax, when the protagonist discovers that her lover has died in the great 'flu epidemic, Till wrapped that discovery in absolute silence, to convey the emptiness she feels.

Pale Horse, Pale Rider is generally agreed to be one of the CBC's best efforts in this first golden age. It is an exercise in the lyrical potential of the medium (unaccountably neglected in the years to follow) with both silences and visual pauses so that viewers can reflect on what they have just seen. Till used Harry Freedman's electronic music to evoke the lovers' almost dreamlike passion and create a surreal mood of distance, fear, and intense anticipation for his protagonist's nightmare. The sound gave the sometimes predictable images life and intensity. Her sick visions of what war would be like were full of dizzying overlaps and dissolves. It is no small task to evoke sentiment without sentimentality, irony without cynicism, and sustained sadness without tearjerking. Porter's lean, evocative prose, Fletcher Markle's dialogue, Joan Hackett's superb performance (ably supported by Keir Dullea), and the sheer beauty of the piece make viewing it a pleasurable experience to this day.

It was shown on the BBC's *Wednesday Play* the following December. Critic John E. Taylor commented in *The Listener* (3/12/64) that "this ambitiously elaborate dream sequence mingled with period reality . . . came off surprisingly well, particularly in the more fantasy treated sequence . . . a refreshing reminder that television can, when it wishes, take wings into fantasy even on a fairly limited budget . . . a considerable tribute to the taste and ingenuity of . . . Eric Till.''

Till's *Private Confessions of a Justified Sinner* (*Festival*, 13/5/64) was the antithesis of *Pale Horse*, though designer Trevor Williams was responsible for the sets in both. Based on a "long neglected Scottish classic" (almost unknown outside of Scotland), it was set in the claustrophobic, evil, sensational, repressed world of Original Sin, the Justified Elect and Predestination that was the Scotland of 1709. The sometimes incoherent adaptation tried for too much and came quite close to achieving it. Basically, the play is a thriller with a psychotic protagonist and a demonic antagonist. The viewer never knows if a pair of hands in a closeup is going to make love to a prostitute or strangle her. Particularly memorable are the eerie, extreme longshots of the nameless Adversary against a bare landscape (shot on location in a Canadian autumn). The play is a not altogether successful juxtaposition of a soundly observed study of fanaticism and a highly subjective account of progressive madness. At times, it is not clear who is who or what is real. Production values are exciting, however, and the suspense never wavers. Of all the drama directed by Till over the years, this is the subject to which he would like to return.

Equally experimental was Paul Almond's version of Clive Exton's *The Close Prisoner* (*Festival*, 11/11/64) which satirizes the medium itself. Henry, a working-class bloke, explains to a television interviewer on "Your Strange Life" (parody of the notorious *This Is Your Life* type of show) how his chest is slowly turning to steel and he has therefore become a British secret weapon. Almond has great fun with in-frame mikes, a frantic floor manager on camera, out-of-focus shots, and an off-screen director whose amplified voice keeps rewriting Henry's life by adding heavy-handed symbolism, sentimental scenes with Henry's parents, and significant music. Since Henry is in hospital, it is also a satire on every Ben Casey-Dr.Kildare cliché in the business. This is more successful Pirandellian comedy, in fact, than the 'seventies adaptation of *Six Characters in Search of an Author*. Henry excises his wife from his strange life by simply running the film of his dramatized life in reverse. His parents ostentatiously mourn the image of his death on the screen, not his dying body. No matter: by now, Henry himself prefers the fiction to the reality. As the steel chokes him to death, he ignores the real experience of his own dying to watch the film of his glamorized death.

The Prisoner (*Playdate*, 16/5/62, adap. Joseph Schull, prod. Leo Orenstein) was very different again. Inspired by Hungarian Cardinal Mindzenty's 1948 arrest, trial, and sentencing, the play by Bridget Boland was made into a subtle, wrenching film starring Alec Guinness and Jack Hawkins. Joseph Schull adapted the play quite differently for producer Leo Orenstein. The film ends with the suicide of the Interrogator, who cannot live with what he has done to break the Cardinal. The CBC version ends with the Interrogator still in command of himself as he condemns the Cardinal to the guilt of freedom and is the more chilling for the change. The camera work is unobtrusive, the sets are massive and oppressive, but it is the acting that is the essence of this adaptation. William Hutt as the Interrogator and Leo Ciceri as the Cardinal gave two of the best performances I have ever seen on television. The complex emotional levels both men found in the script will haunt me for a long time to come.

Filed and Forgettable

These short glimpses by no means exhaust the best of the era, but it would be pointless to pretend that there were not some remarkably bad efforts as well. I reserve my fire for the four that follow, largely because each of them suffered from pretentiousness and preciosity. George Ryga, an experienced playwright who is capable of considerable lyric intensity, passion and powerful imagery, had his cliché-filled script, *Man Alive* (30/3/66) taken with absolute seriousness by producer George Bloomfield and (perforce) designer Rudi Dorn. The theme of "Man's alienation" was buried in documentary-style film footage and slides. Obtrusively "creative" lighting was wasted on numbingly banal choreography through which the *angst* of the suffering common man was signalled by writhing hands. Boy met girl and sang to her, badly. The choral speaking sounded under-rehearsed. Voice collages merely emphasized the ineptitude of the visual imagery. Worst of all, viewers would have been completely baffled about what in the world was going on much of the time. Anyone who did penetrate the fog would not have found the "messages" worth the effort.

The True Bleeding Heart of Martin B. (25/10/66) by M. Charles Cohen, usually a competent playwright, was described in the CBC's press release as "a way out comedy." In fact, the play was heavily didactic. It focused obsessively on an uninteresting person's search for the answer to the questions, "Who am I? Am I a good person?" Again, the camera work was obtrusive, the blocking awkwardly stylized, the lines banal, and the characters stereotyped.

The CBC liked Cohen's *David Chapter II* (20/5/63) so well that they followed it with *David Chapter III* (5/10/66). Both were directed by Harvey Hart. Chapter II was an experiment in what could be accomplished in live to tape production, using a shooting style of quick cuts and apparently fast edits to give the drama some of the look and flexibility of film. In *Chapter II*, David, a young Jewish university graduate, searches for his identity; we visit him again at age thirty-five in *Chapter III*. In both, acting and the camera rhythms were jerky and frenetic. Worse still, the characters presented every conceivable stereotype of Jewish life without depth or insight in either script or performance. Young David's climactic explosion was simply melodramatic. In *David Chapter III* it was that much more difficult to get involved with the anguish of the protagonist's compromised business ethics and dull marriage when even his fantasies were treated with ponderous seriousness: "David pinned to the wall of a long, long corridor," or "David's collapse presented in a film loop" over and over.

Finally there was the NFB-CBC collaboration, *Waiting for Caroline* (29/11/67), directed by Ron Kelly with screenplay by George Robertson and Kelly. The CBC tried to disarm viewers with this introduction: "You may find Caroline, the character and the film, poignant and pitiful, or you may find her and it exasperating and shocking. The film does represent an aspect of our diverse Canadian life, a woman who has not yet found herself. . . . It is a film for adults." However, shots through glass, windows, water, and railings—presumably to create fragmented and distorted views of the self-centred protagonist—frustrated rather than involved this viewer. The worst aspect of the film was its remarkably awful dialogue: stilted, full of pregnant pauses, and predictable.

The 1960s is a decade now much mythologized, when battles about profound social change raged, the music was good, kids were hitting the road and not trusting those over thirty, and the generation gap was widened by drugs, changing sexual manners and mores, radical politics, and weird hair and clothes. Foreign wars, domestic riots and near-revolutions threw the good times into sharp relief. It was also the era when Canadian and British television scrambled to catch up with the times. Indigenous theatre began blooming all over Canada for the first time. Our year-long euphoria of 1967 caught and amplified a wave of confident nationalism. Culturally, it was a good decade for this country, as new artists emerged and older artists found new audiences. CBC television drama in both languages was part of it all, leading the way in the 'fifties, riding the surge through the mid-'sixties, then slowly being left behind as that particular moment headed to its inevitable end in October, 1970.

For thoughtful people in both English and French Canada, our collective sense of self was turned inside out by the 1970 FLQ Crisis. CBC English television drama, like almost all the theatres, backed away from any look at the event. Against all odds, Michel Brault got his brilliant film, *Les Ordres,* into Quebec cinemas within three years. The English section of the NFB struggled with self-censorship when it followed its first examination of the issues, Robin Spry's *Action*, with a year-long delay before release of his *Reaction*. Small dramatized segments were used on *The October Crisis* (1975), the CBC's own belated recognition that something of lasting consequence had happened. Nevertheless, the long silence was symptomatic of the lack of confidence, creativity and concern that beset the drama department from 1968 to 1973. Our culture suffered the gravest shock, yet most of our cultural institutions, particularly television, slammed the window and pulled the curtains on the event and its aftermath. No television play in English has addressed the October Crisis since it occurred.

In my opinion, despite the occasional self-censorship, I do not think there would have been such a failure of nerve in the first fifteen years. That silence also suggests to me one of the most important reasons why we need an unfettered CBC Drama Department, one with high morale and some sense of long-term commitment. Who else, if not the bard in the corner of our living room, is to articulate such seminal experiences, giving them shape and voice and coherence as only fiction can do? When that voice fails to address, as it did from 1968 to 1973, the big questions that plague us as a society, we lose a way of seeing and hearing that no other voice in our culture can replace.

NOTES

1. M.J. Miller, "Canadian Television History 1952-70: Canada's National Theatre," *Theatre History in Canada*, vol. 5, no. 1 (Spring 1984), 51-72.

2. Raymond Williams, *Drama in a Dramatized Society* (Cambridge University Press, Cambridge: 1975). Former CBC president Al Johnson's speech, "Canadian Programming on Television: Do Canadians Want It?" (CBC, 19/1/81), 5, points out that "50 per cent of the viewing time throughout the country is spent in watching drama." The percentage has gone up since then. It is true that VCR timeshifting requires planning and a commitment to find and play the tape at a later date. Drama, however, should benefit from the VCR, because it has a longer shelf life than most sports and current affairs programming for the home viewer.

3. G. Murdock, in *Media, Culture and Society*, 2:2 (April 1980), 151-68.

4. "The Largest Theatre in the World," Fleming Memorial Lecture to the Royal Television Society, 2 April 1981. BBC publication. In 1987, Grenada broadcast a tribute to this remarkable man.

5. CBC press release, undated, on deposit in York television drama archives, York University, Downsvie, Ontario.

6. "A historical and analytical study of the origin, development and impact of the dramatic programmes produced for the English Language Networks of the Canadian Broadcasting Corporation," Unpublished M.A. thesis, Wayne State University, Detroit, 1966, 121. A copy is on deposit in the CBC reference library. It contains valuable interviews with people now gone, but very little programme analysis. It seems unlikely that Jackson actually saw any drama.

7. CBC press release, Ottawa (12/12/54). Alex Barris in *Globe and Mail*, 7 December 1954, called it a "Sort of Canadian 'Omnibus'," and within a month (10/1/55) Bob Blackburn in the Ottawa *Citizen*, already found it too "arty." The twin premises of false analogy to U.S. programming and the assumption that CBC television should serve the mass and not the pluralist audience marred a lot of television criticism in that period—and still does.

8. CBC programme lists from that period. CBC holdings.

9. The *Oxford Companion to Canadian Literature*, in the survey article on Canadian drama, does not mention the television work of Laurendeau or Blais, Ryga or French, and makes no mention of the fact that television drama was the progenitor (or bread and butter) for so many Canadian playwrights, even though radio drama is treated with due emphasis and accuracy. There is no word of Shoub, Schull, Israel, Richler or others. It is clear that, in Canada at least, television scripts are not yet "literature."

10. Collection l'Arbre, vol. g-5, (Editions HMH, Montreal: 1970) 8.

11. *Tube of Plenty: The Evolution of American Television*, by one of the most reputable of American broadcast historians, Eric Barnouw, Oxford University Press, London: 1975, provides an overview of this process that is both detailed and sardonic.

12. *The Golden Age of Television: Notes from the Survivors*, a Delta Special, Dell, New York: 1977, 132-34.

13. See Malcolm Page's annotated bibliography of plays, published by 1977, in *The Theatre Quarterly*, vol. 7, no. 27 and Supplement *Theatre Quarterly*, vol. 8, no. 30, Summer 1978.

14. "The Largest Theatre in the World," The Fleming Memorial Lecture to the Royal Television Society, 2 April 1981, BBC Publications, 5.

15. Quoted in E.A. Weir, *The Struggle for National Broadcasting in Canada*, McClelland and Stewart, Toronto: 1965, 394-95.

16. The *Daily Star*, Toronto, (5/5/66). Cohen goes on to point out how many CBC-trained writers and directors have gone abroad to the U.S. or the U.K., adding that one of the problems of CBC drama is the corporation's failure to develop a formal training workshop for writers. Both points are well taken.

17. Compare Edith Ephron, "Television: America's Timid Giant," (18/5/63), in *TV Guide; The First Twenty-Five Years*, compiled and edited by Jay S. Harris (New American Library, New York: 1978), 75-79, and Eric Barnouw's *Tube of Plenty*, passim.

18. *Ford Startime* was a CBC experiment in a sponsored more upscale anthology. According to Robert Allen and Hugh Gauntlett, the tensions were simply too great for both the Corporation and the sponsor. *Festival*, which was basically *Folio* (by another name) was the result. For another decade, the CBC English television drama department could afford to broadcast sustaining or non-sponsored television drama with all the freedom that involves.

7

ANTHOLOGY:
1968 TO THE PRESENT
"The focus narrows"

Anthology since 1969 has differed in many ways from what went before. The form itself is much less popular: *Festival* disappeared, and with it a place and time when Canadians could count on finding music of all kinds, ballet, theatre, films, television dramas from abroad or originals every week. After *Festival* was terminated the arts became a special event, no longer an aspect of regular CBC programming. For some reason, ambitious and creative storytelling also became a special event, surfacing under various titles a few times every year. In 1970, then, television drama slid into a new phase. In this chapter, I attempt to convey the flavour of the next phase and analyze a handful of programmes, choosing for discussion, as before, the typical, the highly touted and the genuinely excellent.

The years 1969-73 were a time of confusion and low morale in the network drama department. The regions could not get decisions from executives now centred in Ottawa (where no national programmes are made). Interesting programme initiatives came to nothing; they rest in the dead files of the Federal Archives. Many second wave producer-directors had left the corporation to freelance—Paul Almond, Daryl Duke, Harvey Hart, David Gardner, George McGowan, Eric Till. Many of the writers of the 'fifties and 'sixties had retired, gone to Hollywood, or else burnt out. In 1968-69, the network itself was reorganized and a new level of bureaucracy replaced the authority of the one man, Doug Nixon, who had made many of the decisions about the programme mix. He had lost responsibility for public affairs, perhaps because he had helped to bring to life the controversial *This Hour Has Seven Days*. By 1969 he had enough and resigned. Laurel Crosby was appointed to head something called Planning and Production, only to leave eighteen months later. Fletcher Markle, who had made significant contributions to radio drama in Vancouver, worked in television, and was then in Hollywood, became the first Head of TV Drama in 1970. He lasted less than two years in the job. This is the era of *McQueen, Corwin, Delilah*, as the CBC floundered in formula series, ill conceived concepts, and bad scripts.

Markle's decision to pour what resources the department could muster in Toronto into *The Whiteoaks of Jalna*, drained both tape and film drama of budgets and resources. Somehow, Vancouver was spared (overlooked ?) but in the move to centralize and consolidate the energy of Toronto, still the chief drama production centre, was absorbed in the project. *Jalna's* debut was surrounded by more publicity and good will and anticipation than any production until *Anne Of Green Gables* (1986). It also had the potential of an enormous worldwide audience presold by the success of Mazo de la Roche's novels. *Jalna* also had a highly skilled producer-director in John Trent. It should have been a triumph, particularly with skilful writers like Timothy Findley and Grahame Woods.

The problem was that *Jalna* readers, who wanted their old, familiar story, were treated to an ill-conceived experiment in narrative structure complete with flashbacks, multiple plot strands, and intercut time frames, all edited in haste as the air date approached. Of course they were frustrated by this. Viewers, unfamiliar with the novels, were simply confused. The twelve-episode serial-miniseries flopped dismally. It was re-edited into a shortened chronological narrative for sale abroad, but flopped again and morale sagged even more; the CBC Toronto Drama Department was in sad shape during this five-year period. Despite three or four notable exceptions— *Twelve and a Half Cents* (1970), *God's Sparrows* (1970), and *Vicky* (1973, a sequel to *Twelve and a Half Cents*)—much of Toronto's series and anthology output ranged from mediocre to adequate.

DRAMA IN THE EARLY 'SEVENTIES: THE LOOKING-GLASS CLOUDS

A few fairly typical examples will indicate what I mean. *Welcome Stranger* (11/4/73, wr. Kaino Thomas, prod. Ron Weyman) had many twists, some of them improbable. A Swiss doctor who is allowed to lecture but not practise performs some very graphic heart surgery on an accident victim in a farmhouse kitchen. The patient dies and despite the fact that there was really no chance, the doctor is sued. This quite suspenseful introduction prepares us for the main event: his trial. However, the drama never really decides whether to concentrate on the issues of medical ethics or discrimination against immigrants, or on various forms of personal betrayal. The dramatic tension is vitiated by lack of focus.

Like so many others from the early 'seventies, this offering seemed unable to decide whether it was going to be a good old-fashioned melodrama dressed up in trendy clothes or a serious presentation of contemporary issues in dramatic form. The issue of Canadianization of medicine vs. the rights of immigrants came across as an add-on. A reminder that the grieving family is unreconciled, hinting at the possibility of a more complex ending is tossed in at the end. The presiding judge comes up with eloquent moral outrage to balance the books and satisfy melodrama's need for poetic justice. The protagonist stays in the country for the sake of the heart clinic and his humanistic values.

Lighten My Darkness (1973, by CBC regular Charles E. Israel, prod. Ron Weyman), starring Diane Leblanc, worked an old stand-by plot about the effect regaining her sight has on a blind woman and her husband. As is often the case, sight is equated with loss of innocence. Her independence is a threat to the marriage because he needs her need. She goes psychosomatically blind to get him back, he "sees the light," and all is well: interdependence is vision. However, her new determination to be free and independent is never clearly motivated.

Despite the familiarity of the situation and the predictable pop psychology of the plot, this drama did have entertainment value. As the protagonist, Leblanc had a nice quality of openness and freshness. The drama itself was strongest in those scenes, largely without dialogue, in which she discovered the world of vision, for example, seeing her husband but recognizing him only through touching him. Her arrival home and walk in the garden conveyed real joy, not mere stereotypical, sentimental response. Israel also digs a little deeper into the psychology of marriage when he has the husband try keeping her in her dependent role by coaxing or forcing her into having a child.

The dialogue was predictable, however. Husband: "I wish to God you were blind." The resolution is equally predictable. Husband, again: "I realize I've drawn a tight little circle around you . . . I've been terribly unfair to you [pause] but I do love you." Still, the play was not lacking in energy, as were so many in this period. It is upbeat, sensitive in some parts. The flashbacks work. The design is oddly dated in places, but the televisual emphasis is a nice change. The music was unobtrusive. In the early 'seventies, viewers heard very little creative use of music as background, atmosphere, bridge, or counterpoint.

The only reliably good storytelling to be found in television at that time was the half-hour series of short-story adaptations that was first called *Canadian Short Stories* (1970) and then *To See Ourselves* (exec. prod. David Peddie). Part of the idea was to shoot some drama in the regions. "MacIvor's Salvation" (wr. Thomas Raddall, dir. René Bonnière, 24/10/73) is a typical example. The performances and editing set a brisk pace, and the director and actors managed the difficult feat of creating genuinely simple characters: the boss's fair daughter, telegraph operator, MacIvor, and the night telephone operator, whom MacIvor never actually sees. The shy MacIvor is relaxed, even affectionate with his colleague because he cannot see her; eventually, the viewer discovers that she is old, fat, and plain. Her wry and tender look at him as he passes her, oblivious because of her age and appearance, sums up both the comedy and the pathos of the piece. When MacIvor gets up the nerve to give the boss's daughter a watch for Christmas, she is delighted; when he forgets her birthday, however, he is transferred immediately. The little story ends when, seeking consolation, he decides to finally see his telephone operator at last. We are left to imagine what will happen next.

"The Painted Door" (21/11/73), based on a well known short story by Sinclair Ross, has been adapted for television many times. This production is worth mentioning for its excellent sound effects—sensitive, small sounds of wind and paint slap and clock tick which are the substance of the farm wife's stifling isolation as well as her security. The very good cast were supported by a lyrical dramatic structure of flashback and crossfade, the cross cutting of dance and political discussion, the raging blizzard and her soliloquy. Again, there was a good sense of period in the set and costumes. The short story itself provides a very visual symbol of the tensions in the drama. In the depths of the Depression, painting the dark wood yellow was an appropriate symbol of the wife's longing for a more relaxed and brightened life, yet the even slap-slap of the paint also indicates the monotony of the day passing. The sound cue of the utter menace of the storm heard in the background worked. When we saw the blizzard, unfortunately, it seemed less than it should have been.

At one point, direction and dialogue were at odds. Her soliloquy about her husband going out into the storm, her fears for him and feelings of inadequacy were understated and much less passionate than the words themselves. However, the edgy subtext between the wife and his friend, who has taken shelter from the storm, assumes that the husband will-will not make a wished-feared return. Her guilty nightmare and our discovery of the friend in her bed has the undiminished shock of the inevitable. A subjective camera shot lets us see what she sees: a blurred, fragmentary glimpse of her husband in the doorway, hand on the doorframe. In the deceptive calm that follows, the two lovers engage in bits of conversation waiting for his return, thinking her night vision was merely a nightmare. It is the camera that seeks and then finds the frozen body. The last image is of the fresh yellow paint on his mitten.

The early 'seventies also marked the decline of sustaining drama, drama presented without commercials. When one could find truly excellent drama, it was ruined by commercials. Frank Penn in the Ottawa *Citizen* (15/10/73) addressed this issue prompted by what had happened to *Vicky*, a fine study by Grahame Woods of an abused woman, who had gone mad and murdered her own children, starting the long journey back from the fragile sanity achieved in a mental hospital to life in the community again: "Jackie Burroughs wasn't acting Vicky, she was Vicky with her eyes burning with all the fires of her private hell and her words torn from somewhere deep inside, like jagged splinters of pain. And then bad breath. Bad breath for Pete's sake, a commercial about bad breath right smash in the middle of a dramatic mood that television isn't likely to see the equal of in a year of Sundays. . . . So it went. Author Woods's brilliant script, René Bonnière's superb direction, Ronald Weyman's genius as a producer and the beautiful work of a supporting cast that never hit a false note . . . all shot through and through by blurbs for banks and razors . . . bananas, detergents, coffee, jello, floor wax, carpet cleaners and paint washers. The sponsors can't be blamed. In fact, if I were one of them I'd be enraged." His description of the programme's quality is quite accurate. It was that rare event, a worthy sequel to the wrenching *Twelve and a Half Cents*, both directed with great sensitivity and imagination by Bonnière. The protest about what such ineptitude could do to the CBC's audience, creative people and the programmes themselves, went unheeded.

The Play's the Thing—A Better Mousetrap?

The Play's the Thing (January 1974, originating under Fletcher Markle and produced by George Jonas) was to be the jump-start needed by a demoralized drama department on the road back to interesting and even innovative television production. Unfortunately, it was based on the highly questionable premise that novelists of considerable repute should be able to write quality television plays. This does not necessarily follow. As far as I can discover, no one mustered the courage to edit the scripts. Thus this widely touted anthology had more than its share of clinkers, including a nineteenth-century melodrama written by the mistress of the eerily ironic and often surrealistic, Margaret Atwood. Atwood is a superb poet and a very good novelist, but her efforts at television writing, spaced eight or nine years apart (the next was *Snowbird* [8/2/81], from *For the Record*, the last to date, a draft of *Heaven on Earth* [1986]) suggest either that television is not her metier or, more likely, that the CBC cannot provide enough opportunities for gifted writers from other media to master television.

The Servant Girl (7/3/74, pr.-dr. George Jonas) is really pretty bad. Set in the 1870s and filmed on location with excellent period detail, it features a household consisting of the mistress, Hannah Montgomery, an AWOL

Cockney soldier called MacDermott, and an Irish servant girl, Grace Marks. Hannah is harmless, if slightly vindictive; Grace's hatred of her is largely gratuitous. There is no characterization, no atmosphere, no emotional tone; only situation is explored. After Grace and MacDermott have murdered Hannah, his agonizing over the deed is a complete non sequitur, while her taunts about his cowardice are predictable and simplistic. The theme of love-hate potential between servant and mistress is barely touched on. Except for the archetype of a man with an axe poised over a sleeping woman, even the conventional gothic elements are lacking. When MacDermott loses his nerve, they eventually kill Hannah by strangulation. The master returns, MacDermott dies, and Grace gets the last ironic line: "Of all the things I ever wanted in my life, her death was the only thing I ever got."

The Bells of Hell (24/1/74) by Mordecai Richler, who had written competent television scripts for the CBC in the 'fifties, was broadcast after much unnecessary controversy. George Jonas was the producer and director; the music was composed by Harry Freedman. It was thought prudent to have an urbane Gordon Pinsent explain the theme and title for viewers who had forgotten the old song: "The Bells of Hell go ting-a-ling-a-ling for you, but not for me."

With the exception of American actor Henry Morgan as the protagonist, most of the acting was awkward or simply bad. The direction and script telegraphed the largely anecdotal satire: a rapacious wife, druggie son, and cool city secretary who takes her birth control pill in front of our hero, a businessman caught in his mid-life crisis. Even the technical side was inept: the camera is often in medium or long shot with the mike always in close focus; the studio acoustics are sometimes terrible, though this is a common complaint in the period. The controversy which resulted in the programme's postponement for three weeks should not have centred on Richler's satire on "Jewish" life, but on the fact that the jokes were so feeble: for example, the doctor's shakedown of the patient is sophomoric, relying on a needle in the rear end for a punchline. A key scene in the garage was neither clearly fantasy (his) nor terrorizing (hers) nor even simple fun and games (whose?). The music suggested one tone, the voice-over saying, "Beware muggers," another. Throughout, there was no consistent style, attitude, or set of conventions. It shows gays, 'seventies massage parlours, lots of angst, lots of music hype, all noticeably trendy without any bite. Morgan's central performance was very tight and professional, but he was trapped in the middle of a strained farce. For a moment, we do feel badly about his impending nervous breakdown, but then the play stops. It does not end; it stops, much to our relief. Why *Toronto Life* printed the script (January 1974), remains a mystery.

Friends and Relations by Hugh Hood (2/7/74) fared somewhat better.

Directed by Rudi Dorn, it is an effective tract on the need to provide for one's wife and how women can break out of conventional roles. The new widow meets some unexpected obstacles, such as, "Never been a woman realtor in Stuberville before"; "I'll be better than all right. I'm going to have my own life." Episodic, fairly straightforward, the play traces the growth of her personality in a believable way.

Brothers in the Black Art (14/2/74, wr. Robertson Davies, dir. Mario Prizek) opened with an old man's narration over shots of hands at work on an old press. The reporter is also represented by a pair of hands, plus notebook. Again, we have excellent period flavour in a pleasant little play that is rather clumsily structured for commercials. The sound effects of the shop are detailed and unobtrusive. The convention of questions from the unseen reporter, however, which often prompt the flashbacks, is dropped without explanation. Both the protagonist's coarseness and wit are obvious but enjoyable. Her seducer, Phil, commits a suitably gothic suicide using lye; she gives the name to her child. Davies's final twist is that the woman reporter turns out to be that child. The play seems to aspire to tragic farce but, once more, there are problems a good script editor could have remedied: we do not see enough of Phil, the father, to care what happens to him and the reporter scarcely speaks until her perspective is tacked on at the end.

With *Back to Beulah* (21/3/74) we are back in the hands of an experienced television writer, W. O. Mitchell, and director Eric Till. The play, later successfully rewritten for stage, had a very strong cast: Jayne Eastwood, Norma Renault, Moya Fenwick, and Martha Henry as the doctor.

Three female out-patients kidnap the doctor, then drug her with the prescriptions intended to make them functional. The contrast of her hysterics and their calm, apparent sanity is frighteningly persuasive. As more is revealed about the women, we find out that hospital staff "tied off Agnes's tubes when she was sick," and that Betty is a kleptomaniac. Yet their illness is not sentimentalized. They run through a savagely ironic mockery of medical testing on the unconscious doctor, who is tied to her chair. The play reaches a climax when the doctor, humiliation in every line of her body, is made to treat Agnes's doll as a baby. Because she cannot pretend when they order her to breastfeed it, they rip her trouser suit and she screams. In a rage, the leader, Harriet, kills their kitten. After some days, Agnes finally stops them. The doctor has learned many things about her patients, about sanity and insanity. She finally frees herself by forcing Harriet to "kill" her. Harriet, unable to do this because she really does recognize her as a person, attacks the doll instead, nearly destroying Agnes in the process. Except for some rhetorical patches, the play is well written and competently directed, with performances highlighted by nuance and imaginative use of television as a medium. It was the only drama in the anthology to live up to its advance billing.

A Bird in the House

One of the best dramas from those years, and one of the very few to stand up to repeated viewing and analysis, did not enjoy the advantage of the public-relations blitz mounted for *The Play's the Thing*. Allan King's version of *A Bird in the House* (22/10/73, pr. Ron Weyman) was adapted by Patricia Watson from a short story by Margaret Laurence. It is structured in part around the catch phrase, "A bird in the house means a death in the house," chanted by the superstitious, fundamentalist Irish servant girl Noreen during an epidemic that decimates the small town of Manawaka. It is counterpointed in the imagination of the protagonist, twelve-year-old Vanessa McLeod, by another little rhyme, remembered by her austere grandmother from the pandemic of 1920: "I had a little bird. It's name was Enza. I opened the window, and in flew Enza."

The drama works much like the short story, by an accretion of intimate details, objects, and interiors which gain significance as the play unfolds. The snowy landscape entraps the family indoors during the epidemic and is the enemy of the ill, yet frees Vanessa and her father Ewen, the town doctor, to snowshoe away from the mother and wife and talk, without looking at each other, about his experiences in World War I; about death and what it may mean; Ewan's favoured brother Roderick's death in battle; a wider world, "I made some good friends"; duty in Manawaka and dreams of going to sea.

Most of the events in the drama are perceived from the child's perspective, culminating in two shocking changes of pace and mood: one as the trapped bird circles blindly around the room seeking an escape; and the other as Vanessa attacks Noreen wildly after her father's death. As the camera became the bird, so later it becomes Vanessa, wildly futilely swinging at the hired girl. Thus far, the drama has been focused on an imaginative child and her response to Noreen's exciting fundamentalism and back-country superstition, which serves as a bright thread of excitement in the repressive, Scots upper-middle-class household. We see how and why Vanessa loses her father, her faith, her home. However, she finds a new relationship with her mother in their grief.

The last scene is significantly different from that in Margaret Laurence's original. There Vanessa finds a photograph when she is seventeen, which finally releases her tears for her dead father. In the television adaptation, Vanessa is still a young adolescent whose home is being broken up after his death. Waiting for her grandfather to pick up the family, she discovers in her father's apparently empty desk a creased photograph of a lovely dark-haired girl, with an affectionate inscription in French on the back. She also finds a new, barely tested maturity, signified at the close by her burning of this hidden photograph—to share one last secret with her father? protect her mother? out of unrecognized jealousy? as a gesture, to end the old life and begin the new? Unlike the story, which contains Vanessa's adult reflections on her discovery, King's quiet elliptical treatment, and the young-old face of Vanessa looking at the photograph, encourage the audience to draw its own conclusions.

Drama In The Hirsch Years

After 1973, when John Hirsch took the helm, television drama regained some ground by presenting fresh new Canadian drama from alternate theatres (with varying degrees of success) and developing a permanent home for the remnants of anthology surviving the 1975 cutbacks, which amounted to $700,000 and ten hours of drama programming, in the "journalistic" dramas of *For the Record* (1975-86). In 1975, there would be other competent dramas and three quite interesting adaptations: *Baptizing* and *How She Met Her Husband*, both adapted from short stories by Alice Munroe and Eric Till's version of Brian Friel's *Freedom of the City*. The 1974-75 season included *Red Emma*, adapted (and cut by a third) from an original play done at the Toronto Free Theatre; other dramas included *Mothers and Daughters*, *Summer Mornings*, *Last of the Four Letter Words*, *Going Down Slow*, and *The Canary*.

However, as Canadian theatres strengthened nationally during the 1970s, with the four-year term of John Hirsch as Head, television drama broadened its scope to serve Canadian theatre better, even as it began to develop a new wave of original scripts of high quality. In a brief period, CBC television recorded for future theatre historians, and presented to the country at large, Toronto Free Theatre's *Red Emma* (1974-75), the Toronto Workshop Production of *Ten Lost Years* (1975), *Love and Maple Syrup*, Carol Bolt's *One Night Stand* (1978), David French's *Leaving Home, You're Alright, Jamie Boy* (1975), *Of the Fields Lately* (1976), Théâtre Passe Muraille's *The Farm Show* (1975), Rick Salutin's *1837* (1975), the Centaur Theatre's production of David Fennario's *On the Job* (1976), Joanna Glass's *Artichoke* (1978), and the first production in either language on network television of Michel Tremblay's, *Les Belles Sœurs* (1978).

In its 1975 winter "high season," the CBC offered, among other dramas, *Stacey*, a rather muddled adaptation from Margaret Laurence; a competent version of Joe Orton's black absurdist farce *The Good and Faithful Servant*; *The Farm Show*; *Ten Lost Years*, a highly successful theatre docudrama adapted from Barry Broadfoot's oral history by George Luscombe and his Toronto Workshop Productions; *The Trial of Sinyevsky and Daniel*; *Sarah*, adapted from a play about Sarah Bernhardt; *Mandelstaum's Witness*; *Sam Adams* #1 and #2 (an aborted pilot), and *The Betrayal*.

Baptizing

Baptizing (19/1/75, dir. Allan King, prod. David Peddie) was the most interesting of the lot as a television production. It also aroused considerable controversy. Dramatized by Patricia Watson (*A Bird in the House*) from *Lives of Girls and Women*, by Alice Munro, it was presented without commercials. Just before the money squeeze began, the CBC could still afford to take artistic chances even if no sponsor could be found.

King made use of a number of naturalistic film conventions. The pace is quite varied with a lot of quick scenes, cuts from action to reaction shot, very few transitional shots, and a few very leisurely scenes that unfold in a charged atmosphere of longing. Not unexpectedly for a drama about a teenaged girl discovering sex and, perhaps, a little love, there are many closeups. The designers caught the 'fifties ambiance very well, and the director of photography made the rural landscape and small town both pastoral and stifling. Music was used as motif, particularly nineteenth-century opera, which signifies both the mother's escape from her rural life and Dell's dreams of romance and first love that eventually cost her a scholarship—her only hope of escape; there are highly naturalistic and detailed sound effects such as dishes, crickets, steps, clocks, and the Korean War on the news. The director also chose the rather old fashioned but useful device of a narrator for voice-over exposition. An interesting tension develops when we hear a soliloquy in past tense counterpointing the visual present. As the camera closes in on the scene we are caught up in what we see and in the narrator's vivid sense of reliving the event.

The flavour of the period is just right; for example, the "Gay-la" dance, accurate in every detail down to the awful hotel room where "it" is supposed to happen. The fervour and failures of sexual inexperience are reinforced by her encounter with her boyfriend Garnet's revivalist religion. The hell fire and spirit shaking are not sent up as one might expect, but used as counterpoint. We see in the centre foreground their two hands as we hear the stirring phrases of the sermon. Cut to two faces fixed on preacher. The boy looks at her hands. The girl looks down. Two hands meet, finger by finger, gently and quite sensually. He takes her hand. She looks at him. Cut to other faces transported by the preacher's words. Cut to rather a deadpan picture of the Queen—an onlooker. The last scene in the emotional sequence is of Dell in bed caressing her hand and looking at it. The television adaptation retained the original story's emphasis on the physical pain of losing one's virginity and its grotesque comedy. On the whole, the CBC handled teen sex for its mid-'seventies audience largely by implication with a discreet nude scene on the bed, one strategically placed hand, a shot of his hand on her thigh, then her on top of him, and shadows on the wall.

Since this is the 'fifties, freedom is available to a girl primarily through a boyfriend with a truck and contact with a new and more exciting religion. However, dating a grade eight dropout who lives out of town is not acceptable at home. Naturally, this rebellion enlivens the boredom of studying for her finals. The feminist perspective is implied through small details: for example, a scene where Dell is scrubbing while her father and brothers have a beer. They refuse to give her one. In the river sequence, Allan King caught the key transition from sensual nude swimming and Garnet's ducking playfulness to her angry struggle and his desire to dominate by forcing her to undergo a baptism in his river, his religion as a pledge of her promise to marry him. By the time she breaks away from him, the romance is also shattered for her.

In the last scene, she comes in, tall and wet, to hear her mother read the exam results. She has passed without the first class honours needed for a scholarship, so there will be no university. As the mother points out with mingled exasperation and indifference, "You'll have to do what you want." While the opera plays in the background, the camera looks though an open window into a long shot. To an ironic, operatic counterpoint, the narrator, the protagonist as adult, comments on the loss of her fantasy and illusions of love. The last shot is of a 'fifties teenager as she looks up knowingly into the camera from the page she is "reading," disillusionment and rebelliousness etched on her face.

Under the *Performance* title, the dramas shown in the 1975-76 anthology season included *Six War Years* (30/11/75, pr. Robert Sherrin, dir. Allan King). Unlike *Ten Lost Years*, this was taken directly from Barry Broadfoot's oral history. It demonstrated an inventive and evocative use of tape with slow dissolves, direct address to the camera, and overlapping scenes to create the many moods of the soldiers overseas and women at home. It worked just as well when it was rebroadcast ten years later on Remembrance Day, 1985. *The First Night of Pygmalion* was a delightful adaptation of another stage play. *1837* was an interesting but oddly distorted attempt to translate the very successful theatrical collage of Théâtre Passe Muraille's collaboration with Rick Salutin into a series of television images built from back projections and animation, which sought to find and convey the resonance of the original stage metaphors; some scenes worked, others did not. *Kathy Kuruks is a Grizzly Bear* (1976), about the struggles of a long-distance swimmer, was a forerunner of what was first described as "journalistic drama" subtitled *Camera '76*, the forerunner of *For the Record*. Ibsen's *Enemy of the People* was followed by *Horse Latitudes*, yet another tale of mid-life male crisis as protagonist Gordon Pinsent sailed off solo around the world, fudged his data in a desperate bid to win the race, got lost, went mad, and died; I found it unconvincing. *Of the Fields Lately* was adapted from Bill Glassco's Tarragon Theatre production of David French's sequel to *Leaving Home.*

On the Job (27/3/76) introduced Montreal's Pointe Saint Charles playwright David Fennario to English television. It was an interesting contrast with Rick Salutin's realistic script about unions, *Maria*, directed by Allan King. Rita Greer Allen's *Raku Fire* was in many ways the best of the 1976 season—not spectacular or innovative in technique, or soul-catching in theme or performances, but memorable in small ways: a portrait of a woman who lives alone as a potter, exploring a very old and difficult Japanese technique as she works her way through a troubled period with another young woman and a male friend. Other good, entertaining drama included *Now I Am a Fountain Pen*, adapted from and presented as a trilogy about growing up Jewish in Sudbury by Morley Torgov; *The Making of the President 1944*, and an adaptation of Fredelle Maynard's familiar short story, *Raisins and Almonds.*

Ada

One of the finest dramas to come out of the 'seventies was Claude Jutra's *Ada* (2/6/77, wr. Cam Hubert, pr. R. L. Thomas, adap./dir. Claude Jutra), adapted from a story by Margaret Gibson and presented in the *For the Record* series. Janet Amos (Ada) and Anne Anglin (Jenny) were both veterans of Théâtre Passe Muraille's collective creations. Its theme, the personal suffering insanity brings, has been put on film many times, often with melodramatic effect; after all, distorted personal perceptions are wonderful subjects for inventive camera work. *Ada* is very different. It is not, as one might expect, specifically an indictment of institutional care, but it does question the medical misjudgment that prescribes a lobotomy for a brilliant, violently anguished patient, Ada. It also documents how her equally tormented friend, Jenny, responds to this event. The contrast between the Ada who raged against the dying of the light in wonderful poems and the Ada who asks Jenny for a few pennies for some chocolate as a reward for knowing the day of the week makes the viewer want to weep and rage. Kay Hawtrey as the sympathetic nurse, to Jenny after one of her attacks: "Ada was an accident. I'm [tiny pause] sure they didn't mean to do so much. . . . I remember Ada too. I'm as sorry about her as you are." Jenny: "Yeah. Sure. The [forbidden] cigarette is terrific." She butts it out on the floor.

What follows is a painfully awkward visit with her mother (Kate Reid), effusively on the defensive, and her hostile, resentful brother (Miles Potter), who can find nothing to say except the bare truth: "The only thing we have in common is that we all smoke excessively." He then asks her how she has been able to do without sex for the seven years she has been in the institution. The ensuing silence is at last filled with the mother's fantasies about how it will be for her daughter when, one day, she has what all women want, a husband and kids. Jenny interrupts passionately: "I'd like to sleep 'til noon, 'til one, 'til two, even." For her, freedom is a life without schedules.

The drama is not pathetic or sentimental in any way. When a foul-mouthed, cynical patient (Jayne Eastwood) deliberately and slowly crunches down on a mouthful of glass, or when "the Virgin" has hysterical nightmares, the rest of the patients simply live with it, help if they are able, or ostracize the patient if they think it is a bid for attention. They also flirt with the segregated male patients when they cross their path in the hall. Occasionally, they deliberately torment each other. They always try to manipulate the psychiatrists in the group therapy sessions, yet take their drugs with resignation: "Candy time again." Ada needs no drugs since her surgery. She simply needs "pennies for my chocolate" (the drama's working title). The language ranges from the poetic to the colloquial and obscene. As with any good, effective drama, the central characters are also brought to life by clearly sketched, yet individualized secondary characters: patients, nurses and family to whom they can relate.

In many ways, the hospital is normal human society etched into the consciousness more deeply and more painfully by the acid of madness. Corridors can be companionable retreats from common rooms or restraint. The steel mesh shuts patients in, but also protects them from the world outside: patients see the hospital both ways. Compromise and defeat alternate with small victories. The drugs are better than the shock treatments: "Something went ping in my head and I became a lot quieter . . . but it didn't take away the black bird. He just comes less often, that's all." Jenny's terror and humiliation—"What day is it?" "Was I very bad?" after "He" has come—distill the long, drawn out course of her illness very effectively. Her attacks are not sensationalized, but they are terrifying experiences. We see her face take on a frozen expression and hear, rather than see, huge wings swooping in and out of camera range. What we do see is the hallway and vast, carved Victorian stairway through her eyes as the camera swoops in and out on bits of moulding, and harsh daylight pours in through the windows.

This scene is juxtaposed with one of several flashbacks to Ada and Jenny. They are discussing F. Scott Fitzgerald. Ada: "If you're outside writing novels, you're a genius. If you're inside, you're crazy. . . . I am a seagull.

No, that's not it.'' Time winding on, routines punctuated with episodes of madness—how long? what day?—the flashbacks of memory, and Ada's progress, now measured by whether she can remember the words to ''Hickory dickory dock,'' are woven together thematically in a drama that shows time as subjective, elided, and endless all at once.

Ada has a strong plot line. It is an honest reflection of the way life can be for the mentally ill, and it is packed with vivid characters and themes. The piece also has an unusually literate script full of allusions that give pleasure to viewers who connect to them, without diminishing the pleasure of those who will not necessarily recognize or care about the relevance of references to the work of Turgenev, Dylan Thomas, Tolstoy, or Chekhov to the play. Above all, it has varied, believable dialogue. Several long scenes are filmed from a single camera position, forcing us to look at everything in the frame, concentrate on the characters and what they say or do. As with Jutra's earlier films, *Mon Oncle Antoine* and *Kamouraska*, some of the most crucial elements have to be supplied by the viewer. In the last scene, ''the Virgin'' is having a hallucination and the ward is in a shadowy, scurrying uproar when Jenny notices Ada is missing. She goes in search of her to find gentle, child-like Ada choking Alice, the girl who has tormented her throughout the play. Yet we never know whether she has succeeded in killing her. Alice falls silent as the camera slowly closes in on Ada, reciting one of her own poems: ''Slow, weak, awaking from the morning dream/of opening/brings me in contact . . . I am alive/this I.'' There is a reaction shot of Jenny, wordlessly questioning, then a closeup of Ada rejoicing, in tears, as she completes the poem. Jenny: ''Ada, you remember.'' Ada: ''It slowly fades. You won't tell on me, will you?'' (about the attack? The adult hidden in the corner of the brain of ''the child''?) Jenny, an appalled co-conspirator, says, ''I won't tell,'' and walks out of the frame back into the hall. The last shot is of Ada, alone.

Dreamspeaker (23/1/77, wr. Cam Hubert, prod. R.L.Thomas; also made for the *For the Record* anthology) was Jutra's darker and, in some ways, more didactic CBC film for the 'seventies. The healing ways of a Northwest-Coast shaman (George Clutesi) and his mute companion are thwarted by an institution for the care of severely emotionally disturbed children. This is a poetic, sometimes gothic piece, with moments reminiscent of the privileged view of the lad in *Mon Oncle Antoine*. In this film, the runaway boy who had found protection and education in unifying feeling and action in healing ways from the old shaman and his mute friend is brought back to a locked room inside the cage of the institution. The shaman dies and the boy, somehow sensing his death, hangs himself. Cut to the third person in the triad, the other Indian, who places the shaman's body in a tree, then shoots himself. The tragedy of the piece is oddly undercut, however, by the last scene: we see the boy and the man playing in the water while the old man looks on, laughing as his voice-over assures us that they are together in death. Most critics seem to have no difficulty with this ending, but I do. Although the ending is consistent with the shaman's beliefs, the cumulative impact of those deaths and the relentless logic of the forces in society that created them make the ending seem somewhat sentimental to me.

Freedom of the City

Very different indeed, yet equally good is Eric Till's intricately crafted version of Brian Friel's *Freedom of the City* (15/1/75, exec. prod. Robert Allen). The subject is the ongoing tragedy of Northern Ireland. Three people run into the Guildhall (city hall) to escape a riot, then enjoy the mayor's chambers and each other's company. Rumours of IRA gunmen holding the hall foster panic, and troops gather outside. Emerging from the building, hands over their heads, all three are killed. That is all the plot there is. Exploiting television's flexibility and intimacy, Till turns the piece into a study of what electronic television (tape) can do without undercutting or upstaging the ironic impact of the original. There are no commercials, so that the drama can unfold in natural rhythms without superimposed breaks. The story ending is shown from the beginning with very quick subliminal flashes of the three protagonists. Till catches the viewer's attention, then follows the jump-cuts with a slow pan of the empty Guildhall.

Then the drama settles down to show the audience Lily (Florence Patterson), Skinner (Neil Munro), and Mike (Mel Tuck) as they explore the hall and each other. In the beginning, it is a lark, full of gentle fun; a break in routine. Lily, an "earth mother," has a garrulous warmth. Skinner, talking to his bookie, seems to be slightly older, bright, cynical. Mike is friendly and not very forthcoming.

The music is part of the drama's complex texture. First they dance through the huge rooms to a resounding orchestral waltz. "Consider yourself part of the family" marks another phase in their acquaintance. "After the ball is over" ironically counterpoints their glee as they dress up in ceremonial robes. As Skinner dumps a ceremonial hat on Lily's head to cheer her up, they all sing the old music hall clinker, "Where did you get that Hat?" "The Man Who Broke the Bank at Monte Carlo" amplifies the escapist fantasy which the sherry and marble halls have inspired in these working class citizens. Music is also part of the complex time scheme. We hear and see the pub singer turn their tragedy into an Irish ballad even as the kyrie eleison at their funeral is sung on the television set in the corner. Woven over and through those two different time frames are voices from the inquiry into the deaths as well as those of Lily, Mike, and Skinner, commenting on their own deaths from a visual and acoustic limbo.

Visually and aurally, the best example of Till's ability to extend the boundaries of the usual naturalistic conventions of time and space that television so often favours is a sequence set in the pub. Within the framing pub sequence, a setting where Mike, Lily, and Skinner would be comfortable, Till gives us some customers' casual anecdotes. Then we see-hear Michael's dying thoughts in flashbacks (his face in three-quarter profile, in limbo) while the priest administers the last rites. The montage continues as we hear bits of dialogue from the presiding judge, who will exonerate all concerned; a shot of the cathedral with a choir singing the kyrie, then Skinner knowing that going out the door will spell his death: "How seriously they took us, and how unpardonably casually we took them." The flow of images continues with another shot of Michael's body sprawled on the steps and the empty Guildhall, then fragments of dialogue from earlier scenes among the three of them, with slight echo over to distance the viewer from that time and place, followed by a closeup shot of the distinguished visitor's book with their three scrawled signatures. At the end of this montage, we hear the music of the Mass and see, on the same pub TV, a politician expounding on the problems of the poor. Such sophisticated sequence is rare on television, requiring skill and a complex script from which to derive the images and themes. I have never seen a multiple time scheme developed to this degree in television before or since; in fact, the only other place I have seen that kind of complex, temporal collage is in Keith Turnbull's production of James Reaney's Irish Ontario tragedy, *Handcuffs,* the third part of *The Donnelly Trilogy.* Here as there, the multiple perpective gives an audience fragments of truth to set against the lies of the inquiry, the ballad maker, and the newspaper reporter.

Past, present and future are also counterpointed in sequences juxtaposing Lily's story about her mongoloid child (accordion music evokes slow, controlled, deep grief) with a rendition of "The Man Who Broke the Bank at Monte Carlo." Everything coheres when Skinner finally realizes, "They think we're armed." His reaction is to fend off the inevitable with parody and jokes contrasting with Michael's anger, which is deep, real, and articulate. The army's warning on the loud-hailer provokes Skinner into planting a medieval ceremonial sword into the middle of a portrait of a very English lord. This scene is intercut with a scene from the inquiry, which this time provides details on how British army bullets actually work. Skinner, fatalistic but stubborn, over a shot of Michael's hands on the sword's handle, ready to snatch it out of the portrait: "Don't touch that. Allow me my gesture." Lily chimes in for all three: "I never seen a place I went off as quick." One of them (which, exactly, is not clear) crosses out "Freeman of the City'' in the visitor's book, intercut by Till with shots of the frigidly British brigadier readying his troops for a major battle. Lily then offers her hand, and Skinner, recognizing a different kind of gallantry, kisses her. There is harmony all around as well as strain. Skinner sings her out to ''Monte Carlo.'' The three link hands and walk into the bright searchlights. Despite the multiple flashbacks, we have not seen this part of the scene before. They stop with their hands on their heads, pause, then realize that they are surrounded by guns. We have seen each one die before, but now we know how the frozen glimpses of a bloody hand, a tricorn on the steps, a handbag and handkerchief forlornly dropped beside a shoe came to be. It is a visual metaphor for the ''how'' but not the ''why'' of what happens year after year in Northern Ireland. Gunfire. The camera pans slowly around the empty Guildhall, looking for the armed terrorists who were never there. On the soundtrack, we hear the inquiry judge concluding no one is to blame as we see a reprise shot of the guest book. In the last sequence, we both hear and see the Sanctus in the church with the three coffins centred in the frame, then cut to a television shot following the coffins to the grave and the crowds. In a typical and effective elision, the black and white shot of the empty church dissolves into ''real'' present time and space as colour seeps into the image. In the final shot, the credits run on a black background while the music continues. After the tragedy and its aftermath are played out in full, the drama closes with music from Faure's Requiem. The effect is not ironic, nor is it intended to be: the beauty of the music is simply given full value and the viewers left to make of it what they will.

This project began when John Hirsch, with great imagination, went to Eric Till and said, ''What do you want to do?'' Till proposed *Freedom of the City*. Somewhere along the line, Hirsch lost faith in it. Only strong pressure from individuals within and without his own department ensured that the programme was telecast. John Hirsch's four years in the drama department were a very mixed blessing, according to many who worked there at that time. To be sure, he did introduce the country to some of the exciting theatre going on here and abroad. He encouraged the development of *For the Record*, an enduring achievement. It was in his era that *A Gift to Last* was initiated, and *King of Kensington* and *Sidestreet* got their start. He swept out people who seemed burnt out; other creative people left. He was full of ideas, had a mandate (based on a confidential blueprint) from the corporation to put the department back on track, and tried to encourage and to train new directors and writers. When there were cuts in his budget, he made no secret of his displeasure in the papers. In fact, though he could not arrange to see me for an interview, his thoughts on what he was doing at the time are on record in numerous news stories. On the other hand, he had a repeatedly demoralizing effect on many in his team, from story editors to executive producers. He could be withering in his criticism of a rough cut and then forget about it; others could not and the unit's morale would plummet. The image constantly associated with his name throughout my interviews with people who worked with him is that of a "whirlwind."

In a recorded interview with Ross Eamon, the professor who organized the Carlton School of Journalism's CBC Oral History Project (15/10/81), Ronald Weyman sums up Hirsch's legacy: "Tremendous messianic fervour. He put a tremendous amount of effort into making work in bigger and better ways. Both tape and film drama expanded." Weyman mentions, without enthusiasm, a co-production of *The Mother* with Hungary and the training programme for directors, most of them from the alternate theatres who, in the end, "didn't really like television or its limitations." Of *A Gift to Last*, he says: "In Hirsch's view it was rather a pot-boiler. He was interested in getting really significant, exciting and important theatrical productions into the studio. There were some nice ones. . . . *Gift* was a sentimental, soft family show which people loved. He hated *The Beachcombers* for not being important, exciting, inventive."

In 1977-78, the crossover year between Hirsch and John Kennedy, the trend was to rehabilitate a rebel or two, among them the Communist Norman Bethune, who had died forty years earlier in China, and a feminist, Nellie McClung, who left her imprint on our history before World War I. *Riel* followed in 1979. After that, there were no more resuscitations of historic heroes, heroines, or rogues until *Some Honourable Gentleman* (1984-86), an irregularly broadcast anthology portraying political figures like James Whelan, Sam Hughes, and John Grierson; and adaptations from the theatre: *Ma!* (Murray), and *Rexy* (1984) on Mackenzie King.

Bethune and Nellie McClung

It proved easier to bring the already familiar modern hero of the doctor to life—aided by Bethune's having also been an artist, lover, visionary, inventor and loner—than it was to animate a heroine whose early reputation was that of "lady novelist." Certainly, there was more Ethel M. Dell than Margaret E. Atwood in this version of the life of Nellie McClung. No matter that, like Bethune, McClung was a social reformer as well as a strong, gifted woman with a cutting wit, a love of battle, a wicked gift for mimicry, and an eloquent hatred of the sweatshops that financed the Tory bagmen of Manitoba. In this revisionist biogaphy new hats and shrill girlishness were all that defined her character.

What made one dramatized biography succeed and the other fail? One cannot fault the casting in either. Donald Sutherland as Bethune and Kate Reid as McClung are two of our best actors. In each play the designs, costumes, props, and sets were carefully researched and well executed. Nor could one blame a lack of money, or the choice of medium (both were taped), or the episodic narrative structure of each. The success of *Bethune* and the failure of *Nellie McClung* must be attributed to their scripts, the attitudes of their writers, directors, producers and perhaps of the CBC Drama Department at that time.

Both scripts were based on material easily available to any viewer interested in looking up some background. Roderick Stewart's biography, *Bethune*, supplied most of the dramatic incident in the play. He had also prepared a teleplay and there were lawsuits before the script was all finished. McClung's autobiography, *The Stream Runs Fast* (1945), quite clearly shows how she came to be a feminist, a process barely hinted at in the television version. Her novel *Purple Springs* (1921), despite its dated sentimentality, gives a far clearer view of women's fight for the vote. More to the point, the University of Toronto reissue of her collection of essays, *In Times Like These*[1] (originally published in 1915) shows the lady at her best: forthright, funny, and logical. A glance at these sources shows how a good script can bring life to a rather dry work like Stewart's and a bad one can trivialize, even falsify, the record. There is no script credit on the tape of *Nellie McClung*. No one at the CBC, right after broadcast, would answer inquiries about who had written the script. I understand this was because many hands had written pieces of it and no one wished to take the blame.

Under Eric Till's direction, taped entirely in a studio, *Bethune*'s dialogue was crisp, the situations crackled with humour, anger, and above all, energy. Among many memorable incidents were those showing Bethune's ruthlessness, forcing his bride to jump across a deep ravine; his impatience and arrogance as he collapses his own lung (a little poetic licence, here); his sense of mischief when he shocks student nurses with the casual display of his own anatomy; and the eloquence, charisma, and naivety of the man's actual words defending Stalin's Russia to hostile colleagues. Silence and subtext were also allowed to do their work, as in the look between Bethune and a Spanish mother, her child dead in her arms, in the complex shifts of emotions between Bethune and his wife Frances (Kate Nelligan), and in the unspoken rapport between Bethune and Chinese general Neih. The play also worked because it gave this enormously complex man his due; because Sutherland, who had been interested in Bethune for several years, gave a detailed, intense performance as did Kate Nelligan; because the script moved cleanly and swiftly through events now remote, giving them urgency once more, and because the whole work had one focus—the protagonist, the misfit who refused to become a victim.

Nellie McClung was not thus blessed. Wooden dialogue replaced the subject's own vivid phrases. Cozy domestic scenes reassured us that she really was "all woman," but we were afforded no glimpse of the grinding burdens of life on Manitoba farms and in fetid factories. One feisty encounter with the province's premier could not compensate for predictable pastiches of hustings speeches. The long-suffering viewer had to endure the shallow, wilful, improbably girlish characterization of Nellie herself as well as the stereotyped suffragist and the put-upon, but supportive husband. David Gardner, who appeared in both plays and had been just right as Bethune's best friend, was appropriately pained as the stereotypical Edwardian spouse. The play also managed not to use the most effective, entertaining and theatrical piece of suffragist propaganda available to the dramatist(s), "The Women's Parliament," a play skewering the pompous flatulence of the Provincial government, which played to packed houses to raise badly needed funds for the suffragist movement. In the most famous sketch of that piece, Nellie herself played Premier Roblin. Ironically, ACTRA (the performers' union) had kicked up a fuss because originally Kathleen Widdoes, an American, had been cast as this all-Canadian heroine. The CBC gave in and cast Kate Reid, thereby associating this fine actress with the débacle.

So much for what seems to have been a misguided attempt to recognize, belatedly, International Women's Year. The slogan the government chose for that year was: "Why not?" After *Nellie McClung*, we knew.

The first decade of this century was an invigorating time for the women's movement, the stuff of legend, a turning point in our political life; and the CBC managed to make it boring. Most insulting of all, this bit of revisionism stressed the suffragist defeat at the Manitoba polls (even though Roblin fell ten months later) and invented a climax in which Wes McClung forced Nellie to choose between her own cause and himself and his career. Needless to say, in this version she chose him. The scene slanders both parties, as well as undercutting the context and the thrust of the issues that should have vitalized the drama. The nameless scriptwriter or writers rewrote this part of our heritage into a Harlequin Romance. We got neither good history nor good entertainment. It took another five years to get women like Maryke McEwen, Anne Frank, Bonnie Siegel, Anne Wheeler, and Jeannine Locke into the powerful producer's jobs that enabled them to introduce topics of concern to women presented from a woman's perspective.

However, the tone of irony and self-deprecation ("Who us?") characteristic of *Nellie McClung* was more typical of the image of ourselves, both men and women, often found on our television in the 'seventies. A quick scan confirms this impression. *The Newcomers* (Irish, Indian, French, Scots, Danish, Ukrainian) too often talked like textbooks and suffered a lot, although there were a couple of notable exceptions. Formula ethnic clichés, slapstick, and sexism blurred the potential of folk heroes like *King of*

Kensington. In more serious plays, a runaway kid was dumped back into a sterile environment (*Dreamspeaker*), Ada was lobotomized, and an eccentric con-man sailed in circles to suicide (*Horse Latitudes*). Vicky killed her children, but did fight her way back to sanity. Yet among all the losers-survivors, the only large-scale, truly tragic figure was Hedda Gabler, and she was not "ours." The vision had become sharply constricted: by contrast to writers for television, the poets, playwrights and novelists of the 'seventies were not locked into the romantic and ironic modes; they could and did create high comedy and full-scale tragedy. Television did, however, retain the power to terrify and delight—and make myths: witness *Bethune*.

DRAMA UNDER JOHN KENNEDY

Kennedy has presided over a very different period in CBC history. In the early 'eighties, only seventy hours of drama were available in the schedule—and this included sitcoms, copshows, and family-adventure series. By 1984 the schedule began to expand with the use of Telefilm Canada and a shifting emphasis in the corporation itself. Then the drama schedule contracted again during the 1986-87 and following seasons in response to the slashed budgets imposed by Parliament in 1984 and 1986. The reader may wish to look at what Kennedy has to say about this period [Ch. 11] before exploring the programmes themselves.

Cementhead: Constructing or Deconstructing a Myth?

Cementhead (*For the Record*, 18/2/79, wr. Roy MacGregor and R. L. Thomas, dir. R. L. Thomas, prod. David Pears) explored our most enduring myth, the small-town hockey player who makes it to the NHL—although *Hockey Night* (1986) was another fine variation on the theme, with a girl winning acceptance on the boy's team as a goalie. *The Last Season* (1987) on the other hand was a somewhat gothic and overdone look at hockey violence, disastrous love affairs and dark secrets. By no means a simple portrayal of a national obsession, *Cementhead* counterpoints the "impractical" but lovely vision of a young goalie (Martin Short), who prefers building a plane and flying it to being number "17" in the draft for Detroit, with the hopes of his close friend "Bear" Bernier (Tom Butler), who eventually makes the NHL. One of many highlights came when the older Maple Leaf enforcer (played by Eric Nesterenko) whom Bear has to beat in the training camp, is pinned down in a restaurant by an earnest young female quoting Sartre and Genet on the mystique of violence. In response the older man takes out his false teeth, puts them in a glass, and laughs companionably with his young rival at her shocked face.

The humiliation of Canada's loss of the Challenge Cup to the Russians was still fresh in our minds when the programme was first broadcast. At the same time, Rick Salutin's *Les Canadiens*, a play about hockey as a metaphor for Quebec politics as well as a play about the process of myth making, was the play of the year in regional theatres across the country. *Cementhead* and *Les Canadiens* both took a sharp look at where the game comes from, how it is played, and what it costs: hockey is seen as a passport out of layoffs and dead-end jobs, a business, a personal dream, a myth of "manhood," a place to let out the pent-up frustrations of its fans in brawls on and off the ice, and as a surrogate arena for regional and national rivalries and larger territorial wars.

Cementhead was a very good example of how the "journalistic drama" can be stretched into unexpected shapes in skilled hands. There are several strands in the narrative: how and why Bernier dreams of being drafted in the NHL and why his friend, "Weeps," a talented goalie, refuses the draft, as well as how hockey is the only way to escape unemployment in a one-resource town. It portrays the useless hockey violence which Bear hates but must go along with if he is to join any NHL team. It unexpectedly introduces a girlfriend who is unimpressed by the jock image and macho ethic.

This "cementhead" is a man who can sustain a friendship with Weeps, even if he can't understand him, and is confident enough to date a woman like Becky who mocks and questions him. Becky has been established as a young woman who pumps gas to earn a living, is a champion diver and studies accounting at night. One of the funniest scenes in a decade of *For the Record* is when Bear pursues Becky into the women's shower room adjacent to the vacant pool where she has been practising. She calls his bluff by undressing, turning on the shower, and getting in; he calls her bluff by starting to strip and going in after her. Their truce is touching and very funny. *Cementhead* is full of rich characterizations and sustained metaphors in dialogue, setting, and storyline. As the drama progresses, for example, the Sudbury landscape becomes permeated with Bear's mythology: in his daily workouts, each slag tip is an adversary to be beaten, and he names each of the distant rocky hills for a team which could draft him out of this landscape. Yet it also explores a subject close to its audience and their experience. In the end Bear is picked by Toronto and Weeps finds that his home-made plane will actually fly. Like his father, brothers, Weeps, the coach, his rivals for a place on the team, Bear communicates primarily through gestures and images, not words, showing in his case that he is not all jock-talk, a "cementhead." As an exploration beneath one of our most common stereotypes, *Cementhead* had both a fine script and resonant images in an entertaining, innovative production.

In the first scene the camera introduces us to the local hockey team at floor level. We hear the tough-talking clichés of the coach (Kenneth Welsh) stepping between the players, who are stretched out on the floor as the anonymous foot comes down on Bear's shoulder. Bear moves the foot; Weeps gets up and dashes for the toilet, vomiting from nerves. Suddenly, the camera looks down into the narrow cubicle as Bear tries to jolly him out of his perpetual pre-game nerves. In the last scene, Bear takes Becky for a ride on the Zamboni in the same kind of local rink. Protesting, she climbs aboard, and off they go. Behind them, the Zamboni spills a trail of black paint—an ambivalent sign of capitulation and defiance, of serious courtship and pure devilment. *Cementhead* proves that it is possible to infuse a faithfully recreated naturalistic environment and a topical story with all kinds of metaphor, subtext, and good humour. *For the Record* will be discussed further in Chapter 8.

Miniseries did not fare so well in the late 'seventies. One had to wonder what grudge the CBC nursed against the Establishment to produce two very bad drama specials on big business. It is true that they were ahead of the 'eighties trend to glamourize the businessman. As the silver-headed executive in a 1986 airlines advertisement crisply informs his breathless, impeccable subordinates (one of each sex, of course): "Luck is for rabbits." However, the love affair between the corporation and the corporations did get off to a bad start. In the wake of *The Masseys* and then *The Albertans* (both 1979), the captains of industry should have been suing. Each was dominated by portraits of obsessive, humourless workaholics who were also ruthless schemers: in these scripts, the only magnetism belonged to a German con-man; the only machismo was to be found in the radical American Indian movement; and the only strength in rural and aging males. The women were prim, tired or bitchy, and the only excitement was illness, with a little dynamite thrown in. Worst of all, although they had been telegraphed from the beginning, the morals are spelled out for the audience as in *The Masseys*: "We believed that we could change ourselves and so change the world. . . . We tried to be givers, not takers." On second thought, given lines like that, the Masseys themselves might have paid for the whole thing.

Alberta, on the other hand, should have filed grievances with the CRTC for the insulting clichés handed out to native people, ranchers, plucky middle-aged daughters, bored wives, engineers, and self-made men. To catch the previews was to catch the highlights. The three-part series managed to be racist, sexist, reinforce eastern prejudices about western parvenus and be tedious, all at once.

Later, the CBC made some amends by showing the real-life drama of the boardrooms in the current affairs series, *The Canadian Establishment*.

Despite the romance with business going on in the news and current affairs departments, however, the Drama Department did not make real restitution until *Empire Inc.* (1982), and then undid the balance with the stereotypes in *Vanderberg* (1984).

By the late 'seventies all single plays not under the *For the Record* umbrella had become drama "specials." The combination of Claude Jutra and Mordecai Richler should have been special indeed, but *The Wordsmith*, also shown in the 1979 season, turned out to be no more than a mildly satirical portrait of the artist as refugee in a hot Montreal summer of the 1940s. Beautifully designed and photographed, the production was a perfect illustration of Richler's lines: "A Jew is not poor. Having a hard time sometimes, in a strange country always, but never poor." The problem was that plot elements—the housewife's dreams of better things, her crush on the boarder, the child's growing pains, the young writer's attempts to lose his virginity—were over-familiar. The play also suffered from a leisurely, sometimes flabby script. Saul Rubinek's performance as the writer-protagonist lacked energy: the best moments were scenes between Janet Ward, as the wife, and Peter Boretski, who gave a strong, detailed, sympathetic performance as the husband who struggles to normalize his precariously balanced household.

Ma! (1983, wr. Eric Nicol, dir.-prod. Philip Keatley) was a delightful and successful blend of presentational theatrical conventions—sets, visible audiences, direct address, area lighting, actress-narrator—with film realism and touches of surrealistic use of landscape. Philip Keatley achieved what few other directors manage: a television adaptation of a play that is faithful to its theatricality and uses the electronic medium intelligently. Crisp dialogue and a funny, energetic performance by Joy Coghill helped.

Kate Morris V.P. (1984, prods. Anne Frank and Maryke McEwen, wr. John C.W. Saxton, dir. Danielle J. Suissa) was a much better treatment of the glamour and pitfalls of business life than the 1979 studies of tycoons had been. Though caught in a conventional love story for part of this docudrama, Kate Trotter convincingly portrayed brains, beauty, ambition, and an ability to learn the rules and then out-think the business males who make the rules, with the help of a male mentor. If we ignore its Harlequin Romance stereo-typical lover, the drama did have a topical thrust and some nice touches of malicious wit. Although a ninety-minute piece, it also appears in the lists of *For the Record*.

Charlie Grant's War (1985, wr. Anne Sandor, prod. Bill Gough, dir. Martin Lavut) is a two-hour drama special that was warmly received by a large audience. For once, a deprecating and rather unfocused young Canadian was shown as a hero on CBC television. This was not intended to be a docudrama, though Charlie Grant was a real person and the opening noted, "This film is based on a true story." The play combined into one well-told tale, suspense, politics, history, and an attractive protagonist who suffered a great deal but survived. Supported by a good cast, R. H. Thomson, as Charlie, used his full acting range and the result was very moving. It is a favourite around the CBC upper management because it had a hero, suspense, told a gripping story very well and touched off a wave of response in the audience. It was indeed good television.

Gentle Sinners

From the many fine dramas of the first half of the 'eighties, I wish to conclude this survey of the second, often golden age of television drama with a more detailed closeup of *Gentle Sinners* (1985), because it represents that part of drama production that presented a personal rather than socially relevant or topical vision. Adapted by Ed Thomason from a novel by W. D. Valgardson, it was directed by Eric Till with Peter Kelly as executive producer.

Till approached this complex material by treating the foreground characters with detailed psychological realism, eliciting four very strong performances from George Clutesi as Sam, Charlene Seniuk as Melissa, and particularly, Christopher Earle as Eric and Ed McNamara as Uncle Sigfus. At the same time, he treats the background characters as larger-than-life figures of repression and terror. Transition points in the growth of Eric from boy to man are often magical hidden places like Melissa's cave, Sigfus's hidden chapel in the woods, "bought to store grain in" but clearly still used for its original purpose, and the huge, echoing, shadowy hayloft. The acting performances and camera work treat auctioneers Big and Little Tree, Eric's satanic friend-tempter and his grotesque parents in a progressively more exaggerated manner. In the final scenes, the two auctioneers seem like a giant and dwarf out of legend. In a black, rain-wet slicker with a hood and knife, the satanic friend (Todd Stuart) seems like some medieval demon-friar, as well as a dark doppelganger of Eric, whose last gesture is to throw his knife at Eric and Melissa as they escape down the river below him.

In creating these mythological dimensions, Till pushes the background performances confidently towards archetype (skirting perilously close to stereotype), yet he fills in the foreground characters with fine psychological realism and delicately shifting relationships exposed through carefully worked-out, often non-verbal nuances. He also treats the rural landscape, the work around the yard, the auction, and the fishing expedition with respect as well as affection. The visual style juxtaposes the lyrical and realistic modes. Thus the contrasting styles work in rich counterpoint, as metaphors of the inner and outer lives of "Bobby," renamed Eric by Uncle

Sigfus when he washes off the dust of his journey under the pump. After this baptism, Eric starts to earn his new name.

In the climax, all the complex themes are sharply focused down the barrel of Eric's gun as he confronts the parents he hates. Given the psychology of the character and the situation, we are concerned that he might very well shoot. When he cannot, he puts the barrel into his own mouth and presses the trigger. The gun does not fire. Sigfus takes it from him and says he had removed the bullets some time before, having read the hatred and anger in the boy. Uncle Sigfus's intervention is not based on an appeal to sweet reason, or even love and loyalty; it is a tough-minded challenge to Eric to be an adult. Their relationship seems shattered by this, however, as Eric takes off to find Melissa and run away with her

As Eric and Melissa make love for the first, and perhaps last, time, the setting is the romantic gloom and warmth of the hayloft in the rain. Here also there is counterpoint. The lyric tenderness of gesture and touch is modified by Melissa's knowledge that what is coming is inevitable. As she says, the only freedom of choice she has had since the Trees came to live with her mother is to decide who will be the first person to sleep with her. Kid brothers often interrupt romances, but this one, being a Tree, has a gun: the idyll is succeeded by murderous chase and injury, tempest and flight, knife wounds to the hands of fair hero and dark tempter. In the end, like Melissa, Eric accepts the limits of his choices: he decides to go home for four months until he is sixteen, as the social worker had advised. Both Eric and Melissa need to be taken care of until more time has passed.

The last scene of the film is particularly interesting. For the viewer who really wants an idyllic ending, the cues are all there: Melissa reaches safety with Sigfus and his Métis friend, Sam. Eric has written her a note and slipped out, but the old man and Sam, driving the wagon, catch up to him again hitch-hiking on the road. Melissa gives Eric Sigfus's parting gift, his grandfather's watch. Eric promises Melissa he will come back. Then Till gives us a surprise shot of Sigfus himself on the back of the cart as it turns away from the camera. It is a final assurance of protection and education for Melissa and an acknowledgement that Eric has the right not to say goodbye if he doesn't want to.

For other viewers, the key to a more ambivalent reading is to be found in the running *Huckleberry Finn* motif, which Eric loves and shares with Melissa. The archetype of the adolescent grown to manhood includes the loss of innocence through ritual initiation, the necessity of rebellion through the symbolic or actual killing of an animal or human being, preceded by a period of isolation and training by an older mentor who knows the ways of survival, and the choice of returning to civilized ways as a young adult. But the pattern here is a significant variant, as it is in *Huckleberry Finn*, where

Huck does not really settle down, ever. As in so many rites of passage to adulthood, Eric has won his name change and left "Bobby" behind. True, he is returning to the prison of his parents' house, but when he is free in a few months, the Huckleberry Finn motif suggests that he will hit the road again for further initiation into the outside world without the encumbrance or, temporarily, the need of mentor or lover. He is leaving the cart and horses, the lost craftsmanship and long friendships behind for the 'fifties wide-finned chariots and asphalt road. Our last image of him is with his thumb out, standing on the side of the road.

The film's last scene is as rich in multiple levels of meaning as any classic CBC television drama of the past. The film also works for me as a paradigm for the transition of Canadian television drama in the mid-'eighties, from reduced but secure circumstances into an uncertain future. CBC drama reached maturity in the 'fifties. Then our television anthology changed with taste and limits of tolerance of the time to survive into the 'eighties as drama specials, films, adaptations and originals. Indeed, the survival of anthology in Canada, as opposed to its disappearance on American network television (*American Playhouse* on PBS excepted), seems to have pulled many CBC series towards the open-ended, the ambivalent, wide variations in tone, and the quirky in some episodes. Audiences who still expect such qualities from the CBC accept them in series as well, even when they stretch the formula.

No classical or contemporary theatre presentations were adapted for television in the 'eighties, except one or two plays from Stratford or the Shaw Festival, and we are the poorer for it. On the other hand, all the dramatized television scripts in the 'eighties have been by Canadians, as are most of the actors, producers, directors, designers, and directors of photography. Most of all, the subjects are Canadian, whether topical or historical, regional or national, high on society's current social agenda or newly exposed, political or personal: the gaze has turned inwards.

All that may change: co-productions have become necessary, even in the (until now) autonomous United States industry. True, the message to Canadian independent filmmakers from a panel of American broadcasting executives at the Banff Television Festival, 1986, could be accurately summarized this way: "Make it look, sound, and feel American, and you may make a deal." The crowded room was not particularly pleased to get that message, and, in point of fact, clones do not sell very well in the United States. According to Norman Horowitz, who has made some co-productions in Canada, the United States, which he calls the "country where chauvinism is rampant," has yet to put a Canadian co-production on ABC, NBC, or CBS in prime time. I take this as definitive proof that what we do well also differs distinctively from American programming.

The context here, of course, is first the temporary 1985-86, 1986-87 increase from seventy to one hundred and twenty hours of Canadian drama on the CBC, thanks largely to Telefilm Canada; then the twenty-six half-hours of drama produced by Atlantis for Global and TV Ontario, that province's educational network; and finally promises, promises from CTV, which currently translates into a half-hour family series, *The Campbells, Check it Out!,* and one or two specials. The copshow *Night Heat* and the spyshow *Adderley* appear respectively on CTV and Global prime time and were financed initially in the U.S. for CBS late-night viewing. Another factor, unfortunately, has been the Conservative government's apparent determination to starve the CBC of operating funds and not to "inflate the national debt" by supporting a national theatre system with touring performances, a film industry, and indigenous television drama.

I will conclude this summary with mention of yet another 'eighties variant on anthology television drama, a hybrid form of film and miniseries that often satisfies neither film buffs nor television viewers. *The Crimes of Ovide Plouffe* (1985), *Maria Chapdelaine* (1986), and *Joshua Then and Now* (1987), are three examples. CBC participation in the government's Telefilm scheme for independently financed television drama can work to great advantage with two-hour movies like *Bayo* (1986), the very popular *My American Cousin* (wr.-dir. Sandy Wilson, 1986), and the more demanding *Loyalties* (dir. Anne Wheeler, 1987). Additional financing can add significantly to the texture and scope of such films, though it does not follow that good television is the result. At least, however, the forms of "film" and "television" need not be at war (although Mark Medicoff, in an article on "Video Mediation of Film" in *Cinema Canada*[2] suggest otherwise.

The attempt to mate a theatrical film with a miniseries, using an editor as procurer, rarely yields memorable offspring. Both forms suffer, the two-hour cinema version and the six-hour television miniseries, usually in inverse proportion to the complexity and quality of the material. The short form, the movie, is often incoherent, with characters dropped or undeveloped, dangling plot lines, and visual motifs appearing without explanation. If the film is tightly written and directed with a focus and edited for theatrical release, the television version will seem grossly padded with irrelevant detail, rococo subplots, and slower pacing. Movies made from miniseries can work, as Ingmar Bergman demonstrated in the 'seventies with *Scenes from a Marriage* and the stunning *Fanny and Alexander*, and Fassbinder did in the 'eighties with *Alexander Platz*. The process very seldom works for the CBC, however, or the independent producers putting these packages together.

Taking the viewpoint of the movie-making community, Michael Dorland, writing in *Cinema Canada*,[3] comments on the diversion of resources from film into television drama: "In fact the real tragedy of this year's Perspective

Canada [a section of Toronto's Festival of Festivals Film Showcase] is that there was so little in the crop with visual qualities so overwhelmingly filmic, so demanding of a big screen that couldn't run as safely on the small one.'' In a speech printed on the same page, director Ted Kotcheff [who was a producer-director in the early years of CBC TV Drama and TV drama in the U.K.] agrees: ''Films deal in levels of realism that are just not possible on TV. Whereas for me, the system of yoking was inconvenient and a bit wasteful financially [in shooting *Joshua Then and Now*], what is going on now has brought film production to a grinding halt. . . . Let us not pay for the mistakes of the tax-shelter years by dismembering Canadian cinema. Otherwise we'll be back to writers and directors emigrating again to make movies abroad and Canada will be finished as a film country.''

Yet television drama since 1952, not film or theatre, or even radio, has been the dominant Canadian medium for fiction performed for huge audiences. Many individual voices, including a significant number of our cinema directors, were seen and heard, often for the first time, in the thousands of hours of anthology that are part of the CBC heritage. Looking out on a narrower but a clearly more intensely Canadian scene, CBC television in the 'seventies and 'eighties has contributed its fair share to the growing catalogue of permanently valuable and interesting drama in the country. What we have done and what we can do is very good, indeed.

NOTES

1. Number 5 in the *Social History of Canada Series*, ed. Michael Bliss, introd. Veronica Strong-Boag, University of Toronto Press, Toronto and Buffalo: 1972. This edition blazons ''The rise of feminism'' as part of the cover design.
2. No. 136, December, 1986, 13-16.
4. No. 124, November 1985, 10-14.

8

DOCUDRAMA
"To Be Seen Is to Seed"

The debate over what elements actually create a journalistic drama, a docudrama, or a historical drama is a vexed one. Articles and papers on the subject regularly appear in academic journals as diverse as *Screen, Theatre History in Canada, The Journal of Popular Culture, Theatre Quarterly,* and *Cinema Canada.* What is the analyst of Canadian television drama to do with programmes as different as *The Open Grave* (1963), *The October Crisis* (1975), or *Canada's Sweetheart: The Saga of Hal Banks* (1985), all of which mix interviews, a recreation of news footage and dramatization? What about *The Paper People* (1967) in which the values of documentary filmmaking are directly questioned using a purely "fictional" film which is a highly personal statement by writer Timothy Findley and director David Gardner. Thinly disguised headlines were the subjects of many of the 'fifties scripts for the *On Camera* anthology. Twenty years later, in 1975-76, R. L. Thomas devised *For the Record* which he initially labelled "journalistic drama." In between, there were many plays that focused on topical issues. As we have seen, *Festival* and *Quest* examined the whole issue of nuclear war in the early 'sixties, as well as the oppression and displacement of native peoples, and many other issues.

One of the last efforts of *Festival* was *Reddick* (1968), a topical, but entirely fictional, account of a United Church minister's crisis of faith. The climax of the piece is his mock trial by his flock of bikers and hippies which moves from the drop-in centre in the basement into the sanctuary. Forced to answer their searching questions, he reveals a man wracked by doubt. Bitterly denouncing him as a hypocrite, the group dances, fouls the elements set out for communion, and, finally, impales his hand to the communion table. The shocked silence is its own comment. A worried subordinate has already called the police. The film ends with Reddick gathering himself together to defend his flock. Although the examination of the subculture is now very dated, it is still interesting to view thanks to imaginative writing by Munroe Scott, Don Harron's excellent performance and some first class directing by producer/director Mervyn Rosenzweig. *Reddick* was studied in church groups across the country as an aid to trying to understand the hippie revolution and the problems of the shift in mores and values which confronted the established churches. *Reddick II* was produced in 1970. Reddick's personal Golgotha was a late but superb example of *Festival* at its best. *Reddick II,* like the rest of CBC television drama, was weakened by a sense of aftermath after long crisis.

A few drama specials from the early 'seventies have a topical flavour— *Twelve and a Half Cents* (1970) and *Vicky* (1973) are two examples. A concerted effort to produce topical anthology took shape in 1975-76. In the anthology *Performance* (1976), *The Insurance Man from Ingersoll,* portraying the activities of a political fixer, *Kathy Kuruks is a Grizzly Bear,* about long distance swimming, *Nest of Shadows,* about the pressures on a teenage mother, *A Thousand Moons,* about an old Métis woman determined to return to her home, and *What We Have Here Is A People Problem,* about the government's attempt to appropriate an old farmer's land, were all topical dramas. In all but name these dramas, subtitled *Camera '76,* were the first season of *For the Record. For the Record* was the longest running anthology on the CBC (1976-86) concentrating on contemporary issues.

Television demands an endless supply of fodder, and topical material was ideal for live television, where writers could respond relatively quickly and producer-directors get a programme to air before an issue had died. Ironically, modern technology makes that far more difficult, somewhat narrowing the topical subject matter available for drama. Yet this still leaves many issues that remain with us year after year—for example, the ethics of professional hockey as in *Ice on Fire, Big League Goalie, Cementhead,* an episode on *The Way We Are,* in an episode of *Seeing Things,* as well as *Hockey Night* (1984), and *The Last Season* (1986). In fact, a series blending melodrama, soap opera, and sports saga, *He Shoots He Scores* (1986-88), was developed for Radio Canada and the CBC.

With the exception of a few 'eighties issues, like pollution, child abuse, and wife battering, the list of topical concerns from the 'fifties would include much of what is current in the 'eighties. True, we are no longer concerned with alienation in the suburbs, and our definition of conformity has changed; for example, we are not as sure now as we were in the 'fifties about who women are and what they can or should do, though our drama still perpetuates stereotypes taken from the domestic post-war period. In

general, however, from euthanasia to the impact of layoffs, many subjects
for topical drama are recurrent in the thirty-four years of CBC anthology
programming.

If one of these subjects is the focus of a drama, does that make it
docudrama? If *A Flush of Tories* is a conscious displacement, as much a
look at the roots of the still painful Manitoba Schools Question of the late
nineteenth century as it was an account of three completely obscure prime
ministers who followed Sir John A. Macdonald's death, is it therefore a
drama documentary? Is the drama special *Riel* historical drama? Péquiste
propaganda? Barry Pearson's script, *Ambush at Iroquois Point* (14/3/79,
prod. Beverley Roberts, dir. R. L. Thomas, exec. prod. Lister Sinclair), is
certainly a historical drama, just as the *Folio* presentations of *The Brass
Pounder from Illinois* (about Cornelius Van Horne) or *The Trial of
James Whelan* were twenty years earlier. Unlike those two, however, it also
focuses on our love-hate relation with our American neighbours, the
differences that divided us in 1812 and still divide us today. In a less realistic
vein, the romantic photography and lighting, exciting plot and love interest
(she is Canadian, he, American) make it good entertainment. Yet the
symbolically shallow river in back of the farm that forms the international
border gains a more immediate symbolic significance in the 'eighties, as the
free trade debate heats up for the first time since Laurier. *Ambush at
Iroquois Point* recreates a historical moment in closely observed detail. Yet
it blends in a more personalized conflict and refuses to be categorized. It was
to be one of several historical dramas in an anthology called *The
Canadians.* No others were made.

Docudrama? Dramatized documentary? Drama-doc? Faction? Re-enact-
ment? Dramatic reconstruction? No one can even agree on what to call this
apparently familiar genre because there is no real agreement on exactly—
even approximately—what kind of television drama it is, or what it should
do. *Television,* Granada's stimulating, thirteen-part examination of the
medium itself, devoted one hour to a sketch of the form's evolution in which
Sydney Newman's contribution was acknowledged, though not the interplay
of influence between the U.K. and Canada. The programme included
interviews with British playwrights closely identified with influential
docudramas, most of whom had no illusions about their impact on society's
goals or laws. It also included examples of varied uses of the form around
the world. The programme reached no consensus on definitions of the form,
not even a sense of common ground among planners, producers and writers.

Canada has seen no coherent theoretical debate on the aesthetics, ethics,
and social uses of "docudrama" as Britain has, primarily in the shortlived
television journal, *Context* (1960s), and in *Screen* through the 'seventies
and 'eighties. Even so, I think the term itself as I define it below will serve,

since it is commonly used—and misused—to describe one of Canada's favourite dramatic forms, whether on radio, television, or in the theatre.

Television, radio, theatre, and film interwove more complex patterns than I can trace here to establish what I identify as these two main branches of the docudrama family tree: historical and topical. Until recently, Canada's most significant contribution to film was the NFB documentary as defined during the 'forties and early 'fifties by John Grierson and developed by NFB filmmakers. His internationalist philosophy, and his burning vision of a sleeping giant which film could awaken to a sense of national purpose, shaped our film production up to the mid-'fifties. This vision was modified in the 'sixties by a new ideology that attacked traditional documentary for its claim to objectivity. A new generation, including Robin Spry, Donald Brittain, Ron Kelly and Allan King, threw out the conventions of voice-over to guide audiences along predetermined paths, or unobtrusive camerawork which tried to be transparent, music that commented on, or too often inflated the emotional impact, and skilful editing out of *longueurs*. Above all, these filmmakers tried to subvert—or at least inscribe in their films for all to see—the pervasively middle-class point of view which dominated the documentary genre. Yet Grierson himself understood what television could offer: "I think of the unexpected of television. You never know what is going to turn up. . . . There is much more possible in a medium which is God's gift for the operator who commands its relatively private relationships . . . a realism in which . . . you recognize immediately, and as if by providence, the 'naturals'."[1]

In traditional documentary as defined by Grierson, staging the event was the worst of all sins. Such criteria temporarily tarnished reputations of such early filmmakers as Flaherty (*Nanook of the North*) and photographer Edward Curtis, who were born well before the term documentary had been defined. By the 'sixties, documentary had become the filmmaker's personal point of view as expressed by an obtrusive camera. His privileged eye was made evident in rough compositions of hand-held travelling shots and the jump cuts that made the elisions of time and space obvious. The sound was also rough, the lighting "found," and therefore often less than ideal; the filmmaker might well appear personally on camera. There was even recognition of the fact that the camera's presence might alter the event being recorded.

At the time when directors like Truffaut and Godard, leaders of France's *nouvelle vague* movement, were applying the conventions of direct cinema to their fictional films, Canada had produced very few full-length fictions. As producers David Gardner and Mario Prizek have both pointed out, however, many CBC producers in the first fifteen to twenty years were avid film buffs who watched European developments with interest.

Val Gielgud, the Head of BBC Radio Drama, was not singling out documentary drama when he observed to a colleague as early as 1950 that "the fundamental connection between drama and politics is, I think, almost as close as that between drama and religion. . . . When the stage is healthy, its political connection is strong."[2] Well before Arden, Wesker, Livings, Kops, and Osborne revived post-war theatre in Britain, BBC radio had a solid, thirty-year tradition of plays, biographies, historical recreation, and documentary features tackling serious social questions. I heard the work of Lance Sieveking, D.G. Bridson, and Tyrone Guthrie from the 'thirties and Charles Parker's complex 'sixties Radio Ballads when I was doing research on BBC radio drama. British television took a while to catch up to its older sister, "steam radio."

Other factors were at work in the evolution of docudrama's branching tree. The naturalism of American writers Chayevsky, Foote, Rose, and others challenged the stagey drawing-room television of the BBC. Kinescopes of those productions were broadcast in the U.K., and the BBC in turn sent numerous observers to American networks in the 1950s. Critic Raymond Williams, in *Television and Cultural Form* (1974), traced the influence of Americans Chayevsky and Rose on British 'fifties television drama, particularly "their feel for everyday life, the newly respected rhythms of the speech of work and the streets, of authentic privacy"—qualities also found in late-'fifties NFB documentaries and CBC television dramas like Len Petersen's *Ice on Fire.* John Osborne's play, *Look Back In Anger* (1956), first introduced to the London stage an educated working-class protagonist, regional speech, and the gritty realistic detail of the lives of a largely voiceless class that lived in parts of the country unknown to the south of England. Radio drama, then (and still) an important part of the dramatic landscape in Britain, began to feature the regional voices of Bill Naughton, Alun Owen, Rhys Adrian, and Giles Cooper.

Into this growing ferment came Canadian producer Sydney Newman in late 1957 to help stir up British television drama with his emphasis on naturalism—though Clive Exton's surrealistic working-class fantasies also flourished under Newman—topical material and, above all, contemporary relevance. For Newman, the ordinary working families who were beginning to purchase television sets in overwhelming numbers were his natural audience, his untapped resource. He understood that a good story which provoked argument around the dinner table afterwards would gain and hold audiences for anthologies. To succeed with this audience, however, an anthology had to concentrate on the recognizable and the relevant, with a tone that varied from week to week and with enough entertainment value to keep viewers tuned in. British television critic George Brandt, writing in

1962, struggled with the idea of the documentary drama, which he felt "makes an ordered response on the viewer's part extremely difficult."‡ Yet he noted that "a keen feeling for current speech distinguishes a good many television plays," and praised Alun Owen's *Lena, Oh My Lena,* for its "early uninterested cynicism," as well as Rhys Adrian's "semi-expressionist satire." As on the CBC, the range of drama on the BBC was enormous by the early 'sixties, even as the range in the United States was being flattened by series and formula sitcoms.

Canada is not normally an ideologically polarized country: its class wars go largely unrecognized, and class stereotypes are few. The cliché of "peace, order, and good government" is founded on what we believe about our lives in both francophone and anglophone sections of the country. We do not often take to riot, rebellion, or revolution, and we very rarely mythologize the upheavals that have occurred. The notion that we are a peaceable kingdom is much more important to us than we care to admit. Films and plays, even television dramas that tell us otherwise have not fundamentally changed this self-perception, which is shared by most of the rest of the world and founded on the basic truth that the only empires we have ever sought were within our own spacious borders. We may have colonized the North and tried to colonize every native citizen in the land, but we don't hunt for trouble outside our country, we don't start wars, and we don't dangerously glorify nationalism, the flag, or the national anthem.

Yet, we are beginning to recognize and even enjoy the regional variations in our language, although the rich turns of phrase of an eighty-year-old farmer or the distinctly regional idioms that chequer the country are too rarely heard in our drama. In Québec, the heart of politics is language: when Michel Tremblay wrote *Les Belles Soeurs* in *joual*, or street speech, the roof trembled and the walls shook with the outcry. Generally, in fact, the country's francophone popular culture has been far more politicized than its anglophone counterpart, from the early radio soaps to the téléromans. Speaking and writing English is simply the way it is for most of the country; to speak and write French is far more likely, at some point in the life of a Québécois or Acadien, to constitute a political act. The language—that is, dialogue—rather than the image, is the backbone of nearly all television plays. The limited success that has been achieved with dubbing francophone drama into English demonstrates this point. Yet, because of audience resistance to subtitles, a resistance incomprehensible to Europeans and never challenged by consistent exposure here, we do not see much of the very good, often lavishly produced historical docudrama from Radio-Canada such as *D'Iberville* (1970) or *Duplessis* (1980), and we get almost none of the network's other excellent francophone drama.

NEW APPROACHES TO HISTORICAL DOCUDRAMA

A certain kind of stylization in docudrama helps us distinguish fact from opinion, speculation or reconstruction. A viewer stumbling by accident across one of British filmmaker Ken Russell's biographies for television would quickly find out how that premise can be taken to the extreme— sometimes successfully, sometimes not. An example is the narrative structure of his biography of Samuel Taylor Coleridge (1977), built around metaphors and correspondences derived from "The Rime of the Ancient Mariner." We had to sort out for ourselves that Coleridge's courtship of his wife did happen in roughly the fashion recreated by the drama. He did not, however, stab her in the heart with an anchor and trail her body in a row boat, arms outspread like the wings of an albatross. Russell's biographies are not straightforward accounts of his artist-subjects' lives. He drives instead for what he considers to be the essence of the art and the personality, sometimes using outrageous presentational conventions. Several risks are involved here: for viewers unfamiliar with Coleridge's life or work, Russell's somewhat baroque methods were likely first to confuse and then alienate; on the other hand viewers who did have some knowledge of Russell's mode or the artist he was depicting could find the strongly personal vision a stimulating challenge their preconceived ideas about the subject.

Troy Kennedy Martin—the writer who conceived the long-running British hit, *Z Cars*, that propelled the copshow into tougher and morally ambivalent territory—has been one of the most eloquent debaters on the ethics and aesthetics of television. In 1964, he argued that television drama should break away from the theatre and use the camera as freely as film cinematographers did.[4] He thought this would open up, not only the possibilities of more realistic action shot on location, but also new potential for fully stylized narratives and lyric pieces.

In the United Kingdom, the most influential example of a television film that tackled the subgenre of historical docudrama in a way that answered Kennedy—Martin's challenge was Peter Watkins's *Culloden* (1964; shown on CBC *Festival* in 1965). Watkins used some techniques familiar to American and British viewers from the anthology, *You are There* (1953-57), in which host Walter Cronkite reminded viewers each week: "These are the events that alter and illuminate our time—and you are there." We were there while famous historical figures like Joan of Arc and Brutus enacted their crises, all the time answering off-camera questions from "reporters" covering the event. The point of view of the script on the historical figure usually reflected prevailing opinion—or historical myth—although the actors sometimes lifted the protagonists above the superficial sketch. *Culloden* was very different. Watkins's camera interrogated illiterate and terrified peasants, hungry, dirty, and difficult to understand with their thick Highland speech, as well as badly armed, untrained chieftains, battle-hardened British regulars, and even the Bonnie Prince himself, complete with a demythologizing but historically accurate German accent. The techniques of direct cinema also demythologized the battle. Using film in a new way, the historical docudrama had come of age.

As we have seen in the previous two chapters, the early days of CBC television featured the work of the network's skilled radio docudramatists: Lister Sinclair, Tommy Tweed, Joseph Schull, Mavor Moore, and Len Petersen. The fact that they supplied so many early television scripts (often adapted from earlier radio pieces), together with the recruitment of Sydney Newman and Ron Weyman from the NFB, meant continuity for the tradition of challenging topical dramas, less than worshipful biographies, and attempts to show the complexity of our history as well as its colour and high emotion.

Fittingly, it was Alistair Cooke, originator of the weekly radio "letters" that for many years kept BBC listeners in touch with the United States, who defined for both sides of the Atlantic a different taproot of docudrama's family tree. Cooke also hosted the only weekly American network show devoted to "culture"—*Omnibus* in the 'fifties—and is now firmly identified with Britain's main television export to the United States as the host of *Masterpiece Theatre*. On 18 July 1948, Cooke remarked: "Television ful-filled the hungriest, most irresistible of simple human wishes—the wish, when mighty and scandalous deeds are brewing, to be a fly on the wall."[5] How true this has been, from the fascinations of Samuel Pepys and Sandra Gotlieb to *Behind Closed Doors*.

How, then, does one deal with this sprawling subject? The most straightforward approach might be to close the academic journals, momentarily silencing their debates about who should control television, what fiction-faction does to the masses, and how to decode or even deconstruct these "texts," and rely on common-sense observation of what an audience is likely to expect when it chooses these programmes. After all, audience expectation suggests the limitations as well as the assumptions by which any producer, director, or scriptwriter works. To reproduce a for-mula, inflect it with imagination, flop into incoherence, or nurse a hybrid, the writer or programme maker has to know what the audience is likely to expect.

HISTORICAL DRAMA OR HISTORICAL DOCUDRAMA

Let us take two examples: *North and South, Book I* and, thanks to good ratings, *Book II* (American, 1985, 1986) and *Samuel Lount* (Canadian, 1986). Both titles promise historical drama of some kind, though possibly too many Canadians would ask themselves, "Samuel who?"—part of the point of making the programme. *North and South* got maximum publicity and was studded with older Hollywood stars in cameo roles; it was clearly designed as heavy artillery in the ratings wars. Viewers escaping all the hype might have expected a serious treatment of the complex issues of the U.S. Civil War, but these were a minority among those tuning in.

North and South is familiar territory. Other miniseries, one being *The Blue and the Grey* (1983), have treated the same general period with a little more emphasis on historical figures. Viewers who had seen *Shogun, Masada* and others, however, got what they expected: plenty of battles; lots of romance; sumptuous costumes; multiple plotlines clustered around the saga of several generations of a family—here, two families, one for each side. Given American politics and the network's distaste for controversy a viewer would also expect that both sides would be treated sympathetically; no particular effort to reproduce nineteenth-century speech patterns; some effort to reflect the social mores of ante- and post-bellum life in the republic; and, above all, a full ration of melodrama. These are the formula elements. When formula is as well executed as this one is, audience expectations are exceeded and ratings soar.

Does it matter to any but the pedant that Lincoln was actually assassinated several months earlier than was convenient for the writers of this opus to portray? The dramatic structure demanded a later date because the writers (or more probably producers) wanted "best friend in blue" to seek and find "best friend in grey" in the Union military hospital. It follows that it is more dramatic for the two soldiers to weep together about the waste and the tragedy, Lincoln's and their own; the scene is also symbolic of the inevitable reconciliation the genre demands. On closer inspection *North and South* is really historical romance, not historical docudrama.

In the same season as *North and South, Book II*, the CBC presented a co-production, *Samuel Lount*, about one of the two Upper Canadians to be hanged for treason after the tragicomic rebellion of 1837. Would a viewer familiar with other documentary historical dramas mounted by the CBC expect heroic mythologizing as in *Bethune, Charlie Grant's War,* or *Grierson and Gouzenko?* Demythologizing, as in *A Flush of Tories* or *Sam Hughes' War?* Or merely some flesh on the bones of half-forgotten history lessons about the events of that year? Already, a clear definition of the genre based on audience expectations is in trouble; this docudrama could explore any of the three possibilities or even all three together. After all, Brecht and Shakespeare could master all three and did, in *Galileo* and *Henry IV*.

Samuel Lount

Given the events of 1837, *Samuel Lount* should be a tragedy as well as a historical docudrama; some CBC topical docudramas have managed this, including *Twelve and a Half Cents, Rough Justice,* and *Turning to Stone.* If this is too difficult to achieve, then at least the programme should hold the attention by telling an interesting story clearly. Not being a historical romance, it needs credible, if not eloquent, dialogue. Initiated by Elvira Lount, a direct descendant of the protagonist, this co-production had many of the elements crucial to success: very good actors, appropriately cast—R.H.Thompson as Lount, Cedric Smith as rebel leader William Lyon Mackenzie, and Linda Griffiths as Elizabeth Lount; authentic locations, sensitively used, particularly the strange, lovely symmetries of Sharon Temple; and a camera which evoked the pioneer landscape lyrically, yet candidly. What the two-hour drama did not have was a good script, and so it simply lost its way. For one thing, it never made clear why Lount was the person he was. He was written as inarticulate, and even Thompson's expressive face and the director's use of eloquent long-distance shots and silent closeups failed to tell us why this particular peaceful farmer picked up a gun and marched down Yonge Street to Montgomery's Tavern and the gallows.

Because Family Compact members were written and largely performed as caricatures, they were difficult to sort out, despite the presence of veteran actor Donald Davis as Bishop Strachan. The dramatic focus was diffused. We were being told how the impulse to rebellion grew, why the rebellion failed, and why Samuel Lount went to the battle, as well as details of the confused events of those few days in December. Numerous strands of theme and plot can be woven together if the writing is good, the characterization interesting, and the narrative drive hard to resist. This production had several good scenes, but little dialogue of substance, much less subtext; the narrative structure was so episodic that the tonal shifts were clumsy, and no suspense was generated. It was unfortunate that the writer did not treat the material either ironically, as a doomed enterprise changing a nation's history, or else tragically, concentrating on the struggle within one man torn between his sense of justice and his pacifism. Did the pacifist Lount think he was damned even as he resolved to join Mackenzie? As he confronted the dead body of a friend at Montgomery's Tavern? When he looked silently at his last sunrise? Hints are given, but even a second viewing makes no such conflict apparent. The protagonist remains opaque.

The drama closes with a letter to the authorities from his wife Elizabeth (using voice-over) as we watch Lount go to the scaffold, and his wife light the fires of Lount's forge after his death. It sounds like an edited version of an actual document. In the last few minutes, the script delivers a passionate indictment threaded with complex overtones of class conflict, religious conviction, righteous anger, nationalism, grief, and pride that are missing from the rest of the piece.

North and South promised formula and delivered it handsomely. *Samuel Lount,* in the fashion of so much Canadian television drama, promises nothing beforehand except good entertainment wrapped around reasonable instruction about an issue, period, or personalities. Regrettably, it did not succeed in that modest but difficult task. Whether a historical docudrama depicts events on the scale of World War II, the muddle of that forty-eight-hour insurrection outside the town of York, the life of a woman who edited a small-town paper in the Canadian West, or Admiral Lord Nelson,[6] the story has to be good.

Billy Bishop Goes to War

John Gray in his theatrical two-hander *Billy Bishop Goes to War* managed to mythologize and demythologize as well as flesh out for us an aspect of history now largely forgotten, with a fine historical docudrama about a Canadian fighter pilot who won the Victoria Cross in World War I. It is called *Billy Bishop Goes to War*. This musical docudrama was a stage hit across Canada and in several U.S. cities: the equally fine television drama, also starring Eric Peterson with John Gray on the piano, was a BBC co-production with Canada's Primedia television, one of the best Canadian independent houses which, as Nielsen-Ferns, produced a series of immigrant docudramas called *The Newcomers* for Imperial Oil. Canadian broadcast rights were sold to the CBC.

This drama relies on research, skilful writing, unusually creative use of music, and a remarkable performance by Eric Peterson, who has done superb work in alternative theatres across the country as well as a variety of roles for television. The television adaptation combined with surprising success television's ability to elide time and space and to deliver special effects with the presentational conventions of the stage version—bare stage, direct address, one person playing all the parts, a toy aeroplane and later a cutout mockup to help the audience visualize the battles. In this case, television's enhancements ranged from a jarringly realistic reconstruction of a bombed-out trench, to superb chromakey effects placing Bishop inside a colourful set of period line-drawings whenever he returned to Britain on leave. Here, stylization inspired by the original stage show gave viewers a break from the apparent naturalism that is too often *de rigueur* in historical docudrama.

Gray's script, lyrics and music and Peterson's performance explore both the making of the myth of Canada's first fully publicized flying ace and V.C. and the darker human side of that myth. We see Bishop, the callow youth from Owen Sound, Ontario, answer the Empire's call only to be exploited and manipulated as a colonial. Worse still, at the play's turning-point, after Bishop has mastered the skill of killing, he discovers he likes to kill, likes it very much. Mythology centred on personality is quite rare in Canada; demythologizing is common.

Like most good topical docudrama, successful historical docudrama tends to present events from a clearly defined point of view, frequently through the eyes of one person or, sometimes, a small, coherent group. The subgenre is much more identifiable when it centres on a real person, particularly if that person is famous. It is a truism of Canadian studies that, as a people, Canadians prefer the collective to the dominant hero, survival to conquest, and are even afraid of, hostile to, or dominated by the immensity of their landscape. Obviously, this last generalization does not apply to *The National Dream,* a blend of contemporary narrative from Pierre Berton and historical narrative. The dramatized sections had scripts by Timothy Findley and William Whitehead, among others. *The National Dream* celebrates the conquest of the landscape and the invention of a new political entity by some quite distinctive and eccentric individuals, among them Sir John A. Macdonald, Major Rogers, and William Van Horne. Some episodes kept a good balance between the grandeur and obstinancy of our geography and the very human men (there are few details about women in this epic) who drove the railway through it. Other episodes allowed the spectacular scenery or the complicated infighting to overwhelm the narrative thrust. Even so, a handful of images linger in the memory: William Hutt's portrait of the private man behind the charm of Sir John A. Macdonald, for example, as he bends over the bed of his small, invalid daughter; or Colicos as William Van Horne, cigar in teeth, confronting empty prairie to the horizon; the sea of mountains; the distant train on the skyline; track cutting toward the horizon, rail by rail; the canyons that killed dozens of Chinese workers; the fetid bunk houses. *The National Dream,* in its own way, did inscribe the landscape of our not-too-distant past on the imaginations of the audience.

How important are the facts? The issue is not, of course, whether a given incident actually happened in that way, on that day, to that person. Historians are well aware that history is, to a large extent, a construct from available fragments, a pattern derived from constantly revised sets of data, subject to change in methodology and emphasis. Docudrama is not documentary. The issue is not whether the historical or, for that matter, topical event happened, but how knowledgeble the audience is likely to be about the subject. If the producer, director, writer, and designer do not signpost what is thought to be fact and what is known to be fiction, the programme is in danger of doing real harm. We will be watching dramas no different than the "lies like truth" which use the same tactics to sell toothpaste and the American way, or create illusions about gallant fur traders opening up the Northwest all by themselves with no Indians to show them the trails and survival skills they needed to make their discoveries.

Few Canadians would have any real apprehension of what life in a bleak city in the hopeless 'eighties is like for the young and unemployed in Britain. Therefore, for a Canadian audience, the information itself would be most important in *Boys from the Blackstuff* (broadcast in Ontario in 1985). We would see the programme quite differently than British viewers. On the other hand, most Canadian viewers over twenty-five years of age watching *The October Crisis* (1975) would be aware of how little information was given them either in the fall of 1970, or in the years after regarding the never proven "apprehended insurrection." They might well have watched the programme hoping to learn something new that would help make sense of that time.

Shakespeare did not explicitly tell his audience when he had rewritten Holinshed in order to glorify the Tudor dynasty. As the sixteenth-century saw it, history's purpose was to instruct, not inform. We live in another age and should expect to know what we are watching. Is it a fictional analogue to real life, like *Boys from the Blackstuff,* about Thatcher's Britain through the eyes of the permanently unemployed, or the CBC's lightly fictionalized *Oakmount High* (1986), banned in Alberta as too closely resembling the James Keegstra case then on appeal? Is it a reconstruction based on actual events, such as *Canada's Sweetheart: The Saga of Hal Banks* or *Samuel Lount,* with real names, places, and events? Is it centred on the writers' speculation about hidden motives and unrecorded conversations, as in the American docudrama *The Missiles of October* in which J. F. Kennedy debates whether or not to call Khruschev's bluff? *For the Record's* speculation about what went on in the oil industry, *Tar Sands*, used the names of real people; premier Peter Lougheed sued. Does a docudrama extrapolate what is likely to happen, as in *Threads* (1984, U.K.), or stretch to a blend of science-fiction and fact as in *Hide and Seek* (about a sentient computer, CBC, 1984)?

In the case of James Keegstra, most Canadians over age twelve have heard or read that this teacher preached antisemitism unchallenged in a small-town Alberta high school for several years. If a viewer is over age twenty, he or she will remember Lougheed; if over forty years, the dread that filled that October of 1962. This means that a significant proportion of the audience can cross-check the interpretation of events offered by the programme with their own memories and opinions.

As we edge backwards through the life span of the majority of viewers over ten years old, the border between topical and historical blurs for more and more of them. The self-correcting capacity to check a given account with memory (however fallible) diminishes. Indeed, we enter more clearly into the realm of received opinion, legend, even myth. We all know that Henry VIII cut off his wives' heads; why and how many, the average viewer probably does not know, and who he was may also be pretty vague. Edward VII was the randy king of a brief sunset age, wasn't he? Wasn't King Edward VIII's abdication over the "romance of the century?" All three subjects breathe sex and scandal. We may well expect titillation. Thus we are likely to agree with George Brandt[7] that "the BBC's *The Six Wives of Henry VIII* (1970), ATV's *Edward VII* (1975), Thames Television's *Edward and Mrs. Simpson* (1978) and—last but not least—*Upstairs Downstairs* (1970-75)" are what he calls "high class costume drama." He argues that, combined with the advent of colour television, these historical docudramas whetted the public's appetite for lush spectacle and lovingly created detail. He also argues that they satisfied viewer partiality for "backward glances cast at a mythological past [which] spring from deeper insecurities. Nostalgia drama satisfies the need to imagine a dreamland of stable class and other relationships as a consolation for the turbulent present." This theory recalls the standard explanation for the popularity of pastoral Arcadias in urbanized societies from sixth-century Athens on as the desire to escape and have a safe outlet for satire or social comment. Several of the dramas in the historical docudrama anthology *Some Honourable Gentlemen* use incidents from our past to comment cogently on our present; thus they do not qualify as "high-class costume drama."

Brandt's assessment of the three series in question is somewhat problematic, however. Viewers of *Henry VIII* and *Edward and Mrs. Simpson* learned a great deal about the complex political issues that coloured those marriages, and observed many details about the historical period as it was lived by the upper classes, and they met seven very interesting women. He does not cite *Elizabeth R* (BBC, 1971) which was distinguished by particularly fine scripts by John Hale, Rosemary Anne Sisson, Julian Mitchell, Hugh Whitemore, and Ian Rodger. Just how these five writers managed to paint one fascinating, eloquent, consistently convincing character in six episodes under three directors is something of a mystery. Certainly, Glenda Jackson's portrayal of Elizabeth was a major contribution to the coherence of the series. The *Daily Telegraph* critic may well have spoken for the kind of viewer who would tune such a programme in when he called it "history without tears and very addictive and successful into the bargain" (quoted by Halliwell). Neither viewer nor historian can know whether Elizabeth ever truly met Leicester at a church, with or without a secret marriage in mind, and certainly not what they said to one another. But the scene as written is plausible, a reasonable speculation based on gossip current at the time. It also addresses one of the most politically significant mysteries about the very private woman behind the public icon: why she did not marry. The episode which dealt with the romanticized conflict between Mary of Scotland and Elizabeth of England was a fine, complex one-act drama, a classic of its kind.

Both the BBC and ATV thought the makers of *Mountbatten* were guilty of speculation about the man's personal life. Moreover, the script rearranged known historical events. *Mountbatten* was not only a portrait of the man and his wife but also presented an account of the transfer of power from the British Raj to Nehru and his coalition. However, many actual events were omitted, condensed or transposed. Its discreet suggestion that Edwina Lady Mountbatten may have had an affair with the Indian leader particularly disturbed the English programmers. PBS in the U.S. had no qualms about showing the series in 1986. The ethics of historical docudrama are not easy to define.

Interestingly, Brandt includes what he calls the "national institution," of the fictional *Upstairs Downstairs* (London Weekend, 1970-75) in the category of "high-class costume drama," but not the fine series, *A Family at War* (Granada, 1970-72), which depicted the strains and heartaches of a working-class Liverpool family without simplifying or oversentimentalizing the change war brought to their lives. The CBC's *Home Fires* was not as successful, though it had its moments. However, with *Upstairs Downstairs, A Family at War,* and *Home Fires* as fairly representative examples of the type, we are not dealing with docudrama—historical, topical, as defined by viewer expectation, or by any of the subsidiary criteria we have been examining. The effect of public events on private lives may be one of these series' themes, but in each of them and many others like them, characters' private relationships are foregrounded rather than the social issues of the times or the effect on society of major historical events. The past is not being explored with the purpose of helping us understand the problems that plague us now; we are not being asked to change our minds about what we think happened in the past, or think about what should happen now.

Do not mistake my meaning. Too many historical docudramas which try to use the form in these ways have been very worthy and very boring. But other kinds of drama masquerading as docudrama have often been used to ride a trend, titillate a weary audience with gossip purporting to be fact, sensationalize and thus trivialize serious issues, as historical pageant or as fanciful and often sloppy recreation. Since many of the latter have been a hit in the ratings on both sides of the Atlantic, I would hesitate to claim, as does Gus MacDonald, Head of Features at Granada, that "audiences demand to be told the status of the information offered."[8] Until we teach media literacy at the grade-school level, this is a pious hope.

The documentary drama as a genre has considerable range. It has become an accepted truism that Canadian audiences prefer this form to all others, whether the medium be film, radio, television, or theatre. Consequently, authors, directors, actors, and technicians have extended it in many directions. The subgenres developed for CBC television are rooted in the fifty-year old tradition of the National Film Board and thirty-five years of CBC news, features, and magazine programmes like *Explorations*, *This Hour Has Seven Days*, and *the fifth estate*, as well as documentary specials such as Allan King's *Skid Row*; Beryl Fox's on-site documentary of Vietnam, *Mills of the Gods*; the very controversial film about the effects of pollution, *Air of Death*, which named places and companies; *The Ten Thousand Day War*, Michael MacLear's independently produced history of the war in Vietnam, and Gwyn Dyer's remarkable series, *War*.

In this book, I have explored with my reader the CBC response to commonly recognized television genres: sitcom, copshow, professional and family-adventure shows. Thanks to the NFB, the CBC, and independent filmmakers, Canadians are acknowledged innovators in the wide world of documentary. It would be very odd indeed if directors like Denys Arcand, Allan King, Ron Kelly, Robin Spry, and Donald Brittain had not carried those skills over into their drama, or if other producers and directors were not influenced by this tradition. Nevertheless, my overriding concern here has been to analyze CBC television drama, primarily from the aesthetic viewpoint, tracing the context as it affects the aesthetics. I will go no farther, therefore, in debating what the nature of documentary/journalistic/ historical/topical drama is; I leave that to Polonius. The criterion I have used in selecting drama for analysis in this section is fairly straightforward: in a drama which is part of an anthology, series or in a drama special, if information rather than personal relationships is foregrounded, it is likely to turn up here. If, as in the case of *Ada*, *Cementhead*, and *Dreamspeaker*, personal relationships or broad human themes are specifically foregrounded, I have analyzed the drama elsewhere.

FAST FORWARD THROUGH CBC DOCUDRAMA

In the crossover between documentary and docudrama, producers often use fully dramatized episodes based on research into letters, newspaper reports, autobiographies, and the like. One of the CBC's earliest efforts, *The Odds and the Gods* (January, 1955), broke new ground. A look at the previous autumn's Hurricane Hazel, the programme used newsfilm clips, location shooting, narration, and a "dramatic" subplot about whether the rookie weather forecaster should trust his judgment and call a hurricane alert. Pop sociology in a scene set in a trailer camp was followed by shots of two real policemen in a squad car out "in the hurricane." Credits included the Cross Town Car Wash. The dialogue and mix of conventions were fairly clumsy and the special effects crude, but the basic format was established.

The Open Grave

Producer-director Ron Kelly's approach to his subject matter in *The Open Grave* (1963) reflected the programme's origins in the Public Affair's Department. It was to have opened, in fact, with no title at all. J. Frank Willis played the anchorman of the dramatized news coverage and familiar announcers Gil Christie, Percy Saltzman, and Fred Davis appeared as reporters: just as when a news event is breaking, the TV crew also appeared on-camera. The story was that an empty grave had been discovered, that of J. Corbett, railroaded to execution at Toronto's Don Jail because his pacifist beliefs were hurting the arms industry. Much of the dialogue was improvised: only principal actors saw what script there was, and fluffs were retained in the final cut. Toronto locations were varied and authentic, the cast of extras huge. As a final touch, *The Open Grave* was to have been shown on Easter Sunday. Between filming and air time, however, word of its content leaked out, and an Anglican archbishop asked that broadcast be deferred. John Diefenbaker demanded an emergency debate in the House of Commons: fortunately, the minister responsible let the CBC make up its own mind. Just twenty-four hours before air time, CBC brass, including the president of the corporation, previewed and appoved the programme. They also added a disclaimer that threatened the impact of the opening scenes. Nevertheless, when shown *The Open Grave* without explanation, I have found that students are disoriented. They are not sure whether the event "happened" or whether the film is fiction, even though the context is a course in television *drama*.

Subsequent events added to the list of ironies surrounding *The Open Grave*: the play was preempted by a Stanley Cup final and postponed to the following Sunday. Reviews were mixed, but most praised the concept and defended the corporation. Eventually, the BBC, West Germany, and Australia bought the programme. There was some controversy in Britain as well, with some of the audience hostile to both form and content. The final irony came when, in the Fall of 1964, *The Open Grave* won the prestigious City of Genoa prize.

The programme itself remains striking in its use of multiple time frames. The chronological treatment of Corbett's career weaves polyphonically through the chronological treatment of the discovery that his grave has been opened; this complements the multiple focus on the issues of how news is covered, the rights and wrongs of disarmament action and civil disobedience, and a rather too detailed analogy to the life and death of Christ. The documentary flavour is retained through the absence of music, the stumbles and repetitions of improvisation, and the conventions of direct cinema. As drama, it sags in places; as a self-reflexive and self-critical examination of the way a news event is reported, it is quite effective for those who look beyond the self-evident allegory.

In 1975, the CBC used the conventions of dramatized documentary in its belated but detailed examination of the October Crisis. Dramatized segments protected the anonymity of sources, recreated key moments, and occasionally lightened the grim story, as when police, searching *Winnie the Pooh* for clues, decided to stalk a suburban house and found themselves raiding a Tupperware party. These segments were not presented in a neutral way, as simple illustrations or speculations about what could have been said and done. Instead, politicians were shown playing billiards with their ethics in a club; and in a scene staged to look like a macho deodorant commercial a policeman told us how and why he went on strike. These scenes were supported by interviews and news clips that confirmed the basic facts. Much less effective, because overly explicit, was the staging of Pierre Laporte's imprisonment and murder.

Ironically, the programme's most dramatic moments, the ones that tend to stay in memory as the essence of those events, were fragments of recollection by real people, such as justice minister Jérôme Choquette's "the arrests were somewhat widespread and inconsiderate"; the freeze frame of Robert Stanfield's honest, rueful face after he admits, "I would feel easier with myself if I had voted against it"; kidnap victim Jasper Cross, head framed by a wing chair, hands very still, voice carefully controlled, as he tells us: "I composed myself for death. What did this mean, the extinction of Jasper Cross?" *The October Crisis* demonstrates that when fact and fiction are juxtaposed, fact may have a more distinct, though subtle intellectual and emotional impact on an audience than fiction.

Recognizing the renaissance of Canadian theatre in the late-'sixties and 'seventies, the CBC tried to adapt the innovative conventions of stage docudrama to television. *Red Emma*, *1837*, *The Farm Show*, and *Paper Wheat* broke away from the naturalistic style of most television docudrama, while Allan King, in a less obvious experiment, filmed Rick Salutin's *Maria* in what scriptwriter Salutin identified as the style of Frank Capra's upbeat movies of the 'forties. In so doing, King achieved a less naturalistic but somehow "more real" look than documentary. Salutin added that the usual, direct-cinema style "gives you the feeling of what it is like to be a

documentary filmmaker in a factory, but not a worker.''[9] The ability of
fiction to enrich, probe for subtext, use different styles of writing—or
filming, in the case of *Maria* —reflects the hierarchies of the worker's
experience of the factory. According to Salutin, the drama part of
docudrama is what makes the film ''real in the sense of someone actually in
there trying his damnedest to distinguish what of the myriad details in a day
of work is trivial, and what is significant and what must be grasped in order
to be changed,'' another interesting definition of the function of docudrama.

For the Record

Maria (9/1/77, prod. Stephen Patrick) was the first *For the Record*
presentation. It is that rare dramatic programme, a film shot in black and
white: this is a convention that in this context clearly invites the viewer to
compare the film with a documentary treatment of the issue. Director Allan
King uses some direct-cinema conventions such as distant shots of workers
at machines or at distant cold corners waiting for streetcars, their dialogue
carried away by traffic or sewing-machine noise. Rick Salutin's script shows
us the fear-filled, often illegal status of immigrants, the pressure of piece
work, and the petty tyranny of some supervisors. However, this is not
''agitprop;'' he also shows us the boss of the Spadina *sh'mata* factory driven
to the wall by offshore imports, his vision coloured by habits of paternalistic
management and battles fought thirty years ago.

The obtrusive music is 'fifties NFB at its worst. However, the sound
effects are selectively naturalistic and convincing. The plot develops swiftly
through a succession of succinct scenes. Maria (Diane d'Aquila), a feisty,
second-generation Italian garment worker, gets fed up, learns to organize a
union from a sympathetic but realistic union organizer, and then tries to sign
up women workers who are busy sniping at one another because they are
Greek, Italian, Portugese, or West Indian. These seamstresses are betrayed
by the men in the factory, who see them as troublemakers—''We make the
money, we're the ones the bosses fear and respect.'' The men are finally
bought off in a special deal. Maria is also pressured to give up her unladylike
activities by her father, boy friend, and mother, who puts it very simply:
''Papa has a union. . . . Papa is a worker. I am a mother.'' She loses the
first round of her battle only to find out that her father is proud of her and
that she has actually managed to scare the bosses. For Maria, there will be a
next time.

This film has been shown to union locals because it works. It tells the
story simply and effectively, with convincing characterization and dialogue.
It does not cast people as grotesque villains or plaster saints, and it works
through to a realistic but hopeful conclusion without fudging on the

difficulties or costs of a union signup. The still timely issue of whether the garment factory can survive in today's market if the workers are given more humane conditions is left up in the air; so is the question of whether it should survive at all. However, the choice between a bad, exploitive job or no job at all is faced squarely in one of the CBC's few coherent looks at key issues confronting labour and management in this country.

Prophetically, just before the early-'eighties recession, *For the Record* probed the dilemma of the unemployed worker. *The Winnings of Frankie Walls* (1979, wr. Rob Forsyth, prod. Bill Gough, dir. Martin Lavut), starred Al Waxman; his previous role as King of Kensington added pathos to the protagonist's character and his painful readjustment to the realities of retraining, the fact that his wife can earn money when he cannot and, eventually, to a job at a much lower wage. Frankie loses his blue-collar job, then his self-image as sole provider for his family and, thus, his self-respect. The film follows his long climb back through the mazes of Canada Manpower and a retraining scheme. Here, again, there are no simple answers.

Maria and *The Winnings of Frankie Walls* make complex issues intelligible and interesting, helping audiences that know a little learn more, and lending voices to the inarticulate. Many in the audience read or listen to reporters, economists, politicians, but it is through fiction that we come to understand what those thousands of words mean to us as citizens.

For the Record (1977-86) was an anthology of CBC docudramas. At its demise, it was also the only anthology left on CBC television in prime time. The two experienced documentary (not drama) producers who got the series started, Ralph Thomas and Stephen Patrick, told the *Globe and Mail* (n.d.) that they were not interested in "dramatized history"; their focus was to be the "human story," on entertainment not polemics, but also on "our contemporary reality" with "some idea of what's happening in other parts of the country." Later, its objectives were identified by executive producer Sam Levene in a "bible" for writers thinking about creating scripts for the anthology:

> *For the Record* is a series of sixty-minute film dramas dealing with contemporary issues affecting the lives of Canadians. These "journalistic dramas" take current and pressing issues, and through the mode of dramatic fiction, interpret them and give them a human dimension. Each *For the Record* programme strives to entertain, challenge, and inform the audience with a fresh and provocative approach to its subject.
>
> The subjects originate either from within the unit through the input of executive producer, producers, script editor; or from outside the unit in

the form of proposals from writers. Whatever the source of the idea, it is a feature of programme development for *For the Record* that the producer remains the driving force behind the development and production of a programme. This feature is of fundamental importance to the concept of *For the Record*, and demands a great deal from freelance writers and directors as they enter into the spirit of creative collaboration.

Various forms of naturalism and rarely, stylized storytelling have been explored in this anthology. Two *For the Record* dramas illustrate this: *A Far Cry from Home* (1981), a passionate statement about wife battering, and *Rough Justice* (1984), a disturbing look at child abuse and how child abusers can exploit the system. *A Far Cry from Home* (wrs. Helen Weils[10] and Bill Gough, dir. Gordon Pinsent, prod. Anne Frank) was broadcast 1 February 1981 as a ninety-minute special opener to the new season. It had no commercials. The writers laid out an arrow-straight series of events from the wedding day to the wife's slow recovery in a shelter for battered women, defining, without really explaining, both the villain and victim. Its strength was to show, by careful naturalism in design, shooting, and directing, the ordinary middle-class environment in which these events could and do happen. Devices like the subjective camera as the wife is beaten and the fragmented glass reflecting her shattered face seemed almost intrusive in the context of this transparent style. As written, the characters were rather thin, although Mary Anne MacDonald and Richard Monette filled them out. Nor was their background ever sketched in to indicate what would make a middle-class businessman assault his wife, and a middle-class gym teacher endure it. Yet the drama had real impact: its enjoyment index was 81 per cent with an audience of 1.5 millon. In 1981 the subject was barely discussed in the papers or at the breakfast table; *A Far Cry from Home* helped change that.

In *Rough Justice* (25/3/84, wr. Don Truckey), producer Maryke McEwen and director Peter Yalden-Thomas decided to displace audience sympathy repeatedly. The narrative structure showed with equal emphasis the point of view of the defence, prosecution, victim, and molester, the process of plea bargaining, and the trial. Sympathy was basically with the child, of course, but by ironic juxtaposition of scenes, jump-cuts, stylized flashbacks, highly charged cinematic imagery, and complex sound texture, the production continuously undercut or shifted our emotional perspective. As with Brecht, whose alienation technique this narrative strategy resembles, the responsibility is shifted from "bad people" to a system made by fallible human beings like ourselves. Aesthetically, it made a better piece of television than *A Far Cry from Home*, yet the dilemmas of child molesting were no less fully explained here than those of wife beating in the earlier programme. No one style is necessarily ideal for a particular genre.

One hazard of topical docudrama is that it can date very quickly. Yet many of the subjects presented in this series have not, as witnessed by the sale of ten episodes to the American Public Broadcasting Service. They include such perennial issues as euthanasia (*A Question of the Sixth*, 1980), the ethics of pro hockey (*Cementhead*, 1979), the election of the Parti Québécois and its effects on both friendships and an anglophone and francophone marriage (*Don't Forget Je me Souviens*, 1980), and the attempts to assimilate Indian children into white society (*A Thousand Moons*, 1977), and even the threat of the Garrison Dam (*Someday Soon*, 1977, wrs. Rudy Weibe and Barry Pearson) are issues still with us. Each of these examples is also a high quality drama in its own right. The "real" world and the world of *For the Record* have overlapped in various ways. *Maria* has been shown to a union local on strike; *A Far Cry* has been used in courses for social workers. A teenager with Down's Syndrome played the lead in the sensitive and imaginative docudrama, *One of Our Own* (1977). It was a real *tour de force* to show us the world through the eyes of a mentally retarded boy portrayed without condescension or sentimentality. *Man Alive* afterwards made a programme about the boy's experience as an actor.

Some of these docudramas are better than others. In a few, issues are oversimplified and the melodrama a little thick, as in *Certain Practices* (1979), *Reasonable Force* (1983), and *Tools of the Devil* (1985). Sometimes, message and entertainment are weakened by whimsy, slapstick, one-dimensional characters, or sitcom dialogue. But *For the Record* always promised, and usually delivered, good drama with a clearly defined documentary flavour. Subjects were current: union struggles, farmers' fights with Ottawa, concentration of media ownership, the boredom of teenagers in a small town full of indifferent adults, the conviction and later treatment of psychopathic murderers, the ethics of experimental surgery, and acquaintance rape.

For the Record was not often didactic, overdrawn, or a series of events strung out on a tenuous thread of plot. Location shooting, tight editing, and an ear for ordinary speech patterns gave the series much of its impact. In many ways it was the CBC at its consistent best, a blend of good entertainment, reliable information, and new insights about familiar subjects. *For the Record* tried to avoid the obvious traps of the genre. The viewer was not regularly afflicted by clumsy exposition, dialogue stuffed with undigested facts and statistics, or bewildering successions of stereotypes. Several episodes have been excellent television drama, genre and anthology context apart. Even more to the point, *For the Record* very rarely broke the contract true docudrama must make with the viewer, the promise to make clear what is fact and what is fiction.

Whether in print, film, radio, or television, that distinction has some-times been dangerously blurred. Canadians were co-producers of *A Man Called Intrepid* (1979), based on the life of spymaster Sir William Stephenson. Edited and shown as a movie abroad, the miniseries opened with news footage of Churchill's funeral. The camera followed the Queen and the cortege up the steps into St. Paul's Cathedral, where the draped coffin passed down the aisle between Michael York—playing a fictional character who had no counterpart in real life—and David Niven, playing William Stephenson, code name: "Intrepid". On the sound track, the voices of the choir and congregation of St. Paul's mingled with Niven's voice in the singing of the hymn. Already, the director had skilfully blurred what had happened, what was being re-enacted, and what had never happened. Halliwell quotes historian Hugh Trevor-Roper: "Not merely a travesty of fact, it is also a wanton insult to living memory and living people."

It may be argued that with such first-rate gossip as *Edward and Mrs. Simpson* or even *Death of a Princess* (1980), which strained relations between the U.K. and Saudi Arabia, such liberties do not greatly matter. In a series like *For the Record*, however, which shows us our fears, preoccupa-tions, failures, and successes on the record, they matter very much. Whether the programme is dramatized documentary, historical anecdote, impressionistic recreation or topical docudrama, the basic distinction between fact and fiction and the imaginative presentation of both is the essence of good documentary drama.

Ready for Slaughter

Producer Maryke McEwen's *Ready for Slaughter* (1983, wr. Roy MacGregor, dir. Allan King) is another example of helping the audience to understand why the news stories matter to us. McEwen took a considerable risk by opening her film with the graphic birth of a calf that is "turned" in the cow. She knew that the urban majority of her audience would be so alienated at seeing how meat and milk actually reach their tables that the wonder of the birth and its importance to a farm might go unnoticed as viewers simply tuned out. Capturing 1.3 million viewers and an audience enjoyment index of 79, her decision paid off. In any case, the risk was justified because the subject of the drama is the reality of modern beef farming. The title says it all: was it the cattle that were ready for slaughter or the farmers, caught in the credit squeeze of 1981-83?

Yet this was not one of the anthology's more didactic exercises. The programme also focused on more universal themes such as the community vs. the loner, and how an innately conservative society can be radicalized by circumstances and its own mistakes. The drama pointed out that banks encourage farmers into upgrading their operations—through huge loans for labour-saving equipment. Will's latest, an automatic feeder, injures his young son, though the symbolism is not pushed too hard. But it also acknowledged that farmers did willingly sign away their futures, believing prices and interest rates would stay the same forever. Gordon Pinsent played Will, a fourth-generation farmer, and Diane Belshaw his intelligent, competent wife. Other elements which kept the play from being too didactic were scenes looking at how and why their marriage endures under the pressure. The emphasis was on Will as inheritor IV and his son, Jordie, as inheritor V whose heifer, born in the show's opening scenes, is the standard 4-H start to a herd. Later, Will's having to sell her underscores how bad things are. The loss was treated unsentimentally: indeed, Jordie's willingness to sell validated his claim to an inheritance that circumstances, bad judgment, and changing banking practices may yet take away from him.

Other fairly familiar character types link the viewer to the situation, notably the old banker who knows the community (Mavor Moore) and a cold, ambitious young loan officer from the city (Layne Coleman). Another element keeping the drama lively is ironic humour: one of the funniest scenes reaches its climax in the reaction shot of Coleman's face when he confronts the struggling, bawling calf left in his sacrosanct office by the more radical among protesting farmers.

Quick sketches show neighbours who fail and sell out, others who cut back or do not venture, and "survivalist" protest groups. Often in the drama, words fail. Will cannot "explain" to the city banker the difference between a job and farming, except to say "it's my life." There is very little dialogue when Will's neighbours are forced to sell their farm and move; all we see is a rainy day with bits and pieces loaded in a truck, a hug exchanged between wives, and a beer shared by the men.

The drama attempts to convey a sense of farming as a way of life rather than a job, and explain how macro-economics become micro for individuals. The central issue of who "owns" land and resources—pinpointed in the feeling of desperation that drives Will to sell part of his herd for seed money even though he has used it as collateral—might have been a humdrum recap of early-'eighties newspaper headlines were it not for the strong sense of ambience. Working before the 1984 budget cuts, McEwen was able (with some difficulty) to persuade the CBC to find the money to shoot in the Bruce Peninsula, Ontario's premier cattle country; the decision was crucial to the drama's success.

From this location shooting come the convincing details of animals, crops, and machinery as farmers see them. Most people, even in small towns, know nothing of the physical facts of farming, its mixture of hard labour, danger, lack of sentiment, high level skills, and deep satisfactions. The wonder of birth is just that, yet also messy, unromantic, and full of things that can go wrong. Crops are commodities that can be burned, not Keatsian harvests from odes to autumn. Yet the drama, photographed by Brian Hebb under Allan King's direction, is also suffused by a strong feeling of the beauty, danger, necessity, and pride of being able to control and pay for those huge, largely unfamiliar machines which move the manure, wheel in rhythmic patterns at seeding time, and obey the commands of the tightly knit family as they join forces. This scene is counterpointed with the penultimate scene where survivalists and neighbours on their enormous machines come to Will's rescue in his confrontation with the bankers.

Another theme is the mass media; its indifference to the issue until something photogenic happens; and then the farmers' fight to control the message. Here, the media are another useful tool, no more and no less. Oddly enough, real changes in the community's politics were triggered by the shooting of *Ready for Slaughter*, partly because details were reported in the local Owen Sound *Sun Times*. Later in the process, when beef farmers in the Hanover-Durham area saw the programme in preview, the Durham *Chronicle* reported they liked the treatment of their dilemma, except for the beer drinking and rough-housing at the dance.

The process which produced *Ready for Slaughter* reflects both lessons learned over the years and the reasons for success of many of the dramas from *For the Record*. McEwen wins the battle to go on location and selects Lion's Head on the Bruce Peninsula, where a prosperous farm family, the Ceasars, are "cast" for their co-operation and source material. The anthology's staff researchers collect material about the issues from the area, and CBC casting finds local people for scenes at the auction, the square dance, and the bank. The designer adapts what she finds and duplicates the "look" in her costumes. As production begins, the cast closely observe the Ceasars and adapt idioms, inflections, attitudes; they also develop a sense of involvement with the community. The crew, up from Toronto for the shoot, settle into the ambience and quickly become sensitized to the issues addressed by the programme. Perhaps the isolation helps them do better work or, at least, more relaxed yet intense work, to get involved and become a "company".

The only person missing (as always on Canadian sets—in contrast to BBC practice), that is, the scriptwriter, had already recognized that young men can and do brawl and older men may still park in a pickup truck for a romp with their wives, despite the problems presented by the stick shift. He also had a sense of the meetings, dances, auctions, and the other details which fleshed out in production give the programme its credibility.

Obviously, given all this, King needed rapport with his cast, particularly with the less experienced members, but also with the community whose life he was capturing on film. Not everything in *Ready for Slaughter* is "authentic"; the people of Lion's Head had to revive the older type of square dancing because they don't do much of it any more. Yet somehow, all the variables added up on screen to a seamless meld of experienced Gordon Pinsent, new-to-television Diane Belshaw, various good character parts by young, not overly familiar actors like Booth Savage and Layne Coleman, with the whole enterprise balanced by that chameleon Mavor Moore. King achieved a clean, unmannered performance by the completely inexperienced kid who played Jordie as well as convincing one-liners from the people of Lion's Head. There is even a Brechtian "in" joke when the Ceasars bid on Will's cattle at the auction barn. McEwen made the wise decision to use a real auctioneer, and a real caller for the dance. Watching the rough cut, I found myself enjoying the clatter of the machines without music and the less polished sound of the local square dance band more than the lush music and balanced sound effects of the final soundtrack.

Ready for Slaughter, which won a Rockie as the best drama at the Banff International Television Festival that year, is a good example of traditional docudrama technique. It was the result of careful research, location shooting, adding authenticity to the piece; it was also the result of the traditional dramatic virtues of clear narrative structure, good pace, variety of tone, a satisfying and credible climax, and performances of high quality, based on a script which sparked creative responses from director, film unit, editors, and sound mixers alike.

Blind Faith and Pray TV: Two Cultures Look at the Same Subject

Producer Maryke McEwen was also responsible for one of the best examples of how CBC topical drama differs from most American topical movies of the week. *Blind Faith* has its flaws, but it addressed what was then and still is a hot issue—televangelism—within weeks of an American drama on the same subject.

In fact, the CBC's *Blind Faith* and ABC's *Pray TV* (shown on CTV), were both telecast in the 1981-82 season. Each throws some light on the differences between two cultures, sharply reflected in their dramatic treatment of subject, aptly drawn from the medium itself.

Blind Faith (28/3/82, wr. Ian Sutherland, based on a story by Edward Cullen, dir. John Trent, prod. Maryke McEwen) was part of a specific broadcasting matrix which, over a couple of years, had included news clips of Canadian Radio and Telecommunications Commission hearings on the use of new satellite channels by television evangelists and exposes of the worst excesses of television evangelism on the news magazine the *fifth estate*.

During the winter of 1982, the *fifth estate* examined the seamier side of television evangelism, telling Canadians, among other things, that we were sending significantly more money per capita south of the border to the Jim and Tammy Bakkers, Jimmy Swaggarts, and the Oral Roberts than Americans do at home. The CBC mandate specifies that controversial issues be treated with impartiality over the whole broadcasting spectrum, not necessarily in any given programme. So *Blind Faith*, like other *For the Record* dramas could take a specific point of view. The final component in this context, then, is the fact that every day, and on Sundays every hour, on every channel available in Canada, we can find both homegrown and American television evangelists, ranging from the comparatively low-key *100 Huntley Street* to the excesses of *The Ernest Angley Hour*. Note that *Blind Faith* was developed and filmed six years before the 1987 Jim and Tammy Faye Bakker/Oral Roberts scandals errupted into what the media called "holy wars."

Blind Faith's limitations were not the same as the ones that applied to *Pray TV*. The CBC did not have to ask, "Will it sell? Will the affiliates cancel out? Will the Moral Majority boycott the sponsors? the station? the network?" Constraints on *Blind Faith* were derived first from *For the Record* itself and the expectations associated with it, then from the laws of libel and slander. Nor were the contexts similar: Canada does not have an organization comparable to the U.S. Moral Majority as an organized political force, nor do we allow whole channels to be licensed for fundamentalist evangelical programming. Since the 1930s, in fact, the regulatory agencies have refused to allow any group with a specific religious or political affiliation to own a station or claim a wavelength for its exclusive use. Clearly, the context of *Pray TV* as shown on ABC was very different, in regard both to the society addressed and to the commercially oriented broadcasting system itself.

Pray TV was broadcast in Canada two or three months before the CBC showing of *Blind Faith*. It was thus fairly fresh in the minds of people who saw both. Neither programme represents the very best in topical drama that either culture can create, but both dramas may be fairly considered as typical of the better efforts in the genre and the ways the two cultures are likely to respond to very different audience expectations.

There are some striking similarities, in both production techniques and thematic emphasis, between the two programmes. One of the most obvious, and one emphasized particularly by the publicity surrounding *Pray TV*, was the decision to cast against type. In *Pray TV* John Ritter, master of the double-entendre in the successful sitcom *Three's Company*, was cast as Tom, the earnest young pastor, fresh from Bible school, who eventually loses his illusions but not his soul. I assume Ritter was chosen to coax people into watching; also, his puppy dog sexism from *Three's Company* was fairly harmless and not liable to radically undercut his role as a conservative Baptist minister which he played quite well. A different decision was made when Canadian Maryke McEwen chose Rosemary Dunsmore to play protagonist Janet Butler, Holton's convert in *Blind Faith*. Not then a well-known actress, Dunsmore carried no audience preconceptions with her and was thus an appropriate *tabula rasa* on which to inscribe the character.

In *Pray TV*, a handsome pastor is protagonist; in *Blind Faith* a thirtyish, moderately pretty housewife is protagonist. The contrast highlights the two different approaches to the subject and perhaps indicates a difference in attitudes to the audience. On the other hand, the antagonist was cast against type. Heath Lamberts is a gifted comedian, familiar to theatregoers at the Shaw Festival and elsewhere. Normally perceived as everybody's good guy, in *Blind Faith* he was cast as TV evangelist Dr. Paul Holton. Rex Humbard, Oral Roberts and Jerry Falwell are as down-home, earthy, and friendly as is the side of Lambert's *dramatis personae* most familiar to audiences. He had just played a jolly monk in a commercial before this presentation. In *Blind Faith*, the new aura of authority and good sense built into the character, combined with the reassurance he had already projected, gave Lambert's Holton considerable credibility. In this more complex treatment of the subject, the preacher was not necessarily the villain of the piece.

A short plot summary of *Blind Faith* is in order. Janet Butler—in her early 'thirties, mother, wife to an upwardly mobile advertising executive—spends a lot of time alone in her expensive house watching televangelist Dr. Paul Holton. Gradually, we learn that her strict and very religious father had rejected her lifestyle, she has had psychiatric help, she has been searching for a strong male figure to depend on most of her life, and has few close friends or outside interests. She does have lots of household help to look after her small daughter, baby, and David, her workaholic husband. In the film's first half, we see Janet trying to cope with her father's death, David's neglect, and the trendy lifestyle required by his business. Slowly, she becomes involved with Holton, first on television, then as a member of his studio audience, and then even more active as a volunteer and on-camera witness for Jesus. By this point, the focus has shifted to Holton's operation, both on and off-camera. A polyphonic set of images supplies counterpoint. Janet's "live" presence contrasts with Holton's television image. We see his charismatic impact on Janet when he meets her in a scene juxtaposed with her husband David in a viewing room analyzing a sausage ad for flaws.

For a while, David's first cynical, then violent reaction to her growing interest in Holton's work cuts her off from the televangelist's world. In a climactic sequence, however, she makes a decision, goes to the studio, experiences a personal conversion, and withstands her husband's renewed threats to leave if she continues with Holton's organization. In the last scene, despite his "me or them" ultimatum, she returns to the studio; he is left in the foyer, where a monitor showing Holton's programme takes over our screen as he goes out the door.

Both *Pray TV* and *Blind Faith* connected television evangelism to ad-cult. Janet moves from the workaholic, numbers-oriented world of her husband, where love is bought with cheques, to the obsessive, numbers-oriented world of TV evangelism, where salvation can be bought with cheques. Since sausages and souls are regularly equated in both worlds, human relationships in both are subject to images redesigned to fit the demands of the sponsor.

Yet, characteristic of the generally more complex CBC treatment of such things, it's not that simple. Holton's assistant, Peggy, really seems to care for Janet. In a key scene, we see Paul Holton ask his most recent consumer (now user), Janet, whether his current style matches his message. Janet thinks so. Satisfied, he rejects his consultant's suggestions for slicker packaging for his show. Is this out of pride? or out of conscience? or because Janet confirms that his current methods work well? In *Blind Faith*, the ad-motif is pervasive, yet treated ambivalently.

It is equally pervasive in *Pray TV*. The ad-cult connection is central to two of the most important scenes in the film. The close connections between Freddy Stone and conservative American politics build to a rather predictable campaign scene where Jesus and politics, the national anthem and the biblical message are intermingled. The mechanics of handling the money and selling the product are more interesting. Rapid editing, together with television's eternal love affair with any technology that moves, give these sequences a truly seductive quality. Thus, we are as shocked as Tom is to find that the "missionaries" to South America for whom he did a spontaneous appeal on Freddy's show are not people, but channels on a satellite, bought to sell more of Freddy Stone's gospel. The drama's message is as clear as Stone's: two cents a Brazilian soul.[11]

However, the sociological and stylistic links between the consumerism of the secular world and the tactics of the world of the evangelist are never made clear in *Pray TV*. Nor do we ever really discover why people are vulnerable to this kind of message. Instead, the emphasis is on mechanics: specifically, how the message is pumped out. The drama does capture the chilling rhetoric that is part of the television evangelist's stock-in-trade very well. Stone says: "How do we do good to the secularist, the humanist, the pornographer, the feminist, the homosexual, the abortionist? We pray for them, bless them, love them and we vote them out." What we do not discover is what motivates Freddy Stone and his conservative allies—money or power or both. We do not even know the source of his hold over his faithful and ever-growing flock. What we do know, soon after the programme gets under way, is that, in the end, the clean-cut idealist is bound to reject these corrupt and hollow methods of spreading the Gospel. Tom's own brand of religion, conservative evangelical, implies a two-edged decision: it means that the conservative viewer of similar bent will stay with the programme and hear it out, but it also co-opts the doubts expressed in the play, making the issue more a family quarrel than the near civil war it has become within Christianity.

The obligatory romantic interest, in the person of an agnostic young woman, provides dialogue which ensures the audience understands the issue: "Whose path are you on? Gus Keefer's or Freddy's? Up with the top ten, golfing with the President or ministering?" The narrative line even asks us to consider whether the girl friend's widowed mother should give her business to Freddy Stone for an annuity and a blessing. Of course, Tom's true mentor is small-town pastor Gus Keefer, first seen planting flowers around his church. The resolution of Tom's conflict and the answers are plainly a foregone conclusion, though it takes *Pray TV* two hours to get there.

The oddest and most truly unexpected aspect of *Pray TV*, for a Canadian conditioned to CBC treatment of such subjects, is the drastic undercutting of the final confrontation between Tom and Freddy. As the scene progresses, Tom is framed by a huge, empty film screen at the end of a long conference table; Freddy is backlit by windows and surrounded by buttons to push. This is the setting for Tom's "climactic" lines: "You're breaking some moral laws here under the cover of spiritual guidance. It's a betrayal of trust. I see you ending in silence." The audience is compelled to agree with Freddy when he points out that these lines really don't add up to much of a denunciation.

Then, creating confusion rather than ambivalence, the producer decides to have Freddy reply to Tom's accusation by citing passages from the New Testament book of *Revelations* and *Mark* saying that Christ will not come again until the Gospel is preached to all nations. Freddy Stone's organizations must build satellites so that Christ will come again, right now. The result of the confrontation is that Tom and Freddy shake hands and exchange "God Bless You"'s. As some American reviewers rightly observed, the programme's blunt thrust up to this point was largely undermined. The director seemed to realize this, however, for he added a succession of rather gratuitous and obvious symbolic images to close the scene. As Tom leaves, Stone pushes a button to close the curtains so that Tom, framed in light, shuts the door leaving Freddy in darkness. It is at this point that the punning names become clear. The naive disciple has become the doubting Thomas who will eventually be an Apostolic Tom, while Freddy appears to be offering the flock a stone instead of bread. Nevertheless, we are left with the fact that the key confrontation has been significantly diluted.

The drama ends with Tom using a different text from the same Gospel of *St. Mark*. As he preaches on small miracles in Keefer's church, the camera lingers on congregational hands touching one another. The message of the drama then becomes: "Go you [live and in person] to all nations and preach the Gospel." The last sequence is a cross-fade between Tom preaching and another obviously ironic reprise of Freddy's opening gambit, "It's going to be a beautiful day." *Pray TV* is a clear story, told simply, with reasonably good performances, impressive production values, and a very straightforward message which is unfortunately sabotaged by a failure of nerve in the last ten minutes. By the end of the play, the point seems to be that "good guys make house calls." Not a show to raise a storm of controversy and dissent, padded by subplots and rather too long, *Pray TV* was basically what most viewers have come to expect of topical drama on American network television.

Like *Pray TV*, *Blind Faith* did not raise much public controversy. I did find disagreements among students, friends, and colleagues over this piece enlightening. Unexpectedly, *Blind Faith* seemed to be a touchstone for attitudes to religion, women, and marriage, as well as a measure of the viewer's experience (or lack of it) in watching television drama. One of my students, a housewife, identified completely with Janet; what happened to Janet frightened her. Few students defended the theology and politics of television evangelism in all its forms. Some, responding to the only real taboo left in academic circles—that is, almost anything to do with religion—hated the piece for not exposing the hucksters of superstition with appropriate moral outrage. They wanted the piece to adopt a recognizable attitude to its material, not an ambivalent moral stance. In fact, they wanted a version of *Pray TV*.

The feminists in the classes were particularly unhappy because they thought Janet was presented primarily as a victim, and yet she was from the educated, upwardly mobile middle-class rather than safely distanced as lonely, elderly, and poor. Those who had seen the *fifth estate* segment knew that such lonely old women are the most likely dupes of an unscrupulous TV evangelist. The film seemed to threaten their sense of what constituted normal behaviour. Perhaps atypically, few of the students had actually seen TV evangelists of any hue.

My colleague, film critic Maurice Yacowar, thought the film brilliant because of its montage. It is true that self-reflexive use of television as a medium is too rarely seen. Another film colleague, Joan Nicks, saw the influence of Jean-Luc Godard in the rather bland, initially featureless character of Janet. She saw that blandness as an invitation for us all to project our own hopes and fears onto Janet—a point of view my students' varied, violent reactions seem to bear out.

My own initial response was atypical of the rather unscientific sample of reactions I solicited. Perhaps after viewing many hours of CBC docudrama I had seen too many workaholics, neglected wives and empty marriages, too many wide-eyed children pushed into the story for the sake of exposition, too many tyrannical fathers and trendy friends. (I am glad to say that repetitive use of these particular stereotypes ended about that time.) I was also bothered by the tight editing: repeatedly in *Blind Faith,* the build of a scene was frustrated by an abrupt cut. If, in *Pray TV,* some scenes were rather flaccid or overwritten, certain scenes in *Blind Faith* were too elliptical. Yet the narrative repeated the same points of exposition over and over, and I was able to predict several of the plot turns.

The most balanced reaction to *Blind Faith* came from a non-professional, my then sixty-eight-year old mother, who happened to be statistically typical of the largest audience viewing segment for *For the Record* —that is, over fifty and female.[12] She was particularly fascinated by the ease with which Holton precipitated Janet's crisis and then channelled it into a sweet and inflexible serenity, which she perceived as an obvious analogue to Holton's own power trip. Certainly, there are times when one quite likes Holton, and this surprising response underlines the play's major strength: its genuine ambivalence. *Blind Faith* is good topical drama where it particularizes the issue of television evangelism, explores the emotional and intellectual crosscurrents televangelism provokes and creates a desire in viewers to form his or her own judgments about the situation presented. "About six out of every ten viewers," CBC Audience Research found, "are opposed to basic ideas behind the evangelical approach, while four out of every ten were either in agreement or were undecided as to its merits." When enjoyment of each of the characters was indexed, Janet garnered 72, David 64, Al "the crude, success-oriented associate" 44, and Holton 41: "These figures do not imply that the viewers were dissatisfied with how effectively these characters were portrayed. The low enjoyment was simply a measure of the fear or dislike these characters engendered among viewers."[13]

It's worth noting that a number of viewers offered unsolicited comment. In the careful prose of the CBC Audience Research Department, Dr. Paul Holton "made them feel uncomfortable," although "almost seven out of every ten viewers thought the programme was fair in its portrayal of how evangelists work." The *Blind Faith* audience was plainly very different from the mass audience the makers of *Pray TV* wanted to deliver to their sponsors. At a screening of *Blind Faith* at the Second International Conference on Television Drama at Michigan State University, an American producer expressed astonishment that the network or sponsors would allow the character of a Jewish adman to make cynical, irreligious wisecracks to a professed Christian. Had it been made at ABC, three factors would have caused executives to cut the scene: the profanity, the wisecracks themselves, and their attribution to a Jew. Yet my Jewish husband did not even register the scene as problematic, any more than I had reacted as an offended Presbyterian to Janet's bible-thumping father. Moreover, the scene as written and filmed was crucial to the narrative structure.

Where *Blind Faith* is indirect, it works well. Unemphatically, but with a deft touch, the drama makes the ironic point that Janet's open testimony broadcast over television brings nothing but outrage and laughter from friends and associates of her husband; it may even affect his career. The scene itself is treated sympathetically. In fact, her surprise and shyness, warmth and spontaneity open up questions about whether, in her case, the end may justify the means. Certainly, questions immediately arise about whether or not society's laughter is justified. The change this experience brings to her point of view on the manners and mores of her milieu is also sketched swiftly, by inference rather than exposition, at the party and, later, at a dinner David stages to sort things out.

As the film progresses, we scarcely notice that we do not in fact see Dr. Paul Holton in colour and in the flesh until he becomes real to Janet—that is, until she actually comes to the station to see him in person. We hear his voice only and watch her watching him. Even at the station, we first meet his electronic, black-and-white shadow on a monitor over her shoulder before we actually meet Holton. In fact, when we first see this electronic image, the camera is placed where the evangelist would stand to look at her.

The locations and decor also show imagination. The costumes make statements about the characters: for example, Janet's own choices of long sleeves, turtlenecks, and quiet colours contrast with her red party dress, obviously chosen to please her husband. At the end of the drama, her hair is shown in the mid-Edwardian upsweep which earlier signified her role as hostess for her husband; now, it links the hostess role to her new role as counsellor on Holton's show. Other subtle touches include the photograph of her father tucked between the pages of his old family Bible, which tells us all we need to know abut the rigid, paternalistic background that oppressed her.

The climax of *Blind Faith*, Janet's personal crisis, has been as carefully prepared as the confrontation between Tom and Freddy in *Pray TV*. In the two scenes preceding her meeting with Paul Holton, we see Janet sitting outside in the dark, hearing and reacting to the conflicting voices of Holton, child, friends, and husband as voice-overs, each making contradictory demands on her. Cut to a darkened and empty room and a television set tuned to Paul Holton preaching on salvation. In the distance, at the end of the lit corridor, Janet appears; she wanders very slowly down the hall, disappears into another room, comes toward the doorway. There is no image of Holton after the establishing shot; only a voice. After Janet has made an eloquent gesture of confusion and despair, the camera follows her across the room to the couch, where she sits in a closeup of desperation and tears, with the voice-over, the "words of personal salvation," sounding in her ears. In this scene, the writer, director, and author trust the actress and the viewer to do the work. Then, for the first time, we see her make a completely independent decision. Cut to Janet heading through the halls of the studio. She almost demands that Holton see her, hauling him out of the dressing-room business conference. This confrontation does not turn out quite the way we expect it to though not because of a failure of nerve on the network's part. Rather, the drama's ambivalence about the relationship between Holton and Janet Holton as a television evangelist explain the shape of the climax.

First of all, Holton does leave the world of television charisma in response to Janet's demands. They talk in a stairwell, which may be read as the modern escape route in any high-rise building, or as the ancient metaphor for humanity's precarious rung on the ancient ladder poised between Heaven and Hell. Then, in a tense, tight two-shot, with Janet one step above Paul Holton, he uses the gestures and words of the traditional conversion. Her hysteria is observable: so is the genuine truth of her conversion. The conviction is total, the joy is real; the direct hands-on touch (with the camera at a low angle looking up at both) results in a full-scale emotional release. Most unexpectedly, the conversion takes place in private, with no television cameras telecasting the event. The audience is left unsure, not about the effect of the experience on Janet, but two key issues, the lasting power of this kind of experience and Holton's motives. Is the surrender to Jesus or Paul Holton? We know Janet Butler goes on to join his organization, but is that necessarily bad? Holton is not *Pray TV*'s Freddy Stone. He's treated with much more complexity.

Other factors also complicate the issue: for example, what are we to make of the reversal in which Janet defies her husband, an incident that provided a lot of satisfaction for some feminist members in my classes? Are we to take Holton's voice on the monitor, quoting Christ's words about how his followers will be hated and reviled, as counterpoint or as reinforcement to the scene in which Janet exhibits her newly won inner peace in the face of David's sullen resentment? David's refusal to discuss or understand what has happened further frustrates the viewer's impulse to come up with a simplistic answer to the film's questions. Is Janet on a power trip? Is she Holton's puppet? Has she acquired genuine inner strength? *Blind Faith* has the potential for a multi-levelled, open-ended narrative that leaves the viewer to sort things out for her/himself.

Response to *Blind Faith* depends in part on whether a viewer has thought through the questions raised by the television evangelism phenomenon. If he/she has done so, then the viewer will be able to concentrate on the nuance and ambivalence in the drama; if not, the play seems to create a shock of self-recognition, that brings unconscious hostilities to the surface. But it may simply serve as a source of new information about an unfamiliar topic as well as being good entertainment—all legitimate functions of topical docudrama.

Unlike *Pray TV*, *Blind Faith* is not about what happens in TV evangelism, or how; it is about why it happens. In this respect, the programme differs significantly from American docudramas, which are more likely to ride trends than swim against the tide. When *Blind Faith* concentrates on the changing relationship between Paul Holton and Janet Butler, it is subtle, fresh, and unexpected. When it shifts emphasis to a black-and-white cartoon of a workaholic, "claim-his-rights" husband and a neglected housewife, it diminishes its thrust. The obligatory romance in *Pray TV*, which does have some nice dialogue and good performances, also splits the focus for part of the time. In *Blind Faith*, however, it is almost as if we were watching two plays. One is about how a husband should make a more interesting doll's house for his wife to live in, or else the Big Bad Wolf will gobble her up: the other is about a complex and growing phenomenon in modern life, the secularization of society. We see the white wine and Monserrat vacation lifestyle come into conflict with one of the new forms of religion that are part of modern society's answer to pervasive personal disorientation.

In choosing to look at the evangelists per se, *Pray TV* ducked many questions about who supports them and why. It did seem quite clear about who the good guys were: yet, in the last minutes, the bad guys turned out to be not so bad, presumably so that sponsors, the Moral Majority and middle America would stay happy and tuned to ABC. *Blind Faith* made judgment less easy. Its last two minutes, however, which included the credits, nearly undid the complex dramatic effect. Running behind Janet's confrontation with David, her refusal to submit to his emotional blackmail and his despair and doubts, we heard and then saw, on a monitor which slowly filled our screen, Holton in full flight on the "end times," a favourite televangelist's catch phrase. We heard-saw him denounce modern society, threatening organized action with clear allusions to the political values and tastes of Falwell's Moral Majority. Just at the last minute, through hysterical delivery and distorted camera angles, the producer and director forgot to trust the audience and told viewers how to react to their drama.

A final note: the *For the Record* producers seem to feel it important that their subjects not feel betrayed by the presentation of the issue. Since they are, by definition, likely to be strangers to that issue, however interested, it is an impressive accomplishment if those concerned are made to feel they have been depicted fairly. *For the Record* assumed a serious, intelligent, curious audience of varied background drawn largely from the middle-class. It survived for a decade as the reliable source for a range of contemporary, relevant anthology drama because, on the whole, the producers and the three executive producers (R.L. Thomas, Sam Levene, and Sig Gerber) also understood the importance of making an entertaining hour out of their subject. Entertainment, in the sense of particularized characters with lives beyond the topical issues at hand, is what catches and holds the attention of an audience, drawing it into areas that are challenging or unfamiliar. When we care about the woman who is dieting herself into today's anorexic fashion, or the farmer fighting to stop a dam that will threaten his livelihood, then we willingly think about the issues. It is a difficult balance to achieve, but that balance is both the strength and limitation of docudrama, whether contemporary or historical.

DOCUDRAMA OF BOTH KINDS: CHAUTAUQUA GIRL AND TURNING TO STONE

By way of summary, I want to take a look at two mid-'eighties docu-dramas that show what the CBC can do to lift this form above the routine. These are not historical romances, nor earnest recreations of our past in which the central characters somehow fail to come to life. They are not sensationalized, trivialized formula pieces which plug into some faddish social concern. The American and British contexts are not particularly informative for examining either of them and, as with so many other examples in this book, I am not able to do them full justice in the short space available. As with every piece of really good television, we can hear, if we listen for it, a personal voice behind the camera: we cannot, should not try to ignore these quirks in order to place them neatly on the appropriate branch on the docudrama family tree.

The two television films are Jeannine Locke's *Chautauqua Girl* (8/1/84, wr.-pr. Jeannine Locke, dir. Rob Iscove) and John Kastner's *Turning to Stone* (25/2/86, wr. Judith Thompson, dir. Eric Till, prod. John Kastner, exec. prod. Sig Gerber); Note: Just this once I have extended my cut-off date for programme analysis to include *Turning to Stone*. Both have plots that concentrate on tangled human relationships and motivations. Both take as their central focus a relatively unknown area of Canadian life. One belongs to the branch we have been calling historical docudrama based on careful research; the other is absolutely contemporary and also carefully researched. In all other respects, except for the skill and integrity brought to them by everyone involved, they are very different. One ends on a hot summer night in the 'twenties when the Chautauqua circuit breathed life and laughter, fun and uplift into cramped lives on the Prairies. The other contains a fragile moment of rapport as middle-class protagonist Allison and Lina, the "boss" of the cell block, see, feel, but cannot reach the softly lit summer night through the barred windows within an anatomy of a brutal and brutalizing environment—Kingston's maximum-security prison for women.

People all over Canada who are old enough to remember the 'twenties and early-'thirties have fond memories of the annual visits of the Chautauqua circuit. As the drama explained for younger viewers, university students—usually women in Canada, though not in the United States where the idea originated and where some of the Eastern circuits were based—went into small towns to find financial guarantors and organize the logistics of the week-long activities. These young women got the kids involved in parades, projects, cheers, and songs, but their most important function was to work with local people selling tickets. If they had done their job well, by the time the tent and then the performers arrived, everyone for many miles around could hardly wait for the week's events to start. The programmes included speakers on uplifting topics, lectures by well-known figures ranging from famous explorers to British feminist Emmeline Pankhurst, string quartets, bell ringers, baritones, small vaudeville dance acts, and short melodramas which might well involve local actors as extras. Their impact was summed up by one farm woman, isolated for months at a time on a prairie farm and exhausted with overwork, who said simply: "They kept me sane." As Sally, the protagonist, remarks in the film, for many a Prairie child this was the first chance to hear a real violin, as opposed to a homemade fiddle. For many communities across Canada, Chautauqua was the highlight of the year.

This docudrama offered different kinds of pleasure to different segments of its audience. For the small audience of theatre buffs, it was a faithful recreation of an almost forgotten part of our cultural history. For many others, it was pure nostalgia, a happy moment long gone but lovingly and accurately recreated. For some, the script's other focus on how the United Farmers Organization gathered strength and generated real hope for change would provide a new perspective on populist politics. Nearly every viewer would enjoy the Chautauqua entertainment itself, which takes up the last fifteen minutes. Locke chose to show us good-sized samples of these acts, recreating in this sequence the more leisurely rhythms of that time. Yet she and her editor also knew when to cut away to audience reaction or some by-play related to the love story.

Part of every Audience Research Report is verbatim comment from the audience research panel answering the questions "What was there that you found most [or least] appealing?" In the report for *The Chautauqua Girl* my rather speculative conclusions are borne out by the verbatim comments.

The romantic interest is provided by Sally Driscoll (Janet Laine-Greene), a young girl who is trying to escape her own farming background by going to University in the city, campaigning for the first commitment to Chautauqua from the small prairie town and then solving the inevitable crises of organizing Chautauqua performances. Neal MacCallum (Terrence Kelly), who eventually gathers courage to propose to her, is a widowed farmer who battles for fairness to agriculture by running for the new United Farmers' party. There is no doubt that the stumbling, gentle romance which develops between them makes the tone, texture, and issues of life at that time very human and very specific. Pride in the West, a new nationalism, a changing way of life, the sense of a second generation taking hold—all are there.

If one were on the hunt for subtext, it could be found in the sequence of Chautauqua acts which Sally and Neil, separately and then together, watch during the week: "The Treble Clef Four—Musical Merrymakers" sing "Love's Old Sweet Song." The talk on "Acres of Diamonds" urges people to find the riches in their own back yards. The last item, a delightful parody of the inevitably happy ending of this television drama, is the last scene of a melodrama in which "Southern belle" Mellisande is shot to death by "jealous rival" Jerome in the very arms of her true love, Sean—and on her wedding day, no less. If audience expectations define various forms of docudrama, then viewers who wanted a nostalgic trip into the past were satisfied, while they were also given a clear sketch of the rise of the United Farmers' Organization on the Prairies. It had over two million happy viewers, a successful rerun, and lots of foreign sales. It was also very good.

The scene in *Chautauqua Girl* which broke through my own critical detachment was wordless. The drama has introduced us to Mrs. Ferguson (Jackie Burroughs), an old Scots lady, too arthritic to make the trip to town, whose husband is "awfully solitary since the deafness." Her daughter, Nettie, has escaped to a good job in the city and sends money for Chautauqua tickets. According to Mrs. Ferguson, Nettie has told her that "it's like being taken on a flying carpet to London or New York [with a smile] I'd prefer Edinburgh." Nonetheless, the old lady refuses to leave her husband. We pick the scene up initially in silent longshot; then in the flat prairie distance, we hear the putt-putting of a Model T filled with five men in full Scottish evening dress. Cut to Mrs. Ferguson, stooped in her garden, with the bared teeth of a hay rake for a backdrop. We hear a string quartet and accordion begin "Unto the hills." There is no dialogue. We see her come forward, guided by Sally and Neil to a chair, where she settles back listening to a fine baritone sing every verse of the hymn. The camera is not busy; it moves to the rhythms of the music. Then, as the old lady silently struggles with her emotions, again we first hear, then see, a piper lilt into the

Skye Boat Song as he swings up a long furrow beside a field of grain. In this scene, Locke trusts her director, her cast, her own instincts, and the audience when she chooses to savour rather than hurry the moment.

The most harrowing scene in *Turning to Stone* is also wordless and completely dependent on its performances, pace, and build. With the inevitability of good tragedy, the protagonist, young first-timer Allison (in a fine performance by Nicki Guardigni) is raped at knifepoint as an act of vengeance by Sherrin (Anne Anglin, in a chilling yet very human portrait of a psychotic): "You stole my person, then you stole our things." We hear a radio playing loudly as Sherrin terrorizes Allison, cutting her bra off with a knife. Then as the matron is distracted by an incoherent argument staged by Lina's cohorts, there is a cut to Allison's profile on her back in her bunk with Sherrin's knife bisecting her face, millimetres from her eyes. Cut to a brief shot, from Allison's point of view, of her own hands undoing the fly of her jeans. We then see and hear Dunky (a sympathetic character) behind the barred door at the end of the corridor, who notices Allison's jeans tossed on the floor outside her cell. She tries to get action from the guards, who have been distracted by Lina's lieutenants. We see Lina, herself, listening, and three unknown inmates listening outside the open curtained cell door, but when Dunky shouts at them they simply move on.

Next we hear Allison screaming in pain, disgust, terror as the camera closes in on the thin white curtain over the cell door which now protects Sherrin, rather than Allison's privacy. Dissolve image and sound into another white curtain and Allison's profile, showing the bruises of Sherrin's hold on her face, completely isolated from her father. The camera pulls back to show us her figure in a shapeless sweatshirt and pants, arms and legs tightly gathered as she tells her professor father to do and say nothing.

This is not the end of Kastner's fictional probe into Canada's maximum-security prisons. But the image of the opaque curtain hiding the bars and the outside world from the prisoners inside and, by analogy, the prisoners from the outside world gnaws away at the illusions most viewers bring to this subject.

Many of the viewers would have seen Kastner's excellent documentaries on prison life, *The Parole Dance* (1984) and *The Lady and the Lifer* (1985). In fact in 1987, in the third part of his prison trilogy *Prison Mother, Prison Daughter,* he showed the woman on whose case this drama is loosely based. Why, then, did he choose drama as his vehicle for this theme? A note before the final credits tells us that the characters are fictional, but the incidents are based on actual events. The publicity stressed that, though this was Kastner's first try at a fictionalized form, the story could not, in his view, be done as a documentary. I think there is another answer. When, as in *Turning to Stone*, a producer moves from documentary to fiction for a contemporary story, it is for one of two reasons. The first is because access is denied, information withheld or someone might be endangered by being photographed or interviewed on camera (as in this case), or a story is inaccessible to the documentary team for political or logistical reasons. This was the rationale Kastner himself gave. The other is simpler: selecting, shaping, and firing reality in the kiln of fiction can sometimes yield more of the truth.

A few critics were bothered by Kastner's choice of a young middle-class woman protagonist. In the debates about who should own and run the powerful medium of television, particularly those of the Marxist semioticians in Britain, Kastner's decision would be vigorously attacked. Kastner knows, however, that if his viewers are to accept the risk of caring for these characters—every one of them, for there are no villains in this piece, despite the fear and violence—they must be led into this alien world with care. It will be easier to learn how the place actually works through someone with whom they can identify: young, female, reserved, trying her best to survive and not to hurt.

The only problematic detail in the plot is Allison's loyalty to the experienced dealer-lover who coaxed her into a one-time effort at drug smuggling. It could be argued that her refusal is consistent with her determination to take responsibility for her actions even though every prison rule, official and unofficial, should beat that out of her. Judith Thompson wrote a superb script for Kastner with no false eloquence and an ear for the different levels of speech and accent used, for example, by a black woman from the Maritimes, the middle-aged hooker from Toronto who runs the unofficial side of prison life, or a violent, inarticulate, tormented murderer.

Kastner and Till do not present the (unnamed) Kingston prison for women as a sensationalized snake-pit. They make it clear that the pervasive violence, drugs, and alcohol are extensions of the damaged lives that put many of the women in penitentiary—problems intensified by the isolation from family and friends, which is imposed by the fact that there is only one maximum-security prison for women in all of Canada. Kingston, Ontario, might as well be off-planet for most families and friends. Kastner does allude to this indirectly when Merci explains that she has had no visitors in four years. Yet, she also explains that her long letters to her son may be better than visits, because she can say so much of what she wants. Her disjointed soliloquy of self-hate as she slashes her arms with a razor blade is one of the film's most powerful moments.

There is hope and even love among these people, as well as anger and fear. Yes, Dunky (Jackie Richardson) wants to sleep with Allison but, as the film progresses, we see that her love for Allison is both very sexual and quite affecting. Sherrin's obsession with Dunky—"Did you get my poem?" "I missed you when you were away,"—is sad and frightening. Dunky cannot protect Allison from Sherrin, not only because she is not strong enough to restrain her physically (only Lina and her bodyguards can do that), but also because she does have some residual feeling for Sherrin, who is at least loyal and available. Avoiding the "misunderstood" or "heart of gold" stereotypes, Lina (Shirley Douglas) also has her moments: when isolation and weariness show on her carefully made-up face, when she talks about protecting something "warm" (and unspecified) inside her which no one can reach, when she talks to Allison about walking outside on a summer's night, we glimpse the face behind the mask of flip one-liners and queenly confidence. Even the prison supervisor, Mrs. Groves, is treated sympathetically. She is trying to stop Lina (ostensibly the model prisoner, but known to be the "boss") within the constraints of the system, but she is defeated. The only significant male in the whole film is Allison's father, played very skilfully by Bernard Behrens. His refusal to play Lina's game, masking a refusal to believe and understand what his daughter tells him, is the trigger for her rape. He also disarms our disbelief by articulating the "yes, but" responses with which we protect ourselves.

Like the equally fine docudrama on street drugs and prostitution, *They're Drying Up the Streets* (1978), this film does not preach. Simply, it shows, with a relentless pace and a firm grasp, the psychological hell of our maximum-security institutions. "Hell is other people," said existentialist Jean-Paul Sartre in his play, *No Exit*. As a prisoner in segregation points out to Allison, you go crazy, but at least in segregation the demons are inside your own head.

This drama does not end with Allison's death or her complete defeat; it ends with her decision to lay a charge against Lina. For Allison this means strict isolation, loss of privileges, and loss of time towards her parole. Rather than enmesh her father even more in Lina's schemes, however, rather than live with the way the prison actually functions for the general population or with her own self-contempt, she does lay the charge. A different set of pressures may cost her sanity before she is freed, but she has made a choice—or at least, it would be comforting to think so.

Last sequence: a long shot of her going up to segregation with the guard as we hear a voice-over recitation of the rules—no privileges, no school, etc. Allison: "I feel like the kid who had to live in the plastic bubble." "Listen, what are you complaining about. It was your choice." "Was it?" The camera closes in, not on Allison, but on the bars, the corners, the heavy mesh and, finally, the guard watching her on a monitor fed by an overhead camera in her cell. The last shot is a closeup of a flickering black-and-white monitor with a muddy image of Allison looking directly up at us.

I was curious about how audiences reacted to this challenging drama. As one might expect, infrequent viewers of such programmes disliked the film more than frequent viewers. On 25 February 1986, from 8 to 10 p.m.—in prime time during the season when viewing numbers are at their maximum—the average number of viewers, sampled every quarter-hour, was 1.768 million: 2.196 million looked in at some point, and most viewers watched at least 80 per cent of the programme. For some reason, the CBC chose to schedule *Turning To Stone* against the guaranteed crowd pleaser, the Grammy Awards. Even with this competition, however, 18 per cent of all Canadian television viewers chose *Turning to Stone*. Without the Grammy Awards, there might have been more viewers and, perhaps, more protest. On the other hand, such scheduling is inarguably alternative programming.

The enjoyment index was 65, with numerous variables affecting this total. Audience Research found that women tended to be more sympathetic to the issues and characters than men. However, there is a possiblity that some men were watching because the women insisted on tuning in and staying with the drama. Not surprisingly, the audience found it "demanding" rather than "relaxing" (76 per cent), and that "it made you think" (89 per cent). Audience Research points out that when a play makes the viewer think without involving him or her in the lives of the characters or issues, you have a formula for failure, at least when it comes to enjoyment indices, something worth remembering whenever the enjoyment index of docudrama is assessed.

The audience enjoyment index for the story line showed a very different set of perceptions between the women in the audience and the men: the women's E.I. was 67 per cent, the men's only 53. Reacting to individual characters, slightly more men liked Lina than did women (56 to 52 per cent). Allison had an overall enjoyment index of 68 per cent.[14] This percentage is later broken down into women, 72 per cent, men, 63, with women over fifty being a little more sympathetic. Despite the warmth of her character, Dunky obtained an enjoyment index of only 60 per cent. I wonder whether this is because black, lesbian, petty criminals cannot as yet get sympathy of 40 per cent of a CBC audience, or whether it is because she fails to protect the protagonist, or for some other reason altogether. Overall, the audience research report suggests, viewers found the programme believable and learned about the difficulties of prison life. Although 53 per cent of them felt little sympathy for Allison in the beginning, only 38 per cent were unsympathetic to the end. 82 per cent agreed that, "It was a programme that you get involved with." Compare the figures for *Chautauqua Girl* where 42 per cent agreed that it was exciting, 77 per cent that it was a programme you get involved with. Not surprisingly, only 28 per cent of the audience for

Turning to Stone found it relaxing. 72 per cent found *Chautauqua Girl* relaxing. It should be noted that the questions on which these findings were based were prepared in consultation with the producers John Kastner and Jeanine Locke.

The most interesting finding was that the enjoyment index which specifically measured the play's ending was quite low: 47 per cent overall, with the 50^a and up age group dipping down to 41 per cent. Many people liked the programme more than the ending. It is known that older viewers watch more television and seem to have a higher tolerance for anthology as a form. It is possible that they may still be conditioned by their experience of most available television to expect a closed rather than an open ending. In this case, they may have expected to see Alison released, transferred, or even killed.

Let me conclude with two opposing voices. "It's either news or drama but obviously [*Tar Sands*] shows that mixing the two doesn't work," according to Roderick McLennan, lawyer for Peter Lougheed, who negotiated an out-of-court settlement in his suit against the CBC. Film academic Seth Feldman, on the other hand, argues that "like the folktale, the docudrama is an exercise in social integration, a form which self-conciously encapsulates primary social or psychological lessons within the most widely accessible entertainment medium."[15] The article relates the television form under discussion to cinematic form, concentrating chiefly on *Maria* and *Running Man* (1980, wr. Anna Sandor, dir. Donald Brittain, prod. Bill Gough). I disagree with both points of view. Shows mixing drama and news do work if the audience is clearly told which is which as the story unfolds. Nor do I think that the teller of tales in the corner necessarily weaves docudramas that integrate audiences into society, or reaffirm socially acceptable norms. In my own view most of the best Canadian docudramas are as likely to disrupt or subvert as they are to assist social integration.

Many docudramas work as investigations of situations which trouble viewers or as exposure to problems they have not considered before: *Final Edition, A Place of One's Own, A Far Cry From Home, Running Man*. A few function more as entertainment: *High Card, Snowbirds*. The best do both. Among the latter, *Dreamspeaker, A Thousand Moons, A Question of the Sixth, Every Person is Guilty, Don't Forget, Je Me Souviens, Anne's Story, Out of Sight, Out of Mind, One of our Own*, and *I Love a Man In Uniform*. In the same category come the ninety-minute drama specials *They're Drying Up the Streets, Crossbar, Tyler*. Overall, of the six or eight topical dramas in each season of *For the Record*, the CBC got one or two really superb pieces, several competent dramas, and one or two efforts that remain dull or confused. As you can see, the standard of historical and topical docudrama has been high. The 1984-85 season was the last for this long-running programme. Three topical dramas were broadcast in the fall of 1985 including *Oakmount High*, but not under the *For the Record* title. This kind of drama will still be made every season as it has been for the last three decades, simply because the CBC does it very well, audiences like it, and the form is essential to finding out where "here" is, and who lives here, and sometimes, how we might better live together.

NOTES

Part of this material appeared in *The Journal of Popular Culture* 20, no. 1 (summer 1986): 127-40.

1. "Grierson On Television," in *Contrast*, 2 (Summer 1963), 22-23.
2. Fontana Technosphere series. Fontana/Collins, London, 58.
3. "The Domestic Playwright: Some Thoughts about Television Drama," in *The Review of English Literature*, vol.3, 17-28, 22.
4. Quoted by British broadcasting historian Asa Briggs in "Arts and Techniques," in *Sound and Vision*, vol. 4 of his *History of British Broadcasting*, Oxford University Press, London: 1979, 686.
5. *Encore* vol.11, no.2, 21-33.
6. Briggs quotes his weekly *Letter from America* (BBC, 8/7/48).
7. A British version of the life and death of Admiral Lord Nelson, *I Remember Nelson*, was another simple yet successful approach to the subgenre of historical biography: six half-hour vignettes told from the respective points of view of seaman, wife, mistress, best friend, etc., composing a chronological yet complex portrait of a fascinating man.
8. *British Television Drama*. Cambridge University Press, Cambridge: 1981, 21.
9. "Documentary Style: The Curse of Canadian Culture," *Marginal Notes, Challenges to the Mainstream*, Lester and Orpen Dennys, Toronto: 1984, 86.
10. I have been told that Helen Weils is a pen name for Cam Hubert/Ann Cameron (*Ada, Dreamspeaker* and a fine radio piece on wife battering, called *In the Belly of Old Women,* 1986).
11. On 2 February 1982 the St. Catharines *Standard* picked up an AP wire-services report that "the distinction between art and reality is proving too much for TV audiences. When a non-working telephone number flashed on screen as part of ABC's movie *Pray TV* Monday night, an estimated fifteen-thousand people tried to call. In the drama Ned Beatty, playing a television evangelist, said : 'If you have a problem of any kind—financial, medical, spiritual—call the number you now see on the bottom of your screen and let us help you to be born again by the spirit of God'."
12. 41 per cent according to "Audiences and Their Reaction to *For the Record*: 1981-82 Season," CBC Research, Toronto: May 1982.
13. The Enjoyment Index is a measure of viewer satisfaction with a show's content . . . based on a five-point Likert scale that ranges from one (I enjoyed

the show very much) to five (I did not enjoy it at all). . . . Most shows range between fifty-five and eighty-five. The mean is about seventy-nine. Main drama characters who obtain EI's above seventy are generally regarded as having successfully established a fairly strong emotional link with viewers''—a letter to MJM from Oley Iwanyshyn, Senior Research Officer, Audience Research, 14 Aug. 1986.

14. "In the case of STONE'S Allison, her 68 is probably the result of an admixture of some less than heroic qualities plus the fact that, in the end her role takes on heroic proportions. In that sense, her EI can be regarded as consistent with the ambivalent nature of her role." Ibid.

15. "The Electronic Fable: Aspects of the Docudrama in Canada" in *Canadian Drama,* vol.9, no.1 (1983), 40.

IV

THE PERIPHERAL IS CENTRAL

9

EXPERIMENTAL
"The Seedbed"

In television drama, the word "experimental" is often used to describe two quite different, though related kinds of programmes. In one, the director and/or writer have decided to experiment with the conventions of television: in non-linear narrative using flashback, or flash forward; fantasy, or direct address to the camera; a stationary camera shot sustained throughout a long scene—although Eric Till's extensive use of longshots to convey the isolation of figures in the bleak landscape of *The Private Confessions of a Justified Sinner* is also expressive of the director's personal vision. Another favourite technique, particularly during the 'sixties, drew on the theatre of the absurd to subtly but continuously disrupt the surface realism of domestic situations and environments. In the 'sixties, *Festival* introduced viewers to this style with scripts by James Hanley, Clive Exton, and Pinter. Jacques Languirand's *Grand Exits* was a superbly effective Canadian example.

To confuse the issue, however, the term "experimental" is also commonly used by journalists and publicists when the subject matter is new to television. Yet today's controversial subjects—gay rights, abortion, animal rights activism, Québec separatism—are usually next year's inside page and, often, the following decade's commonplace. The "experimental" subject may still yield fresh and fascinating drama, but it is more likely to move into the main stream of daily concerns. Thus, such "experiments" in form as the jump-cuts of the 'sixties become the commonplace visual conventions of advertisements in the 'eighties, and once avant-garde subjects like "unmarried mothers" in the 'fifties, or "living together" in the 'sixties appear casually, as mere facts of life, in late-'seventies sitcoms. My working definition is that experimental television combines unusual subject matter with innovative form.

The 1950s

The need for experimental drama as favourable soil and climate for the development of quality and originality within the traditional drama anthologies was recognized early in the history of the CBC. A press release dated 15 May 1953 announced *Playbill*, a summer series of half-hour plays, shown at the family viewing time of 8:30 p.m. Under Robert Allen as CBLT's drama producer, *Playbill* would be "based as far as possible on the original writings of Canadian authors. Additionally, producers will be given a free hand to develop and employ experimental techniques which will improve the dramatic qualities of the productions." The first-season anthology included "Metronome," a psychological drama about a woman haunted by memories of her dead stepson, and Len Petersen's "They Are All Afraid" (1953), a play which had aroused strong reactions in its original radio version.[1]

In the next season, supervising producer Sydney Newman announced "plans to introduce many new Canadian actors to the television screen in the half-hour weekly series. . . . *Playbill* will be operated on the workshop principle. However, the new series will not confine itself purely to 'experimental' matters, but will offer the CBC TV viewer a wide variety of solid dramatic fare" (press release, 18/5/54). New writers included George Salverson and Leslie MacFarlane, both of whom became prolific contributors to CBC television, in addition to the already established writers Joseph Schull, Ted Allan, and Melwyn Breen. The series' various producers included drama regulars Henry Kaplan, David Greene, and Silvio Narizzano, all of whom were central to the first phase of drama production. Others were borrowed from "Space Command" (1953, Murray Chercover), "Fighting Words" (Arthur Hiller), variety (Harvey Hart) and "Ad Lib" (Leo Orenstein). *Playbill* was an environment where fledgling producers could learn to "fly" the complex controls of live drama. The word "experimental" in this context seems to cover the try-out of new producers, actors, and writers rather than the reshaping of television forms or the inclusion of challenging content. It would not be the first time the word was used to disarm criticism rather than describe a programme's content and style accurately.

By 1959, *Playbill* originated in various production centres across Canada and was scheduled on Thursdays at 10:00 p.m. This was not a prime-time slot, but the freedom to tackle more adult forms of drama may have compensated for the inevitable loss of audience. In fact, right up to 1976, the CBC continued to protect at least one slot in the schedule for experimental drama. Moreover, most of the popular anthologies occasionally experimented with form. As we have seen, as the 'sixties turned into the 'seventies, even series television occasionally presented scripts and production techniques which expanded on the ordinary conventions of TV storytelling. In this chapter, however, I intend to confine my discussion of the experimental form to time slots and programmes that actively encouraged viewers to expect the unexpected. I will begin by describing each anthology, and how it differed from the others.

Q for Quest

The CBC did not clearly set aside an "experimental" time slot until 1960. *Q for Quest*, produced for a thirty-minute later evening time period, was supervised by Ross McLean in its first year, and in the '61-62, '62-63, '63-64 seasons by Daryl Duke, a Vancouver producer who directed numerous episodes from the first season on. According to the Toronto *Star* (24/2/62), Duke castigated the rest of TV drama for its use of stale conventions and fear of risk. He seems to have resented the emphasis on big business, the middle-class, and suburban angst which, he felt, characterized most anthology drama of the early-'sixties.[2] He also claimed that Parliament was threatening to censor *Quest*. In fact, E. Brunsden (PC, Medicine Hat) had described it (Ottawa *Citizen*, 15/2/62) as "depraved, disgusting, absolute garbage, a rank violation of the sanctity of the Canadian home and family."

Yet in an interview with *Time* (16/3/62), Duke seemed to deny *Quest*'s role as a forum for experiment: "I am not trying to be avant-garde. Drama ought to be as accurate as reporting. I want my shows to be as contemporary as a news bulletin." It was a theme he reiterated in many interviews. By 26 September 1962, however, he was proudly telling Ralph Thomas on a television interview that *Quest* held the CBC record for the most irate viewer letters received. Duke portrayed the programme as focusing on "social purpose dealing with matters that relate to contemporary society and with problems entirely new to television," using "a more ragged, tough kind of shooting and cutting; more closeups, no equivocal medium [shots], and a greater emphasis on savage lighting techniques. I want to fill that half-hour tight."

The next season (1962-63) provided a fair sample of the programme's range: "The Trial of Lady Chatterley"; a jazz ballet called "District Storyville"; "The Evolution of the Blues," and George Ryga's superb half-hour play, "Indian." This kind of activity attracted an offer from the BBC (Nathan Cohen in the Toronto *Star*, 20/4/63) that Duke turned down to continue with *Quest*, now defined (*Star*, 28/12/63) as "a kind of commando task-force in Canadian television running around with sten-guns shooting out the lights." Despite the rhetoric—very un-CBC and, at this time, very un-Canadian—Duke plainly viewed his programme as a place, not only for technical innovation and topical, controversial content, but also for what he bluntly termed, "moral debate." Certainly, *Quest* did provoke some hard thinking on the loaded issues of the day; some programmes would still be provocative twenty-two years later. One response to this kind of provocation was a "Declaration of Canadian Women" by the WCTU, Women's Catholic League, and others (27/10/64) condemning the "misuse of the CBC to spread propaganda for perversion, pornography, free love, blasphemy, dope, violence and crime"—an outburst provoked by *Festival* as well as *Quest*. Across the country, newspaper editorials were divided on the declaration.

By January 1964, Duke had decided to go south for *The Steve Allen Show*. Before he left, he triggered one final burst (*Star*, 30/1/64) when he accused the CBC of elitism, of taking its tone from the *Massey Report*'s "government by garden-party philosophy." In his view, the corporation thought "serious drama should be encouraged and the vulgar arts should not. This attitude has been outpaced by the social changes in our time." Instead, the CBC schedule "should be a grenade explosion aiming to wake up the country about where it's going and why." His farewell was a prophetic but certainly not a nationalistic statement: a half-hour of Bob Dylan. The Toronto *Telegram*, in a tone often used by those left behind in Canada, commented sardonically: "It can't be frustration or a search for different opportunities. It has to be money." It was announced on 23 September 1964 that supervising producer Robert Allen had cancelled *Quest*.

The following are brief analyses of programmes in the *Q for Quest* anthology which I managed to see: they are a reasonably representative selection of the drama presented.

1960-61: Executive Producer, Ross McLean

"Burlap Bags" (wr. Len Petersen, prod.-dir. Harvey Hart) was a world-famous radio play adapted for television. The abstract impressionistic script had made brilliant use of radio's flexibility and fantasy to help the listener imagine a world where disembodied hands and voices inhabit piano boxes, and people are either transformed into their machines or wear burlap bags on their heads. The play was staged for television on a set attempting to portray the "no place" that radio can evoke with such ease. A wavy scrim, gobbos, or cutout patterns filtering the lights, and a painted floor left actors free to move in a rather inadequate representation of limbo. The music and sound effects, however, over-did the themes of conformity and despair. Though a brisk pace and energetic performances, particularly that of Percy

Rodriguez as Tannahill, lent the piece intensity, the nightmare fragmentation of consciousness and distortions of perception which had made "Burlap Bags" such an innovative radio masterpiece were blunted by visual representation.

Plays which excel as radio, in fact, rarely succeed in television. Reuben Ship's classic satire, *The Investigator* (CBC *Stage '55*), was a huge underground hit in the United States even though the offical recording was not allowed on American airwaves for fear of the McCarthyism it dissected so effectively. It was not subjected to TV adaptation. Mavor Moore's *Come Away* was, and some of its delicate aural transitions and collages of images suffered in consequence. On the other hand, Joseph Schull's *The Concert* (6/2/58) actually gained in ironic impact since the audience knew by mid-play that the blind heroine's new friend was black.

1961-62: Executive Producer, Daryl Duke

The play list for this season still exists, running from 8/10/61 to 27/5/62, with twenty-three programmes and eleven preemptions. Even in this earlier period, viewing habits were hard to form thanks to interrupted scheduling, but anthology was—and is—much more vulnerable to preemption than series television. The season was a mixture of plays, music, a conversation with controversial novelist Henry Miller, dance, a film, and two adaptations from the theatre.

"Dreams" (18/2/62) was an unscripted, unrehearsed, lyrical film by Allan King exploring the relationship between two young people. Robert Whitehead's introduction described the programme with dubious accuracy as a look at "the strange, fleeting, brutal transition between youth and maturity." What we saw was a restless youth leaving his girl because she wants to get married. An experienced documentary filmmaker (*Skidrow*, 1956; *Rickshaw*, 1960), King was finding his feet here in a new form of documentary, enhanced or fictionalized by editing, that would flower in a few years with *Warrendale* and *A Married Couple*. In "Dreams," the self-conscious conversations, the contrived transitions between scenes, and the choice of a humourless protagonist created a sentimental and rather banal, slow-moving piece.

"The Neutron and the Olive" (3/11/62, wr. Rudi Dorn, prod.-dir. Paul Almond) presented a man obsessed with the threat of nuclear war who has a drunken, lecherous, haunted confrontation with a young woman at a cocktail party. Almond used unusually subjective, first-person camera angles to contrast the surface chatter with the interior agony of the characters. Bernard Hayes as the man played the rather rhetorical dialogue with force and intelligence. Sharon Acker played his confused sounding board with warmth and charm.

Jules Ffeiffer's "Crawling Arnold" (4/2/62), though it unleashed the current of invective that rained down on *Q for Quest* over the next two years, seems to have been simply a provocative piece of contemporary *angst*. I have not managed to see it.

1962-63: Executive Producer, Daryl Duke

"The Wounded Soldier" (3/3/63, prod.-dir. Mario Prizek), adapted by Jack Kuper from a short story by George Garrett, was an anti-war fable about a disfigured soldier who, refusing to hide, decides to become a freak in a circus. As the world's funniest face, he attracts a woman; losing his novelty and her, he returns to hospital and gets a new face. Prizek began the play with ominous music, a ticking clock without hands, and a bandaged face seen through bars which are revealed as the side of a hospital bed. The political satire was a touch predictable but the exposure of the nurse's mixed motives and needs was a satisfying touch. The dread of what will emerge from the bandages is an old Hollywood cliché that still works. We never do see the protagonist's face from the front, but can paint whatever horror we please based on scars on his neck and ears. The public's fascination with and revulsion toward the man are conveyed by a double exposure of the protagonist in a mask, walking over a stills of photographs of crowds.

"Eulogy" (14/4/63, dir. Harvey Hart) actually presented two contrasting eulogies: one taken from *Another Country* by black American novelist James Baldwin and the other by a rabbi at a boy's funeral, which was expanded into "An Early Grave" (1963) by Wallace Markfield. Baldwin's sharp eyes and ears, and his passionate pride in being black, combined with Percy Rodriguez's restrained performance gave the words extra power, intensified by the songs, "Freedom" and "Come go with me." The second eulogy, with its swift speech rhythms, stripped setting, and different idiom focused not so much on the dead person as how to grieve. The two were linked by the eulogies of rebellious, gifted young men in minority cultures, and each created a complementary but intense emotional effect.

1963-64

"Night of Admission" (19/11/63) was another new departure in programming. Shot partly on location at Sunnybrook Hospital, the play (wr. Frank Freedman, prod.-dir. George McGowan) showed what it was like from the patient's point of view to be admitted in the middle of the night. It also showed how the experience was distinctly different from the melodrama of "doctor" shows like *Dr. Kildare* or *Medical Centre*, which adhered to a fairly strict formula at that time. Don Francks gave a first-class performance as the patient whose anguish and illness intensified his incomprehension and apprehensiveness. Francks was often seen in closeup, with intimate atmospheric music heightening the range of emotions. The rest of the hospital was seen almost entirely from his point of view, using flat camera angles and foreshortened or fragmentary glimpses.

"The Bathroom" (14/1/64, prod.-dir. George McGowan) was that all too rare event in CBC prestige anthologies, a comedy by the well-known west-coast humorist, Eric Nicol. The plot was simple: outside a bathroom in an old people's home, various characters line up to take their turn to encourage a potential suicide to reconsider. The set was a clever mix of clean, institutional, comfortable mismatch. In this as in other television drama of this period, the pace is more leisurely than a viewer of the 1980s is accustomed to, but it suits the wry dialogue and relaxed performances by Mervyn Blake, Eric House, and Ruth Springford. The subtext of neglect and boredom, of lives in reprise, is reinforced by a lightly ironic counterpoint of phrases from old songs, apparently meandering on in the background.

"Bedlam Galore for Two or More" (7/1/64, prod.-dir. Mario Prizek) is fairly typical Ionesco: "The exits you find are never any use." Trapped indoors for years, two people try to ignore the storm and war raging outside; the freshness here is in the tension between naturalistic acting and dialogue, spiked with non sequiturs, and the absurdist situation and set. Jack Creley as "He" and Norma Renault as "She" gave good, rapidly paced, crisp performances. The design was fascinating: a huge, nineteenth-century drawing room full of fragments—among them a Venus de Milo with arms, a nude statue of Liberty, bits of bodies, radiators, and dolls. The basic action consisted in the progressive demolition of this environment, a substantial challenge for live studio drama.

"Two Soldiers" (8/10/63) marked the return of George Ryga, whose script, directed by George McGowan, represented a major technical inno-vation because it was completely taped on location. Two soldiers, friends on leave, are walking the last mile or two home down a country road. They talk, then get involved in horseplay which gradually becomes a sharp look at the prospect of nuclear war and its aftermath. One is a born soldier, a leader; the other is ambivalent, sensitive, something of a clown. Both are alienated from their working-class, small-town families, have conservative values, and project machismo. Yet their common sense, working away at the almost unimaginable issues, humanizes without trivializing the whole question and society's inadequate answers to it. The natural light, long tracking shots, and sense of landscape and distance give the play a very fresh look, and reinforce the perception that this issue involves everyone.

"The Brig" (21/1/64), part of a play by Kenneth Brown which had been shut down by the New York Police, continued the *Quest* tradition of stirring it up. George McGowan presented a shortened, somewhat expurgated script, but it was still a very powerful production. The set was huge, cold, a cage, more realistic than the chalk lines of the original production, but every bit as effective. The play shows a navy "glass house" or military prison brutalizing, through sheer mental torture, guards and prisoners alike. Every word is shouted, every move made at right angles and on the double; the pressure is relentless, the isolation complete. Ed McNamara, Len Birman, Gordon Pinsent, Jonathan White, James Peddie, and Louis Negin, among others, gave it all they had, creating a memorable half-hour that felt like three days.

"Kim" (27/1/63, dir. Eric Till), adapted by Hugh Webster from the letters of Kim Malthe Bruun, a young Canadian who died in the Resistance at the hands of the Gestapo, was one of the best scripts in the four seasons of *Quest*. Characteristic of many of Till's pieces, the structure is entirely contrapuntal. The framing narrative shows us Kim's fiancée and mother reading letters full of love of adventure, life, and hope which are cross-cut with events in Kim's cell and during his interrogation. As Kim, Garrick Hagon gave a detailed and quite moving performance. The effect of the piece exploring the inner strength and love of a young man under brutal pressure was achieved largely by indirection: viewers were left to infer for themselves the relationships between fear and pain, the glimpses of beauty inside and outside the cell, and the never-quenched hope.

The following extract from an interview with actor and adaptor Hugh Webster[3] gives some insight into how a complex programme like this developed. *Kim* might have been a standard, "talking heads" treatment of a book of letters; it could have been sentimental, self-righteous, or melodramatic. But it was none of these things. Webster also adapted *The Offshore Island* (12/3/62), and *Private Confessions of a Justified Sinner* (13/3/64) for Till. The partnership was clearly creative for both men.

> Hugh Webster: Eric gave me the book, *The Heroic Heart*. I read it and I was very moved by the whole thing. I was also struck with the central idea, which is of a young man finding a spiritual peace while his body is being broken. At the very point when the clubs are landing, he has managed to move his mind on to another plane where he doesn't feel the pain. So we wanted to do this point and counterpoint of the brutal reality of what his body was going through and at the same time, try to get the fact that—for lack of a better word—his soul was moving on a completely different level.

MJM: What made you include the mother and sweetheart? There is a lot of intercutting as well—for example, the wordless encounter between the girl friend and the German soldier on the stair during a bombing; there are many moments which are nonverbal.

Webster: I can't remember what our thinking was on that. I do remember us having long conversations, Eric and I, about [the two women]. We had to have that element because it was so much a part of him.

MJM: Those scenes also served as a bridge for the audience from ordinary to extraordinary—And the design, too [by Rudi Dorn] depended, both thematically and in technical terms as a studio piece, on the interplay of light and shadow. Always with such material there is the aesthetic question of explicit *vs.* implicit violence; *Kim* had only a few, carefully selected violent moments.

Webster: That's exactly what we wanted to do. For instance, when the machine gun started executing Kim and two others in a cellar, the sound was a bouncing ball coming down the stairs. Remember? Bang, bang, bang—

MJM: It was a superb and simple effect because those stairs had been used over and over again. Current television is often so literal, so deadly in its emphasis on realism. . . .

Webster: Sure. There's nothing for the imagination at all.

MJM: I am developing the notion that this has something to do with the increasing dependence on film and on location shooting. . . .

Webster: Except when films are properly made. I remember Eric and I were very influenced by films. I was really struggling with *Kim* after our conversations, knowing we wanted to do something very special with this little half-hour. So I was really struggling. One night when I was in New York, Eric came down and we sat around and talked about it. I told him, "I've only been able to write one shot, but I think it's the one that will enable me to get going on the whole thing. It's the last shot," —which is the way that I always wanted to work in those days, to get that very last shot. [In this case] it was the sand pouring out of the sand bags and covering Kim's face as he sank down to the bottom of the post after he was executed. I told Eric what the shot was. He said, "That's great. That's great." After that, we bashed away and bashed away.

Eric and I used to feed each other. He'd throw out some idea about a scene that I might just have finished writing. It made a mess of what I had done. But at the same time, when the idea is as powerful as a lot of his ideas are, then you've got to say to yourself, "Wait a minute, now let's see if a marriage is possible here." I would try to work it in so that it dovetailed with what went before and what came afterward. And so there's this constant play back and forth."

This conversation summarizes part of the process of creating experiment in television forms. In the more flexible first fifteen years of television drama, designers, directors, writers, and, more rarely, actors and cameramen, could try for all kinds of visual effects, using all sorts of narrative structures on all kinds of subjects. Since the CBC has never had the luxury of trying out a dish just for the family before it put a dish on the menu, it has served up some unpalatable messes, as well as many wonderful meals. The menu only becomes bland fast food when the drama department refuses to take risks; to "experiment."

Prizek's Eyeopener

Q for Quest was to be replaced by *Eyeopener* (in the winter of 1965). Its producer, Mario Prizek, had wide-ranging interests and credentials in both experimental content and television formal innovation. Prizek's statement of intention, to Gordon Hinch, Director of Network Programming, described the profile of the series. "I hope that the material assembled will present a sharp, hard-hitting picture of what it is like to live in the contemporary world. It will eschew the wish-fulfilment images usually flashed upon the scene in cliché formulae as supposed 'slices' of life. It will, I hope, make full use of the television's power to disturb, in order to make people sharply aware of things around them that they take for granted. But the shock [sic] it might provoke would be the 'shock of recognition,' not the shock of titivation or provocation.'"[4] He goes on to argue that range is absolutely necessary in drama, and that giving people what they want ignores the fact that public taste can be modified, expanded, and developed—an axiom for all *Festival* and *Quest* producers, of whom he had been one. Every programme of the initial twelve was to be issue-oriented, including the three revues. Issues were not to be merely topical but, rather, relevant for a long time to come. "Formal experiments will be strictly governed by the demands of the subject matter": there would be no period pieces or changes in intensity of presentation. In other words, Prizek was not about to ease up on the viewer after a particularly challenging show by programming light satire, revue or a personality profile. Yet despite his statement, Prizek did vary his anthology with comedy, psychological drama, satire, absurdist drama, and gentle naturalism; however, he did not wait for public affairs to test the waters on sensitive subjects.

It is evident from the files on deposit in the Federal Archives that Prizek saw *Eyeopener* as satisfying some of the same needs as *Q for Quest*. He read all the scripts himself, deliberately avoiding the use of script readers in an attempt to retain a more immediate contact between writers and producer. Writing to a New York agent, he defined what he was looking for as "strong punchy drama about contemporary issues. I want to avoid period pieces or way-out 'coterie' drama. I hope to touch many controversial issues in the series—segregation, overpopulation and birth control, anti-semitism, the dehumanization of modern society, the Bomb, etc. . . . The sort of play I am looking for will look at the issue obliquely and even ironically, making the viewer aware of the full ramifications of the problem by implication while looking intensely at some point on the periphery . . . a small concentrated action suggesting a large issue"—in twenty-eight minutes and ten seconds.[5] This declaration of intentions should be read in the context of both the history of CBC experiment to date and the excitement generated at the time by *This Hour Has Seven Days*. Information about *Eyeopener* is rather fragmentary.

The Eyeopener, Tuesdays at 10:30-11:00 p.m., opened with Leslie Macfarlane's *The Blind Eye and the Deaf Ear* (dir. Melwyn Breen). Other scripts broadcast were *The Trial of Josef Brodsky*, arranged from transcripts by Stan Jacobson; *Sara and the Sax* by Lewis John Carlino, *The Golden Bull of Boredom*, a jazz opera about tabloid sex by Lorees Yerby, and Jules Pfeiffer's *The Lonely Machine*. *The Black Madonna* (dir. George McGowan) was adapted from Muriel Spark's short story, and *The Guard* (dir. David Gardner) was a short story adapted by Jack Gray. *Hear Me Talking to Ya* (dir. Paddy Sampson) had text by Don Francks and music by Ron Collier. *Uh-huh?* consisted of Pinter sketches arranged by George Bloomfield. *The Tulip Garden*, by George Ryga, told of a man who so deeply loved his wife that he was prepared to confront neighbours about burying her in the garden as she wished—"an outspoken commentary on our disparate approaches [to] the idea of death," and a gently crafted drama that has been redone at least twice. Finally, there were Anthony Burgess's *The Wanting Seed* (dir. Stan Jacobson) and *The Dutchman*, a trenchant play about race relations by Leroi Jones; it is unclear whether or not the latter was actually televised. To the experiment-starved 'eighties this sounds like a feast for the mind and spirit. After *Eyeopener* was not renewed, six years passed before the CBC decided to set aside another experimental slot.

CENSORSHIP AGAIN

Eyeopener was responsible for the promulgation of "Prizek's Law," which is, according to Hugh Gauntlett, that "'you must not screen a producer's show [before it is aired] without the producer being present—which is, on the whole, a sensible rule." Gauntlett and Prizek agree on the facts.[6] At 5 p.m. on the day the first show of the anthology was to be broadcast, Prizek ran into a publicity man who asked him what show was replacing the one cancelled for that night. Prizek knew nothing of this. It transpired that the network management had cold feet over "A Borderline Case," a satire on Québec separatism featuring the then largely unknown Second City troupe. Presented as a documentary, it took as its premise the fact of a nuclear explosion. Everyone denies knowledge of it. The superpowers are accusing each other, and war is near when "Lawrentia" speaks up to say "we did it" and all hell breaks loose. Prizek: "They regarded it as dangerous political propaganda." "They" included Robert Allen, Hugh Gauntlett, and a new administration up the line beginning to twitch about the controversies surrounding *This Hour Has Seven Days*. "They" pulled the show. Gauntlett tried unsuccessfully to phone Prizek as events took their course. Prizek's response to "this league of frightened men" was, as he recalled: "Gentlemen, I cannot predict the vagaries of the CBC far enough ahead to do a thirteen-week series." Thereupon, he resigned.

After this, amid ugly controversy, came the cancellation of the provocative, satirical, current affairs show *This Hour Has Seven Days* (October 1964-May 1966), despite its enormous and vocal audiences. English-speaking CBC producers saw this as blatant censorship and threatened to strike; on-air hosts Patrick Watson and Laurier La Pierre and producer Richard Leiterman even managed to involve friends in government. Meanwhile, Cabinet ministers hostile to the show were leaning on the corporation. CBC upper-management, which had been battling against the inclusion of specific items, was especially disturbed by the refusal of the "Seven Days" team to go through normal channels, thus eluding management control of what went out. They were also genuinely worried about the editorial judgment which sometimes favoured making news rather than investigating stories. Prizek's problems foreshadowed the final crisis in confidence between producers and management. *This Hour* died a very public death which precipitated the exodus of some of the best producer-directors in CBC television.

The country-wide controversy surrounding the cancellation of *This Hour Has Seven Days* and its effect on morale at the CBC were still reverberating when Daryl Duke was called back to fill the breach with *Sunday*, the purpose of which was clearly to reassure the country that the energy and commitment had not disappeared with Leiterman, Watson, and company. In the *Telegram* (August 1966), Bob Blackburn reported that *Sunday* would replace Leiterman's *This Hour Has Seven Days* and Duke wanted to use Leiterman and other "Seven-Day" people. His full-time return coincided with the return of Ross McLean and Mario Prizek.

Duke intended to emphasize more human interest and less hard news than his predecessors: "*Sunday* will be an attempt at total journalism using a combination not only of politics and other significant current events, but also drama, music and the arts" (CBC press release, 19/10/66). He also announced that he would be using Robert Hoyt as a producer and Larry Zolf as a producer-story editor (both from *Seven Days*), as well as Peter Reilly, a journalist and producer-story editor, and Ian Tyson and Leonard Cohen as "regular on-camera personnel": "One week I might use all five and the next week only two or three." Motivating the "new theatricality" that he envisaged for public affairs was "the new experience in mixed media and total environment now being carried out in the United States, especially New York."

Duke's rhetoric has to be read in the context of the super heated late-'sixties: "*Sunday* is a show which will run the gamut of human experience, mixing raw journalism with the emotional intensity of a medieval bear-pit or bullfight arena. With *Sunday* I hope to give a mind-expanding full view of life . . . a view which has all the intensity and chaotic clarity of an L.S.D. trip. . . . Since the world is exploding and changing in new ways, the inevitable desk and the neat studio look seem sterile to me and incapable of conveying to the viewer the true impact of the exciting but irrational world. . . . In short, I am trying to approach a format which, in its ultimate form, could provide the first experience in psychedelic television." *Sunday* would be produced before a live audience.

The word "psychedelic" connoted mind-expansion in that era, but the striking thing now about Daryl Duke's vision is his idea that coverage of the world as it flowed past television's window-looking glass did not need to be categorized into "features," "documentaries," "current affairs," "variety," or "drama." There could be a single programme in which all those perspectives and all those tools could be seen working on the stream of events. What Raymond Williams would later call "flow" would be contained within a single framework. The obvious danger was and still is, that news and information might become entertainment packages, complete with jolts every thirty seconds. Moreover, Duke's vision of a flow of information, analysis, music, drama, and graphics in one programme could not be sustained on a daily or perhaps even weekly basis; yet it makes the "high-tech," hard-news style which now prevails seem very limited indeed. Twenty years later, the *Journal* briefly tried dramatized scenes and short skits, but the lack of conviction and talented writing was obvious; these elements largely faded from view after its first season.

The new programme was launched with a theme song by Ian and Sylvia Tyson and plans for a satirical version of *Macbeth*. Within three weeks, it was on the front pages of the papers, and CBC President Alphonse Ouimet was apologizing for the inclusion of a "sex film" made by the BBC. Former Ottawa *Journal* editor Senator Grattan O'Leary came out in vigorous protest at both cost and content on 18 March 1967: he qualified the show as third-rate, and was quoted as saying that part of the trouble with the CBC was that "a number of young producers . . . appear to have just discovered sex and they appear to have discovered that it is wonderful." "Free speech is not an absolute," thundered O'Leary. Sound familiar?

Sunday was less than four months old when it was cancelled. Nathan Cohen commented that its perpetrators "had the wrong talents for the show's need. . . . Daryl Duke is, by instinct and cultivation, of the minority in his tastes. That of itself made him the wrong choice for a programme designed to be topical in its widest sense and to persuade the general public that the programme was in their interest and image." (CBC press clippings file, no date.)

Program X

When *Eyeopener* left the air, *Festival* was the only place where viewers could find formal experiment until 1970. Then *Program X* (1970-73) took up the task, beginning on Thursdays at 9:00 p.m. under the eye of executive producer Paddy Sampson with fifteen shows; the associate producer was George Jonas. According to an interview with Bob Blackburn (*Toronto Week*, 14/11/70), the budget was only $5,000 a week and the majority of the shows were to be shot in studio, on tape, with performers on ACTRA scale. There would even be a few low-budget films. Sampson said that the series was intended, among other things, "to start the script factory again": he then admitted that it was easier to work with scripts from established writers.

In view of the kinds of plays subsequently presented, it is interesting that Sampson should have stressed as non-experimental *Program X*'s character. Could this have been because the ratings drive and the more conservative cast of the corporation as a whole in the 1968-73 period made it politic to avoid emphasis on the elements of startled recognition or surprise which usually characterizes successful experimental anthology? Well before the series opened, a CBC press release of 23 November 1970 quoted George Jonas as saying: "*Program X* is only experimental in the sense that we try to explore, sometimes using different techniques, sometimes new writing and performances. The real experiments never reach the screen and what does is past the experimental stage. The finished product is something we think works.''

Could that have been the way it really was? Not reaching acceptable broadcast standards was given as the reason for shelving a film by David Acomba, although there was dark speculation in the papers that the CBC had bowed to pressure from the Ontario Provincial Police. Yet I have the reiterated evidence from many people that programmes had to be aired regardless of quality for budget reasons.

The assumption underlying the producer's comments seems to have been that the word ''experimental'' either had pejorative overtones for the general public in the 'seventies, or was synonymous with ''in process,'' ''unfinished,'' ''unprofessional,'' the ''laboratory version,'' or ''not the final saleable package.'' If that is how the term was (and still is) generally understood, it was reasonable to adopt this defensive tone. ''Training'' productions, not meant for broadcast, are certainly not the same as experiments. Genuine experiment breaks new ground with audiences and/or performers and technicians through the use of new conventions, fresh or often controversial subject matter, and new technology. Experiment and innovation can also be the result of the very personal and idiosyncratic vision of one person: a director, writer, performer, cameraman, or designer; it should not necessarily connote esoteric, unfinished, or inexperienced. In any case, it was deemed necessary to have introductions and explanations by Charles Oberdorf before each drama on *Program X*. Perhaps Sampson thought his rather eclectic anthology needed a continuing personality; perhaps he felt the audience needed help. In my experience, presenters—as the British call them (a more honest word than ''host'')—are a superfluous annoyance. The only exception I can recall was when Gordon Pinsent had to explain to an audience in 1980 that, yes indeed, the play they were about to see really was going to air ''live.'' By 1980, live television drama had become the ultimate in ''experimental'' television.

Program X got off to a shaky start because the brass, despite protests on all sides, decided to put *Canadian Short Stories*, written and edited for commercial sale and possible export, in the same time slot as Sampson's *Project* (the working title), although it was a wholly different bouillabaisse. They were going to call both kinds of drama *Theatre Canada*. The fact that both were a half-hour long and on tape seems to have been behind the scheduling and the meaningless title. Fortunately, when the venerable Dominion Drama Festival, then in the process of becoming Theatre Canada, protested that its title had been preempted, the corporation changed its mind and permitted each anthology to seek an identity under a separate title.

The 1970-71 season of *Program X* included *Openhouse*, a suspense play; *Ashes for Easter*, a dramatic monologue; *Musical Chairs*, a film fantasy by

Warren Collins; *The Couch*; *An Evening with Kate Reid*; *The System*, a
satirical play by Eric Koch and Frank McEnaney; *Sniper*, written by Rudi
Dorn, about a young soldier who sees his own life as a surrealistic fantasy
while chasing a sniper at a campus demonstration; *One's a Heifer*; *Boss*, in
which a middle-aged insurance salesman kills his dog, Boss, perhaps to
punish himself; and *Banana Peel*, a science-fiction piece starring Billy
Van—he slips and his life flashes before his eyes. *A Day That Didn't Happen*
showed how prolonged absence changes two former sweethearts. *Charlie
Who?* was a piece of theatre of the absurd in which a husband shrinks until
his wife throws him out with the trash. *Parallel 68*, by John Reeves, was a
drama documentary about the invasion of Melos by the Athenian League in
416 BC and its parallels with the suppression of the Czechoslovakian
"Spring" of 1968.[7] In *Generation Game*, young students and then adults
responded to hypothetical situations. *Nothing to Declare*, by Norman
Snider and David Cronenberg, was a film about a young man faced with
choosing between "Establishment" and revolution in a North American
Civil War of 1982-87. *Lemonade*, by James Prideaux, featured well-known
American stage actresses Eileen Herlie and Martha Scott. Rod Coneybeare's
Gerber's Girls was described in the press releases as "an ironic drama in
documentary form" about a feminist consciousness-raising group being
filmed by a television director (Coneybeare) with an all-male camera crew;
intense personal conflict arose among the suburban housewives, the
director, and a psychiatrist.

One of the most highly praised of the half-hours was *One's a Heifer*,
adapted by "that formidable stylist, Rudi Dorn" from a story by Sinclair
Ross, described by Bob Blackburn in the *Telegram* (25/3/71) as "a boy's
nightmarish experience with a demented old farmer. It's an extremely
effective mood piece. It has nothing to say but if it doesn't get your nerves
on edge, nothing will." *One's a Heifer* starred veteran television actor Ed
McNamara and his fifteen-year-old son, Miles. This is another of the
handful of short drama pieces which have been redone over the years
—along with *The Painted Door* and *Coronet at Night* —most recently on
Bell Global Playhouse.

Concerto for Television (1971), a musical and visual fantasy by Norman
Symonds that combined music and sound effects with film projections to
convey an impression of a wind storm, provoked Bob Blackburn (*Telegram*,
19/3/71) to this outburst: "Blind, dumb, mindless fuddle-duddle . . . an
exercise in pretentious self-indulgence that used so often, justifiably, to
enrage the public whose money is squandered on it and which the CBC
seemed to be getting out of its system—until last night when its switchboard,
predictably and understandably, was ringing with complaints." Apparently,
the CBC could still hit a nerve with a critic and, hearteningly, with an

anthology audience. *Program X* chose to show another "Concerto for Television" in the next season.

From lyric experiment to deft and acerbic Noel Coward, from thrillers to poetry readings, topical and docudrama scripts, from star turns to efforts by unknown writers, *Program X* presented a varied menu, mixing the serious and light, conventional and unconventional, recalling its precursor on late night television in Montreal (1959-69), *Shoestring Theatre—Teleplay.*

Warren Collins's colour film, *The Double-Jointed Turned-On Picnic*, was based on a real picnic and a real marijuana bust. Even in black-and-white kinescope, it catches the easy, "groovy," relaxed tone of the period. Collins achieved this through jump-cuts, a hand-held camera, the use of double speed, glimpses of half-familiar faces from alternative theatre including those of Martin Lavut and Marilyn Lightstone, an exaggerated soundtrack, and the overall style of a home movie. Two detached European voices on the soundtrack, plus the casual style and the inclusion of popular culture images from "the mainstream," cast a slightly sardonic shadow on the hedonism it purported to celebrate. The attempts at surrealism are not particularly successful, but the ironic conclusion is fun. Predictably the picnic is busted, but the "grass" which supposedly created all these playful, subjective visual effects turns out not to be the real thing. The police have no case and the turn-on remains in Collins's camera.

One of the more intriguing offerings in the following season was a one-woman thriller called *The Tape Recorder*, by Australian Pat Flower (dir. Mario Prizek), a script which the BBC had chosen to produce as their first colour drama. Patricia Collins plays a chic, but rather frigid, very self-controlled secretary who moonlights typing manuscripts and is drawn into a terrifying cat-and-mouse game by a demented writer who communicates by tape recording. Melodramatic as the plot sounds, director Mario Prizek did not see it as such: "I felt that the situation was very delicately realistic, that there would be no intrusion by gimmicky camera work. I wanted to focus on the character and the situation and this is one of the reasons why it was shot in continuity—live on tape as a continuous performance [so rare a process that it could almost be considered experi! mental]—rather than by the usual method in film and TV of shooting in fragments. A television play allows, in addition to the verbal text, silent moments. We try, when possible, to have those moments of silence—they're relevant to the character and the strange situation. The chief interest is, I feel, the character. So it was directed as a character piece—not as a thriller" (CBC press release, 24/2/72). What Prizek says he tried for, he accomplished; the taut half-hour was indeed a character study first and a thriller second.

The next two seasons saw a change in time slot for *Programme X* to Friday at 10:00 p.m. The anthology continued largely as before, with *Relax, Old Man*, a comedy by Anthony L. Flanders; an evening featuring Gordon Pinsent; *Hitch*, the story of a lonely hitch-hiker; *Outport* by Angus Braid, shot on location in Newfoundland, and a Canadian gothic horror story by theatre writer James W. Nicol called *The Ballad of Willie and Rose*. *Corsican Justice* was adapted from a short story by the nineteenth-century master, Prosper Merimée. Skilfully edited and very funny extracts from comedienne Barbara Hamilton's successful stage show made up *That Hamilton Woman*. There was also an "abstract story, rather like Kafka's writing," called *The Last Man*, as well as *A Predicament*, a comedy by the experienced television writer and novelist Morley Callaghan.

Mavor Moore, who oversaw the earliest days of television drama, contributed scripts: *The Store*[8] and *Getting In* —the latter a two-hander, starring Gerard Parkes and Colin Fox, about a man who, being interviewed for admission into an unnamed organization that is desperately important to him, is unable to find out the rules of membership. It was described by Blaik Kirby as a "small but flawless gem" (*Globe and Mail*, 28/5/71). He also adapted his delicate radio play *Come Away*, a wonderfully evocative and multi-levelled mood piece, full of gentle comedy and bitter memory, centring on the reflections of an old man as he answers the questions of a little girl gathering leaves in a park. Her companionship gives him new hope, but in the end he discovers that her name, "Morte," is fateful: she is Death, come because, "it's time." The radio version, using the special intimacy and flexibility of that medium, created vivid, fragmentary pictures of people and conversations in the man's past using a few phrases or a pause, which the listener filled with her/his own imagery and significance. Television can be rather too literal a medium for such a play, but here the strong script guided the viewer through the old man's dreams and memories. The man was played by George Waight, an experienced Winnipeg television actor and bank executive who was in his mid-'eighties at the time of production.

Sid Adilman, entertainment columnist for the Toronto *Daily Star* (n.d.) noted the anthology's passing: "*Program X*, under executive producer Paddy Sampson, during its three years on the air has given many emerging talents a chance to have their works presented. Each *Program X* this season costs $7,000, considerably less than its successor [which he mistakenly identifies as *The Play's the Thing*] . . . a prestige series ordered by Eugene Hallman, vice-president of the English language network." Later in the column, Adilman more correctly equates the new anthology with its true predecessor, a hodgepodge anthology of comedy, series, and single plays called *Sunday at Nine*. According to Adilman, *Program X* was cancelled in 1973 to "free up the tape facilities for *The Play's the Thing*." Yet there were no real savings in cash or even studio facilities. When he defended the programme, Sampson had no axe to grind, since he had already requested a replacement; as he pointed out, it was just beginning to pay off in terms of new contributors and internal talent development. He placed *Program X*, accurately, in the tradition of *Quest*, *Eyeopener*, and the satirical revue *Night Cap* (1973) as a training and proving ground for new talent—a commodity badly needed in the early 'seventies, when the corporation had been drained of talent and morale and the Drama Department was at a very low ebb. It seems, however, that the constraints of priorities, production facilities, and scheduling came ahead of the development of actors, writers, and directors, at least until John Hirsch was firmly enough in control of the department to try again.

The Last of Its Kind: Peepshow

Peepshow (1975), initiated in the fall of 1975, was the last truly experimental anthology on the network. Its avowed purpose was to train new writers and directors, all but one of them men, whose roots were in the regional and underground theatre. In part, this reflected the CBC's belated recognition that interesting things were happening all over the country in the theatrical renaissance of the late 'sixties and 'seventies. Under producer George Bloomfield, *Peepshow* helped pave the way for adaptations of all kinds of non-naturalistic Canadian stage plays for television by more experienced producers and directors. If *Peepshow* served in part as a lab for these drama specials, then very briefly, the symbiotic relationship between *Quest* and *Festival* was recreated.

Viewers who sampled this anthology series saw among other things a series of "microdramas," playlets three or four minutes long. *A Country Fable* was a surrealistic piece about a farmer who falls in love with Mary Tyler Moore. *Herringbone* is Tom Cone's fantasy about a child whose parents will him into adulthood. Codco's Newfoundland satire, *Festering Forefathers and Running Sons* (20/11/75) spoofed *Riders to the Sea*.[9] *Peepshow* also presented *Death*, by Larry Fineberg, and a number of other films and tapes, including two quite skilful films by David Cronnberg, *The Italian Machine* which forshadows his obsession with the hazy distinction between "art" and "machine" and *The Victim*. Directors in the training programme included Stephen Katz, Paul Thompson, and Eric Steiner.

Coda: Catalyst Television

Since 1976, the viewer willing to try something different from the largely naturalistic anthology and series drama which dominates the CBC has had to look to the regions. *Last Call* (see Chapter 10) originated in Vancouver;

Wonderful Grand Band, a mixture of music and comedy situations, and *Up at Ours*, a comedy series set in a boarding house full of wildly original characters, came from Newfoundland. In Alberta, *Catalyst Television* grew out of the theatre company of the same name that began in 1977, under Professor David Barnett, as an interventionist or "interactive" forms; plays were commissioned from Catalyst by organizations as diverse as the Alberta Drug and Alcohol Abuse Foundation, the Association for the Mentally Handicapped, and the Alberta Law Association. The philosophy and field work progressed under Barnett and his colleague, Janet Selman, in spaces as various as living rooms, jails, and town halls.

Certain basic relationships were developed between actors and audience. The intention of the theatrical event itself was to build scenarios based on research into challenging social issues like drug abuse, the rights of the mentally handicapped, and life in prison; these were performed for the people directly involved in the problem and the point of the performance was to get them thinking about solutions. To accomplish this, actors would provoke audiences into asking questions, which they answered in character; then they would try to get suggestions for other scenes to be improvised. Finally, the audience switched places with the actors as a sort of laboratory or rehearsal for action in the "real" world, they were encouraged to act out several alternative solutions for the problems being explored.

Barnett and Selman were experimenting in a modern theatrical tradition of intervention and interaction that extends from Brecht, 'sixties street theatre, and political theatre like *teatro campesino* to Augusto Boal's work in the barrios of Brazil, Québec political theatre of the 'seventies and Chris Brookes' Mummers' theatre in the outports of Newfoundland. Catalyst Theatre casts the spectator as protagonist in a drama to which she or he can contribute. The audience is urged to create the outcome itself as a first step to changing their situations after the players have gone home.[10]

Here is the context that accounts for the atypical look and feel of *Catalyst Television*. Taped for television, and shown across the country, the programmes cannot be fully interactive, since viewers cannot create new scenarios themselves. Instead, the television audience saw actors on platforms with props, performing scenes for people who had experienced teen suicide in their families or were mentally handicapped. These people were free to add their comments; others asked questions, then the actors went on with more scenes. Progammes had originally finished with a phone-in that was not televised. Barnett supervised all of the television presentations.

Despite the phone-in, this kind of presentation cannot duplicate the live performances for specialized audiences of prisoners or alcoholics. Most of the television audience will not have a direct stake in changing the rules of the society of which they are a part. But this is not standard docudrama, either: the television versions grow organically out of an intense and intimate interaction between the people who live the problems and the actors trying to help them articulate their concerns. Despite the much more formal conventions of lights, cameras, performers, and the occasions when the audience feels constrained in the presence of invisible viewers, some of the intimacy between performers and audience survives. Moreover, the drama is often to be found in the unscripted comments or arguments from the audience.

Given the circumstances and the medium, information and not interaction is in the foreground. Though the process and purpose of the television version differ from the original, the distillation of that process of interaction still gives *Catalyst Television* a charged, authentic environment, a thrust and energy which will catch the eye and ear of the viewer: a message gets through.

As for the 1980s and into the 'nineties—we wait. Experiment and innovation do occasionally appear in anthologies, specials, and even series. Nevertheless, experiment is now a luxury, like most other kinds of risk taking. Experimental film feeds the mainstream with talent and ideas. "Fringe" or "off-Broadway," or "alternate" theatres nourish the mainstage of London, New York, and our own major population centres; so too all other arts. In Britain, Channel 4 keeps both ITV and the BBC alert. In contrast, American network television shows the effects of being robbed of its few experimental programmes in the 'sixties. If the money and commitment can be found, the choice between reviving a past model, or our more restricted present, is obvious. We, the viewers, wait.

NOTES

1. In the *Stage 44* anthology (24/2/44) and rebroadcast six times.
2. A characterization which also fitted the American context, according to Kenneth Hay in "Marty: Aesthetics vs. Medium in Early Television," in *American History/American Television: Interpreting the Video Past*, ed. John E. O'Connor. Frederick Unger Publishing Co., New York: 1983, 94-113. Note *Marty*, like many other early television successes, became a movie. Thus, *Requiem for a Heavyweight*, *Judgement at Nuremburg*, and others took on second lives in the imagination of audiences.
3. Interview, June 1984. The late Hugh Webster was one of the most reliable television actors on CBC television.

4. CBC documents collection, P.A.C.

5. Ibid. Frank Cooper Associates Agency, New York, 27 April 1964.

6. Interviews with Hugh Gauntlett, November 1983 and January 1984. Interview with Mario Prizek, September 1983.

7. Author and producer of CBC radio's Prix Italia winning *Beach of Strangers* and *Triptych*.

8. One of the few television scripts to be published; Moore's *Come Away* is another. See bibliography.

9. Members of the same troupe appeared later in the CBC comedy series *Up at Ours*, *The Wonderful Grand Band*, and *The S and M Comic Book*.

10. For a more detailed account of the interactive side of *Catalyst Theatre1*, see Alan Filewood, ''The Inter-active Documentary in Canada: Catalyst Theatre's *It's about Time*,'' *Theatre History in Canada*, vol. 6, no. 2 (Fall 1985), 133-47.

10

REGIONAL
or
"What Toronto Doesn't Know!"

Endless words have been spent on the "problem"—the correct term should be "asset"—of regionalism in this country. When regionalism at its most xenophobic is not tearing at the national fabric, it is one of the major strengths of a nation not yet fully mature in a cultural sense. On the one hand, there is the producer in Newfoundland who absolutely would not give room or notice to CBC in-house professionals in contemporary sound and lighting methods if they came from Toronto or, worse still, Halifax; in the end, the corporation had to bring in BBC technicians—and no one seems to have found this mélange of regional pride and leftover colonialism at all unprofessional. On the other hand, the highly original characters found in the boarding house of *Up at Ours*, which originated in St. John's, could have come from nowhere but the salt and rock and roll sensibility of Newfoundland.

We Canadians spend resources we are never sure we can afford to create two complete television systems in two languages with two bureaucracies. We build well-designed production facilities in every region but Toronto, where the majority of English language television is made, only to find with a change of policy or priorities or political masters that there is no money for regional programming. The country's communities lose their chance to project particular points of view, sets of images and ideas about their activities and attitudes. The rest of us are robbed of the fresh infusion of talent and ingenuity regional programming has contributed to the network whenever it was—tolerated? allowed? occasionally encouraged?

The winds from Toronto, then Ottawa, blow, the programmes ebb and flow. Regional drama seems to flourish whenever someone on the spot takes an interest in it. Winnipeg, the home of vigorous indigenous theatre, was a logical place to generate drama with a regional difference. I have found a tantalizing list of thirteen half-hour plays produced in Winnipeg, irregularly broadcast in a 10:30-11 p.m., or 11:15-11:45 p.m. slot from 1959 to 1961; the production record resumes in the late 'sixties with the work of Don Williams. Jack Philips, a regional executive with persistence, managed, after four years, to find a bit of money for Don Williams to make a couple of modest specials each year for the Manitoba audience. Therefore, some quite good drama went out from CBC Winnipeg for local and national audiences for close to a decade until a new regional manager, in the late-'seventies, decided there should be no more of it. In most cases, then, regional drama has depended on the will and talent of one or two local producers and the determination of an executive far enough up the ladder to get them money. In Halifax, and even in Edmonton and Calgary, there has been very little regional drama, although more is in the planning stages; in Regina until the mid-'eighties there has been none.

St. John's is another story. A very successful family show called *Tales from Pigeon Inlet* began as a radio anthology with anecdotal scripts by Ted Russell. Some of these were adapted by Tom Cahill, the researcher, animateur and producer of much of Newfoundland's drama output. *Tales from Pigeon Inlet* began in 1978 with a pilot called "The Hang Ashore," and has appeared as seven more tales at various times around the country as part of regional exchange. In one way, it resembles *Cariboo Country*: the same characters thread through these stories, but no single one is star. Several episodes were showcased on *The Way We Are* (1985-, exec. prod. Robert Allen), a regional anthology shown at 7:30 p.m., but not carried by CBC affiliates. *Up at Ours* was another local show using Codco performers. In 1980, seven episodes from two seasons' worth of shows appeared on Metronet, stations owned and operated by CBC English Television. The wild and wonderful humour of this collection of characters living in a boarding house left me laughing, though with nothing very meaningful to say about them. *Joey* (1983), an adaptation of a St. John's stage hit, and a rather lifeless portrait of Joey Smallwood, was not a Newfoundland production, being directed by Ted Followes and produced by the network. The last drama on the network to come from St. John's before *The Way We Are* was *Sir Humphrey Gilbert*, appearing on the five programme regional anthology, *The Undaunted* (1984). Written, produced, and directed by Cahill, it juxtaposed Gilbert in Elizabethan dress with modern scenes around the city and countryside in a wry entertaining drama about how he came to explore the nearby waters and what he might make of contemporary Newfoundland.

Other purely local productions not broadcast on the network (which I have not seen) were *As Loved Our Fathers* (on Confederation), *Yesterday's Heroes* (1984), and *Where Once They Stood* (1986), an anthology reenacting incidents from Newfoundland history in eighteen episodes. Television drama in Newfoundland seems to have started only in the last decade or so with a little help from Philip Keatley and his training programme, but the quantity produced has been phenomenal. Determination, a dash of pride and the confidence that there was an audience for material targeted for Newfoundlanders has made up for scarce dollars.

Montreal is a special case: English drama originated here from the beginning, in 1952, when the single channel was used by both language groups. A lot of local drama and some network drama originated there in the 'fifties and 'sixties, but output since the demise of *Teleplay* has been very sporadic. I will discuss programming from Montreal and Winnipeg later in the chapter. I begin with Vancouver because over the three decades of television drama in Canada, Vancouver has been the true counterweight in quality and quantity to a Toronto-based vision of this country and its concerns.

In reconstructing the history of television drama in Vancouver, Winnipeg, Montreal, and Toronto, I have not been able to detect any consistent broadcasting policy at either the governmental level or within the corporation concerning the role of regional drama in either the regional or the national programme mix. Since 1975, we have seen the money squeeze from the Liberal government and the body blow of severe budget cuts in 1984 and 1986 by the Conservatives resulting in the near collapse of these distinctive voices—even though there was a regional anthology in the 1985-86 season for the first time in fifteen years. The perception of this anthology in Vancouver, at least, was that their ideas, their style, their flavour—which is, after all, the point—had been run through the sieve in Toronto so many times that producers have given up and withdrawn or else made baby food as ordered. This is not a fair judgement of the anthology, *The Way We Are*, but the perception itself is what is so damaging.

The distrust that exists between producers and area heads and "the administration" was exacerbated in 1985 by the clumsy way in which old and valued servants of the corporation were declared redundant and the widely held belief that the first round of cuts was applied almost at random as far as numbers and names were concerned. This was what I heard, at least, in Toronto, Montreal, Vancouver, and Ottawa in the spring and summer of 1985. Since no one person or group of executives would admit to deciding how many or whom, there was no one to reason with or blame. Internally, the people who make the programmes have seen their numbers and crucial support staff shrink, while administrative personnel do not appear to change in numbers or strata. I have not met one person connected with the making of drama who does not express deep frustation at the gap between themselves, their priorities and the decisions which affect what they do, and the executives down the street, across the country, and especially, up in Ottawa. The second round of cuts has deepened the sense of uncertainty and drift.

A more creative, if frustrating, source of tension in the regions is common to all countries. In this corner, "the boonies"; in the other, "arrogant, ignorant snobs," to cite the sort of invective that is likely to flow. Both are balderdash, of course. Fighting words are common to every clash between regions and network when scarce programming resources are being allocated. Will a given programme be "up to network standards?"—whatever they are. Answers vary. When Don Williams in Winnipeg made his film, *Death of a Nobody* (1967), based on an actual racist incident against an Indian on a reserve, it was not shown on the network until it had been awarded the best programme prize in the CBC's internal "Wilderness Awards," predecessor of the Aniks. Suddenly, it was up to network standards and scheduled for national broadcast. Something good could come out of Galilee, or Winnipeg. On the other hand, though it had its moments, the three pilots shown locally for a sitcom called *Mayonnaise* (Vancouver, 1984, based on a stage play) were not up to network standards, if those standards exclude *Delilah* and *Flappers*. The failure of *Mayonnaise* had nothing to do with lack of technical expertise or even being too "regional" in tone or subject. The basic situation centres around a klutz of a cartoon writer and his loyal, handsome friend who come from Toronto to seek their fortunes in Lotus-land; the concept was defeated by flawed writing and dramatic construction as well as character stereotyping.

As in so many other areas of our political and cultural lives, the creative people in the regions are often paranoid and those at the "centre" either negative or flippant. I have heard in passing: "Why, after twenty years, do we have to keep proving ourselves?" And I have also heard, "There's no way they can make stuff for network out there." "Out there" is all points east, west, or north of Toronto, including Hamilton, where privately owned CHCH, quietly and often in a very entertaining way, produced a two-hour drama special on the War of 1812 and *Niagara Repertory Company*, an anthology that is sold to other private stations across the country. "Out there" also includes privately owned CKND in Winnipeg, where once or twice a year, high quality dramas by Alan Kroeker appear. Four were picked up by the CBC for national telecast: *Hunting Season, Reunion*, and *In the Fall* (1982-83). *Tramp at the Door* (1985), the well-told story of how a prairie hobo enters the life of a Manitoba family, was shown New Year's Day, 1987. *The Prodigal* (1984), though it has not been shown nationally, is an effective study of an artist's love-hate relationship with his dying father. Note that neither CHCH nor CKND is a "major population centre."

Nonetheless, the 'fifties tradition of a local station putting on a play has vanished.

America has no regional drama. Cincinnati, Minneapolis, Maine, even New York are often created on Hollywood sets, and they look and sound like it. One exception is the hit of the mid-'eighties, *Miami Vice*. Being shot on location does not automatically make it good television, despite the fresh visual setting, the contemporary music and two compelling performers dressed in pastels. Nevertheless, I sometimes wonder if the success of *Miami Vice* and *Magnum P.I.* (shot in Hawaii) are directly proportionate to their distance from Hollywood executives.

Here it is on the politically mobilized West Coast that rebellion against centralism has been most clearly articulated. Part of the intervention at the 1974 CRTC hearings on behalf of the B.C. Committee on the CBC centred on the importance of regional and local broadcasting. Like many others on that occasion, the committee deplored the dearth of distinctive Canadian content, indeed of distinctive content of any kind, and the corporation's central Canada-Toronto mindset. By 11 March 1976, "The B.C. Committee to Reform the CBC" was publishing a manifesto in the Vancouver *Sun* "protesting against the steady erosion of regional content in CBC broadcasting, against the steady Americanization of the corporation's offerings, against the steady decline in quality of both television and radio programming."[1] It had then grown to include viewers and listeners, artists, CBC staff members, the Canadian Wire Services Guild, the local branches of ACTRA and the Writers' Union of Canada, the regional associations of CBC radio and television producers, representatives of the Communications Studies Department of Simon Fraser University, and thousands of individuals, among them three hundred citizens from the small town of Canal Flats. The proposals relevant to television were that the CBC restore programming and financial autonomy to the regions, leave mass entertainment to the private stations, and, instead, provide a truly alternative service, an echo of the forthright, largely futile 1974 recommendations of the CRTC. The B.C. Committee recommended further that the CBC develop an interprovincial exchange of programmes—this happened, although for some reason Southern Ontario seems to see very few of these—and that it redress the balance between staff salaries (51 per cent) and the amount spent on talent (11 per cent), as revealed in the CBC's 1976 annual report. Six months later, in a November speech to the Men's Canadian Club called "Reflecting Canada and the Regions" (10/11/76), CBC President Al Johnson promised the regions an extra half-hour a week on provincial affairs, and "that the CBC, over a five-year period, would increase significantly the regional production for drama, music and variety—whichever programmes are most appropriate and indigenous in the several regions of Canada.''[2]

Two problems, still unresolved, arise from that sort of answer. One is that, as Johnson admitted, "this [would] involve a substantial increase in the allocation of funds."[3] Therefore no such increase in regional programming could be achieved then, or in the nine years since. Perhaps more important was Johnson's silence on the central question: who should retain artistic and financial control, the region or Toronto? Who decides what programmes are "most appropriate" or "indigenous"? Among those who choose to remain in the regions, there seems to be a conviction, complicating this discussion, that people who go to Toronto soon think like Torontonians. Yet, surely, one of the better things to happen in broadcasting is that "expatriates" retain their sense of being outsiders and observers; audiences can profit from these outsiders' sharp, imaginative looks at dilemmas, habits, and characters in drama, wherever the scene is set. On the other hand, there is also strong evidence to suggest that little has changed since Gail Martin of Simon Fraser University pointed out that "regional television producers have no control over their own budgets, no decision-making power over program ideas, script submissions or frequently over major production decisions. Everything is decided from Toronto."

In the 'eighties, regional drama can survive to serve its special audiences and reach out to the rest of the country only when there is a head of drama and, even more important, programme planners and executives who will make room for the cream of regional production in prime time. If and when that happens, as it has happened before, the whole country is the richer. We see diversity: new faces, landscapes, attitudes. We see cheeky and, occasionally, challenging probes of the collective psyche, different ways to laugh at and with ourselves. When the regions are gutted of people and resources, we all lose. What a visiting Ontarian like myself sees is buildings without programmes being made, good programming without a place on the national network, and a national network in which a miniseries—*Vanderberg,* about Calgary oilmen—was filmed partly on location after a fierce struggle by the producer who was eventually parachuted in with his crew. It is also worth noting that *Vanderberg* was conceived, produced, processed, and edited in Toronto.

Britain, which has a both fiercely regional and unswervingly nationalistic culture, is served by BBC-1 and BBC-2, with a programme mix that includes strongly regional dramas and serials and special programmes for Scotland or Wales or the North. The independent licences are deliberately given to consortia based in the Midlands, Yorkshire, Tyneside-Tees, or the South-west. In return for their slice of the Independent television channel assigned, they are expected to provide regional voices as well as international hits like *Brideshead Revisited*. So far, this sharing of responsibility has worked because there has been the political will to make the system function and reform it when it failed. The political will to sustain and strengthen our cultural fabric is missing in Canada right now. Herschel Hardin's *Closed Circuits: The Sellout of Canadian Television* (1986) points out how the Board of Broadcast Governors (BBG) and its successor, the CRTC, have refused to enforce the regulations that already exist for responding to new technologies or to act on its own (1974) impeccably thoughtful, nationalist analysis[4] of our situation. This political will may well return, but the talented people who leave will not. The generation that now thinks good programming must come from the United States will continue to grow up in ignorance about its own country. In March 1987, the CRTC did, at last, insist that the collection of affiliates who call themselves CTV spend a significant portion of their revenues to achieve a good four-and-a-half hours of television drama a week and twenty-two hours of miniseries and specials by 1992. They even set a timetable for arriving at that modest target as a further condition. The cries of pain were loud and long, but there was a grudging promise to comply. Given the civil war within the organization between BATON (Bassett-Eaton) and the other affiliates who have equal voting rights, the network itself may not survive to implement the order, nor is it clear that the CRTC will act to enforce it if necessary. We shall see.

In the West, any Ontarian or Quebecker who will listen is going to hear about the perfidy and blindness of the East. Quebeckers add language to their list of grievances, Maritimers see the whole country to the west of Quebec as Upper Canada and *terra incognita,* and the majority of Ontarians living outside Toronto will complain fervently about being ignored by both Ottawa and Toronto. But Toronto cannot help being the cultural centre of English Canada any more than New York, Montreal, Rome, or London can help being the financial and artistic centres of their cultures. Toronto is both a major population and financial centre. The point is that we can have centralized programmes for the whole country and regional programmes to keep us honest, give us a balanced view of ourselves, if enough of us want this to happen. In this book, I have been directly critical of the CBC's lack of pride in its considerable accomplishments, its loss of a sense of what it has achieved and how, and by whom—that is, its strange lack of a sense of tradition. The viewers, with their short memories, are no better.

The relationship between drama and its audience has always been symbiotic. It is a matter, not of ratings, but of the service the public expects from the public and private sectors in return for the privilege of using its air waves. Yet, even with the public subsidy of Telefilm money, neither CTV nor its affiliates will invest in regional drama. In 1985, when CTV in co-production with Scotland financed a much-touted (and quite enjoyable) series about a Scottish settler family, *The Campbells* did not evoke the

Selkirk settlement or Cape Breton, but, again, southern Ontario. When viewers are silent or don't bother to think about the tunnel vision of our television, we get, not the programming we may want, but the programming we deserve.

While usually silent on the marked differences in quality among the American television dramas they enjoy, Canadian viewers are noticeably and sometimes outspokenly hostile to the similar differences in quality in CBC TV drama. Often, the criticism has not been justified; too much good film and tape, too many scripts from all over the country have battled their way on to the network. But there have also been arid periods. There has never been a sustained, thoughtful overview of the relationship between regional and centralized storytelling at the CBC. Yet it is a basic thesis of this book that the stories we tell and the myths we make are far more important to our lives than the ephemera of sports and news.

Vancouver

Vancouver is a long, long way from Ottawa and CBC headquarters.[5] It is also far from Toronto's executives and the Drama Department. Television drama from Vancouver has been a distinctive voice from a distinctive region of the country, suspicious of Toronto centralism and the hated and envied East, which often seems to begin at the Manitoba-Ontario border and end at the frontier of Québec. Vancouver has frequently been ignored and underfunded, yet it has enjoyed the freedom to go its own way, relatively free of interference, out of sight and out of mind. The feeling seems to be that those who choose to stay in Vancouver—as Andrew Allan, Robert Allen, Mario Prizek, and so many writers and actors, editors and cameramen did not—somehow have to keep proving themselves, time after time after time. Too often, around the English network corridors, people say without much thought, certainly without knowledge of what has gone on over the years, that the regions cannot make good drama—at all, ever. Of course, this is nonsense.

There are several reasons for Vancouver's preeminence. First, a great many talented contributors chose to live in B.C. It had a CBC radio production centre, an NFB unit, and active theatre groups to nourish local television in the early days. Thirty years later, Vancouver still supplies drama with a "made in B.C." stamp. Vancouver was also at the end of the line before there could be simultaneous or even delayed transmissions. When kinescopes had to be "bicycled" across the country, the tendency was for each region to make programmes with three or four cameras in some sort of studio that were not prohibitively expensive. As the days of studio anthology waned, Vancouver was an invaluable help in the transition, thanks to its persistence in hanging on to its own film unit. Another factor was that until 1969 there was no formal head of TV drama to veto projects, though major expenditures for series in particular, had to be sold to Toronto-centred network executives. The consequence was that, in the period from 1968 to 1973, when network drama had reached its lowest point in quality and coherent vision and Montreal regional drama had been closed down, there was a group of producers in Vancouver happily, and creatively, making film and studio television drama. This regional drama department was partially dismantled during John Hirsch's era in one of the periodic swings toward centralizing production.

CBC television drama in Vancouver has had a long and honourable history, starting with good children's programming like *Hidden Pages,* on which directors learned their craft and writers learned how to dramatize prose, tell a story swiftly and clearly, and build characters. By the late 'fifties, Vancouver was producing its own anthology, *Studio Pacific*, as well as contributing to *Playbill* (1953-64), the network anthology of regional drama. Also in the late-'fifties, producer-director Ron Kelly was trying his hand at documentaries (*Spanish Village*) and television drama: *Seeds* (n. d.) was one result, and another was a beautiful mood piece called *Object Matrimony* (1958) for *Vancouver Playbill*. This range of experience stood him in very good stead when he devised the excellent and controversial *Open Grave* in 1963. Throughout this period of growth and change, producers and film crews were not frozen into naturalism as the only mode of storytelling—and are not to this day, if some of the episodes of *The Beachcombers* or Don S. Williams's *Last Call* (1983) are taken into account.

The 1950s reveals a regional focus on the Anglo-Saxon attitudes of those who are either expatriate or lovingly imitative of Empire. There are also realistic and unselfconscious portrayals of older citizens as protagonists and subsidiary characters. The question of the place of the first peoples in history and contemporary life recurs again and again as a matter of course, both centre stage and in the background of the stories. Vancouver as "Lotus-land," the place where the 'sixties drifters washed up after their cross-country treks along with their values, sensuality, drugs, and general aimlessness comes to the forefront in the drama of the late-'sixties and early-'seventies. The whole country shared this preoccupation, but the problem was particularly acute for the West Coast, and the north-south influence of the laid-back, tolerant style, together with the weird and wonderful feudal politics of B.C. lend the clash of generations a different flavour from, say, *Reddick*. The major premise of *The Manipulators* is that of the good parole officer: listen and let clients make their own decisions. The comparison here is with the Canadian and American series dedicated to actively investigating, fixing, power-broking, and interfering in the style of *Wojeck*, or *Quentin Durgens, M.P.*, or *McQueen*. The individualism of both the interior (*Cariboo Country*) and the coast of British Columbia (*The Manipulators*, *The Beachcombers*) contrasts strongly with the "fix-it, now" mentality of much of central Canada. Especially in *The Beachcombers*, too much interference usually has unintentionally ludicrous or embarrassing results.

As film becomes a major medium for television drama, this city, with its mountains, sea, lush growth, elaborate gardens, oddly shaped clapboard houses, houseboats and bush planes, log-booms in the harbours and windswept beaches in the city itself, with its casual dress and prosperous, middle-class populations of Japanese, Chinese, and native peoples, its larger numbers of elderly and kids hanging out in the temperate climate, the bicycles and walking malls—this city so different from the rest of Canada has been recorded on film for television from the 'fifties to the 'eighties. What particularly strikes this inlander's eye is the extensive, unselfconscious use of the mountains, the great, green rain forests and, especially, the sea in all its moods. Vancouver becomes a special visual presence in drama as a counterbalance to Toronto.

In Vancouver, television drama began as it did everywhere else in the 'fifties, with producers learning virtually on the job on technically simple programmes, then graduating to half-hour local studio anthologies with titles like *Vancouver Playbill* (1958-59), directed by Philip Keatley and others with supervising producer Frank Goodship, *Vancouver Television Theatre* (1960), and two incarnations of *Studio Pacific*: the first in 1959, with Philip Keatley and John Thorne alternately directing, and two more short anthologies in 1966-67 and 1967-68, also called *Studio Pacific* and directed by Len Lauk, Elie Savoie, John Winther, and others.

The Early Days

Ron Kelly used a pick-up crew (accustomed to documentary and news production to do one of the first film dramas of the 'fifties: "Object Matrimony," which he wrote, produced, and directed. It is a wry and carefully observed story with a meticulously detailed pair of sets, photographed with a depth of field rare in television at that time that reads clearly even on the faded kine. It was beautifully acted by the two principals. The story is simple: eager with anticipation, a bachelor in his sixties goes to meet a "good Christian widow" with a laughing face. The suspense revolves around, first, whether a sour-faced look-alike is the lady he has come to meet, and then, when the right one does come in, whether he will put his identifying rose in his lapel and declare himself. His agony of indecision and her growing pain as no one introduces himself-herself transforms the vigorous little comedy of the first ten minutes into a small but moving tragedy. There is no dialogue as we watch closeups of him twisting the rose on his lap intended for her, then letting it fall as he loses his nerve. The largely visual piece works very well indeed.

Kelly did two other films: "Bit o' Bark," now missing, and "Seeds," about the violence of a gang of young kids to a small child; some correspondence in the Public Archives expresses the doubts of some of the brass concerning its subject matter. I have not seen it. Dennis Duffy of the B.C. Provincial Archives did see it, and thought that it did not have particularly gratuitous violence. Edward Bond, one of Britain's master playwrights provoked similar reactions three or four years later with *Saved*, but that play did get on stage despite the censorship by the Lord Chamberlain. Kelly lost his battle, but went on to win a different one when the CBC decided to air *The Open Grave*.

Studio Pacific

After a hiatus of eight years during which Vancouver was producing *Cariboo Country*, a commercially sponsored anthology went to air called *Studio Pacific*. I saw four: "How Beautiful with Shoes," " The Treasure," "Some Are So Lucky" (which had been presented live twice on the network, in 1956 and 1960) and "The Clubman" (n. d.). As with most other anthologies they vary in quality. "The Treasure" was a melodrama in design, lighting, makeup, acting, and photography, with most of its suspense quite spurious. "Some Are So Lucky," is a working-class drama typical of its writer, Hugh Garner. A bored sales clerk (Lilian Carlson) trapped in a dull marriage meets an old flame, then gets progressively, embarrassingly, and belligerently drunk; the last shot is of her sitting alone. The characterization is skilful, the character rather difficult to spend a half-hour with.

Lilian Carlson appears in many West Coast dramas with Barbara Tremain, Merv Campone, Terence Kelly, Frank Wade, Robert Clothier, Rae Brown, Lee Taylor, George Clutesi, and many others. As there is a sort of CBC Toronto acting company and another in Montreal until 1970, so in Vancouver certain skilled actors appear again and again.

"How Beautiful with Shoes" (wr. Wilbur Daniel Steele, prod.-dir. Elie Savoie) was an uneven, yet interesting mixture of melodrama and intricate character study. A psychopath captures a restless, pretty, rather vapid young farm girl named Amarantha, who is plagued with inchoate longings for some beauty in her life. The posse's hunt for the psychopath and the clichés of dialogue, clearly derived from Hollywood Westerns, set up this formula thriller. However, there is a ten-minute interval after he has kidnapped the girl in which his sensuality and poetry alternately repel and ensnare her imagination and ours; she is also moved by his need, his worship, and his touch. Through flashbacks, we begin to get threatening glimpses of his disorder. He attacks her: she stops him from choking her by rejecting his romantic image and screaming "I ain't Amarantha, I'm Mary, Mary." "Mary, Mother of God," he says, as he seeks comfort in her lap and goes to sleep. Eventually, they are found, she wrenches herself away, and he is shot. In the last scene, after she has pushed away the pawing puppy love of her steady boy friend, we see her carefully laying herself down on the bed, hearing his voice in her head, remembering his strangeness and kindness, not sure, finally, whether she is Amarantha or Mary.

"The Clubman," written, produced, and directed by Len Lauk, is the most interesting of the four. An epitaph for the old colonial values and a "ruling class" that had graces and standards as well as oligarchic power over the province's affairs, it is also an experimental film using extensive flashbacks. It begins with Colonel Geddes's graphic and fatal heart attack. We then see what led up to his death. First, he meets with old friends—a doctor, a minister and a businessman—to decry the loud business types who are now members of his club. Then he begins to think that everyone is laughing at all he stands for: the boy playing in the park (himself as a child); the young man home from World War I who joyfully greets his young wife (himself as a young man); the contemporary lovers in the park; and the nouveaux riches at the club.

There are no villains. The driven, rather loud businessman, Hepburn, actually has respect and affection for the Colonel despite a previous snub—a nicely observed moment of ice-cold courtesy in which the timing and closeups say it all. Later, the Colonel happens to overhear Hepburn's casual remark that the memorial plaque on the fireplace lists only officers and no "other ranks." Enraged, he fells Hepburn with his cane. To his mystified friends, he explains: "Someone had to shout, 'Follow me' . . . with those ridiculous whistles. They're all dead, you see." In his old age, he has struck a blow for his generation against all the laughter he thinks he hears. Ironically, because he has upheld the standards and honour he believes to be under attack, the club committee is compelled to expel him for violence against another member. The play does not, as cliché would demand, circle back to the heart attack. Instead, remembering himself as a child, the Colonel runs into a flock of geese. The last shot shows him walking down the park path towards a very modern highrise, upright and satisfied—a member of a club glimpsed in this play before the "club" values truly disappeared. Despite some old-duffer dialogue and the rather obtrusive distortions of a fish-eye lens, plus an echo-mike on the laughter, the piece is fully engaging. The visual imagery and settings are evocative and the character interplay complex. "The Clubman" is a non-standard thirty-five minutes long. Perhaps it could have been trimmed, perhaps not. What interests me here is that, as late as 1967 in Vancouver, a play was still allowed to find its optimal form.

Where the Action Is—A Centennial Anthology

In 1967, Vancouver also did an anthology in which writers speculated on the circumstances and actions surrounding particular historical incidents. The anthology, called *Where the Action Is*, was under executive producer Philip Keatley. Again, I saw four: "Spud Murphy's Hill" by Eric Nicol, "Tumult With Indians" by Betty Reid, "Cataline and the Judge Boy" by Paul St. Pierre, and "The Kamloops Incident" by George Ryga.[6]
"Spud Murphy" (prod.-dir. John Thorne) is based on an RCMP report about a local moonshiner. Nicol's version is delightful entertainment; it was photographed with a tremendous feeling for landscape (and the seductive curves of the still) by John Seale. "Tumult with Indians," its title drawn from a Hudson's Bay Company journal and also produced and directed by John Thorne, simply does not work at all. With its black-and-white cardboard cutouts and clumsy exposition, the confrontation between James Douglas, then a young Hudson's Bay clerk (later Governor), and the Carrier Indians, whose customs he violates to prove some point, looked and sounded like a popcorn matinée. In unconsciously ironic contrast, it closed on a completely different note of dignity and truthfulness as Switzer read from the chief's epitaph, which states quite simply that he "had the life of the late Sir James Douglas in his hand and was too much of a gentleman to take it."
"The Kamloops Incident" and "Cataline and the Judge Boy" are both Westerns of a different stamp in which, respectively, American cowboys are prevented from lynching a suspected thief in frontier-justice style and a land speculator is stopped from getting some pasture from a well-respected but illiterate Basque pack-train master named Cataline. The language is rough and colloquial because both Ryga and St. Pierre have an ear for the regional rhythms and idioms of actual speech, and because the regional executives did not demand washed and ironed language from dramatic characters. Both are very entertaining.
Leo and Me (1970s) was a comedy series conceived originally for Chuck Shamata with Michael J. Fox. It was shelved, then revived and given a regional airing in a dozen taped episodes. Since all tapes but the pilot have been wiped, there is no way of judging whether the CBC was justified in not putting the series on the full network. In the 1970s, Vancouver's energy went

into series: *The Manipulators* (1970-71) and, in 1971, the longest-running, most popular and most successful export of them all, *The Beachcombers*.

Drama Specials

Stacey (1975), *Kaleshnikoff* (1974, wr. Paul St. Pierre), *A Gun, A Grand, and A Girl* (1976), and *Ma!* (1983) were among the few drama specials which came from Vancouver after 1975. *Sacrament*, for *Performance* (wr. W. O. Mitchell, prod. Philip Keatley, dir. John Davies) is one of the most interesting. Its best moments are very good indeed, chiefly in the emotional confrontation between Keith, the "kid" (a fine performance by Ian Tracy), whose father has left and the Sea Captain (Murray Westgate), who has been building a small sea-going ship in the basement. When that dream dies because of the old man's health, Keith confronts him on the ferry boat ride which the Captain tries to offer Keith as a substitute. To the boy, his idol is now just another lying grownup in his life . His scream of anguish and anger echoes back to the old man as he scoots down the ladders of the ferry. Running after him, the old man falls and dies of a heart attack.

The rest of the play concentrates on how the boy learns to cope with this second loss. In the end, the close relationship of mother and son reasserts itself. The truthfulness of the dialogue, direction and central performances creates some very affecting moments indeed. The drama ends when the mother (Mary Pirie) picks up the heavy electric saw with which young Keith had first tried to finish the boat and starts to work on the skeleton of the boat with her son.

However, the structure of the play is too episodic, and weakened by stereotypes of a nosey old maid and a swish homosexual. The plot meanders somewhat and the climax comes early. Worse still, it is nearly ruined by saccharine music which undercuts the eloquence of the old man confronting implacable reality and the agony of the child coping with inevitable disillusionment. The basic situation is, of course, a variant on Mitchell's explorations of boyhood and the wisdom of surrogate fathers, which goes back to *Jake and the Kid*. According to Keatley, this programme was thrown away in the schedule; it seems to have been broadcast on New Year's Day.

Ma! (wr. Eric Nicol, prod.-dir. Philip Keatley) is a very interesting as well as entertaining adaptation of Nicol's stage play, successfully combining the conventions of live studio drama (unit sets, area lighting, direct address to the camera), taped studio drama, and location shooting in a lively, funny, and fast portrait of newspaperwoman Ma Murray. Joy Coghill's performance was a pleasure to watch.

In 1980, Don S. Williams had taken over as executive producer of *The Beachcombers* after being a one-man drama department in Winnipeg. He joined other drama producers Elie Savoie, Hugh Beard, Jack Thorne, Brian McKeown, and Neil Sutherland; Sutherland developed Vancouver's latest shot at a series, *Red Serge* (1986, 1987). It began as an idea Williams had for an anthology about the experiences of RCMP spouses over the decades. Two unrelated stories were made as pilots. The rest were about three daughters and their rascal of a moonshining father, all Americans, who stalk and become engaged to three Mounties. Both the pilots and the connected stories are pleasant entertainment, reminiscent of the less serious episodes of *Hatch's Mill*. The running characters were likable, funny stereotypes; one an ingenue, one raised to be a refined but motherly young woman, and one an awkward, fearless tomboy. The Mounties were even less differentiated. Four episodes of *Red Serge Wives* (its initial title) were initially broadcast all in one week to the B.C. region in March 1985 to test the market, all well advertised. In 1986, six episodes were seen on the network as part of the regional anthology, *The Way We Are*. In 1987, six more episodes appeared together under the title *Red Serge*. They continue to be good family entertainment with some very funny double entendres for adults. However, the death of the fine actor, Ed McNamara (who has appeared in CBC TV drama since 1952 and who played the father) may put the series back into limbo.

Vancouver also produced *Nellie, Emma, Daniel and Ben* in the 'seventies, and, in co-production, two family adventure shows: one with West German television, the kidult *Ritter's Cove* (1979-81), artistic control remaining with the CBC, and *Danger Bay* (1984-). In the latter case, though the CBC is one of the partners and guarantor of the network prime time in Canada that ensures the Telefilm money, either the CBC has virtually no artistic input into the series—John Kennedy says it has, but the Vancouver people say there is very little CBC control—or the standards for good family adventure have changed since *Beachcombers* days. As argued in Chapter 3, I find *Danger Bay* derivative and improbable. However, the ratings and overseas sales for *Danger Bay* are good.

Two other pilots for series were developed with regional funding in Vancouver in the 'eighties. One was the atypical sitcom, *Mayonnaise* (exec. prod.-dir. Don S. Williams and Brian McKeown), written by John Ibbotson, the author of the stage play on which it was based, and Edmund Heeley. The elements in *Mayonnaise* are egg-in-the-face slapstick, the vinegar of a cynical view of urban life, with the salt of sexual farce and the pepper of very urban, Vancouver one-liners and the soothing oil of stereotyped situations and gags, backed by a really funny, detailed apartment set. Unfortunately, they do not blend; as in an unsuccessful salad dressing, they separate into layers. Despite the fact that this was a sitcom with a memory and some good wry secondary characters, the show did not become a network offering and there was no more regional money for further development to get it right.

Constable, Constable (wrs. Michael Mercer and Lyal Brown) exists as four episodes at the time of writing. The executive producer for the first two (1984) was Don S. Williams and, for the next (1985), Robert Frederick, both of whom are credited with creating the series. Jackson Davies, as Constable John Constable from *The Beachcombers*, got very warm responses from audiences and is an experienced actor with a considerable comic talent: it thus seemed logical to the CBC to create a "vehicle" for his talents. The basic situation is that Constable John has been transferred to the ultimate dead-end of the RCMP, the detachment on the UBC Campus and the comfortable housing development on the university endowment lands. The detachment's Sergeant thinks that "out in that lawless wilderness there are 20,000 students," a campus full of disorder, potential riot, and "Commie" Russian chess players. The other characters are Danny, a handsome, funny guy and an old friend of John's who lives on a houseboat, drives a Corvette, and runs charters, and Constable Roxanne Douglas, a competent, warm, young officer who checks Constable John's well-meant protective stance in the first episode. The cast is rounded out by John's sister and niece; a female clerk with the manners of a tank, and Corporal Lamontagne, a wry francophone who had an affair with his sergeant's wife and is in exile at UBC. There is a perceptible change in tone between the two pairs of episodes. The first two are primarily comic, with overtones of adventure, while the second are closer to copshows than comedies: all are entertaining in a somewhat more adult key than *The Beachcombers* or *Danger Bay*. Regrettably, the series has not been developed further.

One of the most innovative experiments in television forms and formulae in the 'eighties appearing on the CBC was made in Vancouver. Morris Panych's musical, *Last Call* (1983, wr. Morris Panych, prod./dir. Don S. Williams) is a "post-nuclear holocaust cabaret," adapted from a theatre cabaret devised for Vancouver's Tamanhous theatre. By rights it should be analyzed in the chapter on experimental drama. However, it has never been broadcast east of the prairies.

"*Last Call*" is performed by the two remaining survivors of a nuclear war in a wrecked television studio in the futile hope that someone else may exist to watch. It is rare in contemporary television, and very refreshing, to see a programme both imaginative in its conventions and so accessible: my university students are always "knocked out" by it. Most have never seen a writer or a director play with the medium itself in this kind of self-reflexive, biting way, much less in the service of an acerbic look at the causes and effects of nuclear war.

The music is the key to persuading an audience to stay with this painful subject. The switching, inserts, time-tape loops, and certainly the use of cameras as analogous to weapons do not really detach viewers from what they see. The convention that no one but the two on-camera characters is present in the studio is essential to the premise of the plot. However, the reassuring implication that there may be "someone else" behind the camera, also manipulating the image, becomes evident for the attentive viewer once the dramatic situation is established. This simple fact helps protect the audience from the relentless thrust of the material. The audience is thus more likely to consent to become deeply involved with the questions of why there has been a war, why everyone else has died, to follow the chronology of feeling in the songs, maintain interest in the autobiographical details and, above all, to deal with the relentless logic of the play's last moments.

"Gross" is more convincing than "Morose" as a performer, except when blind Morose is singing on his own and whenever he is confronted by the gun Gross uses to control him. His middle-class bathrobe contrasts with the ill-fitting, tattered clothes (with one tell-tale bullet hole in the back) worn by the "rubbie" Gross. The economy of detail in the costume changes— sweatbands, hats, umbrellas—also assumes the viewers' willingness to accept a common convention of theatre and old television comedy that one person may play many people. Gross's Jekyll and Hyde makeup, a product of the nuclear logic in that half his face is ruined, works well. The designer's use of the cables and broken masonry breaks up the monotony of the set, creates interesting blocking, and gives texture and shape to the composition. The high tech, now in ruins, produces a nice contrast to the bits of "sets" recycled from variety and game shows used in the various musical numbers. The pools of light and shadow also allow for surprising images and imaginative blocking.

In *Last Call* Panych inverts the logic of Beckett's *Endgame*. Blind Morose is not the master: the mobile Gross, overtly psychopathic, holds the gun and can see. The stories within the story have a much more explicit, if less resonant, significance for contemporary North America. Most important of all, the mutual tormenting and mutual interdependence lead not to "something taking its course," but to death. Where *Endgame* is a richly resonant parable that can be read in many ways, *Last Call* is sometimes too specific in its satirical, despairing thrusts. In fact, the mixture of Brechtian song-commentary has overtones of Weimar cabaret. According to Williams, there were real commercials in the programme, but the play's structure must surely have subverted them. The behind-the-scenes convention of visible television apparatus with out-of-focus cameras, ragged dolly shots, buttons pushed randomly to produce commercials or the CBC logo, the mixture of parable and polemic create the multiple perspectives of a play by Pirandello.[7]

Yet *Last Call* is not derivative. It does belong to a certain modern theatre tradition which the average viewer does not have to know anything about, placing the very local and topical allusions that make it so vivid to a West Coast audience in a wider context. Most of all, the play's structure, specifically the false ending with lyrics hinting that brotherhood can ensure that "we'll be here tomorrow," bubbles connoting fragility and balloons that burst floating under the credits, worked beautifully to lull us into a Lawrence Welk "champagne" finish, which is then superbly undercut. One naturally thinks that when the credits start to roll, the play has ended. Only on a second viewing of the tape did I discover the truth. Gross and Morose do not make friends in their dying over a glass of beer. They do not even coexist, as their prototypes do in Beckett's *Endgame*. For no "reason," Gross kills Morose . . . and that is how the programme ends.

Both the archetype of the "odd couple" and the eternal optimism of most television programmes reinforce the shock and surprise of the play's last few seconds. Audience expectation reflects a deep desire for tolerable endings. Even the BBC's very different but harrowing two-hour film, *Threads*, shows us survivors, although it does end with a child giving birth to a—we don't know; the last frame is her face frozen in a scream. In fact, some sort of survival is implicit in most science fiction about post-nuclear societies. Such an assumption is necessary to get on air a programme like the heavily publicized ABC ratings-blockbuster, *The Day After* (1983). In its surreal, absurdist, highly stylized conventions, *Last Call* is fundamentally more truthful and thus, paradoxically, more realistic than either the documentary style used in *Threads* or the elaborate special effects of *The Last Day*.

It is seven or eight minutes too long. The scene of transformations, where Gross becomes a judge, a clergyman, and so forth, is redundant and slows down the pace. The relentless drive of the numbers becomes repetitive. The numbers which either tell us a story or give us a glimpse of these two before the event works best, as does the post nuclear lullaby near the end. But overall, the basic form, a television cabaret of stylized song and dance, works as a satire about society as flashback, as a summary of arguments about the politics and sociology of the arms race, on satire on television forms, and television's narcotic effect. It is also a drama about a primal conflict: the relentless struggle for power and survival between two men.

Alone, a spate of programmes cannot change public opinion or persuade those who have not made up their minds. The CBC's producers chose to work on a whole series of anti-war plays in the early 'sixties, but Canada did not "ban the bomb." People do not make up their minds on the basis of movies like *On the Beach* or *War Games*, or even the NFB's Oscar-winning film, *If You Love This Planet*, but by what mass media psychologists call "the two-step process": 1. a broadcast, or screening, and 2. live (not televised) discussion by credible people known to the viewer. The famous case of *Cathy Come Home* (1965, prod. Tony Garnett, dir. Ken Loach, wr. Jeremy Sandford), also broadcast by *Festival*, is a case in point.

Nevertheless, I think *Last Call* should be shown on the network. Although it is accessible to most viewers, there would be some hostile responses to it since it does not look like ordinary television. *Last Call* and the CBC would be accused of being flippant, pathological, despairing, revolting, too complicated, to frenetic, too stylized. Fine. The CBC needs more explorations of form, more innovations, more risks and more *ideas* as opposed to issues in its television drama right now. The subject will not date. For those who do watch, it will engage the mind as well as the emotions, with perhaps a more lasting effect than a two-hour ABC movie set in Kansas, which copped out about the most fundamental question of all: "can the species survive?" The American answer was "yes, although the suffering will be appalling." The British answer was, "yes, but it will probably not be worth it." *Last Call* says "no." However, I reiterate my argument; that it should be broadcast on the network is not primarily because of the subject matter, but because it is interesting television.

It would be inappropriate to assume that experimental television will never be seen again, or to mourn the passing of experimental drama and the ways its techniques and conventions animate more conventional television. Yet without its yeast, the daily dough of television drama is far less likely to rise. If experimental drama disappears altogether, audiences will either tire of the resulting blandness or seek their "sliced white" on readily available American channels. Programme slots where directors, designers and, most important of all, writers can try new ideas are essential to the health of television drama as a whole. Risks are essential to creative work, even if only 75,000 or 100,000 of us watch the results at 10:30 on a Sunday night. At present, experiments in television form are rare on the CBC and, of course, almost unknown on the other networks. Independents seeking foreign sales do not show much inclination to experiment either. This leaves the CBC the source of our only alternative programming, where the range of styles and subject in our drama has been too narrow for too long. The regions are a logical seedbed for such activity. Regional as well as national experimental CBC drama will reappear. It must, or the daily bread of Canadian TV drama will grow stale, flat and, inevitably, unprofitable.

WINNIPEG

Marilyn MacKinnon, in the *Winnipeg Free Press* (1/1/82), interviewed Don S. Williams, who had been responsible for most CBC Winnipeg's TV drama. Williams told her that he had left for Vancouver "with a mandate to get local and network drama going, through innovative or, if necessary, subversive means," very much in the twenty-year-long Vancouver tradition. What Williams did not say about the move was that he had been told quite clearly by the new Winnipeg manager that all drama in Winnipeg would be closed down forthwith, for reasons not specified. Local managers are free to call their own priorities: sports, variety, current affairs, or (rarely) drama. Yet, as Williams pointed out to MacKinnon, up to that point CBWT had been a traditional source of work and support for the local performing and creative community.

Williams commented to the reporter that there had been a drastic decline in Winnipeg output from 1977 to 1980: "It's sad. One of the things we had going for us was the typical Winnipeg spirit. We had a lot of things against us, like limited services. . . . Vancouver seems to be filled with ex-Winnipeg TV people from the CBC and the private stations." When Williams left, CBC Winnipeg English drama died. In 1983, the Winnipeg film unit was almost phased out. By 1985, only two of twelve film editors remained. It should be noted, however, that TV drama continued to exist on the French service. Since 1981, the CBC French division in Winnipeg has also been making drama for the Radio-Canada network on a regular basis, including *Pierrot* (1980) and *Anne Gaboury* (1985), produced and directed by Leo Foucault, an executive producer at CBWFT.

Meanwhile, CKY, the local CTV affiliate, taped the Manitoba Theatre Centre's production of *The Tempest* with the help of the Department of Education, and CKND initiated a drama series in 1981 called *Front Row* as well as commissioning an original work entitled *Autumn Lady* from the Royal Winnipeg Ballet. CKND's *The Prodigal* won the first conference award ever given for foreign professional programming at the Fourth International Conference on Television Drama at Michigan State University.

It was shot on location in Winnipeg, written and directed by Allan Kroeker, adapted from a short story by Guy Vanderhaeghe, and starred Ed McNamara as the father, blind and dying of cancer, and August Schellenberg as Tom, his alcoholic and estranged son. *The Prodigal* explores the anger and misunderstanding between the two men and how communication is finally achieved when the son decides to draw his father. Tom, who runs away from his family and emotional commitments, always paints his father from the back; yet in one picture, off in the distance, we also see a boy running toward the square, strong figure. The paintings and drawings were superbly done by the designer. From the father's perspective, however, they have been accomplished "by sucking on the federal tit." Tom's alcoholism is another barrier between them. Yet they join in resisting the "good" daughter who acts as peace-keeper. Fortunately, cliché is avoided. Her exasperation and militant middle-class values are tempered by real love for this irascible old man and his self-destructive son. Nevertheless, "Snookums," as her father calls this mother of teenagers whose husband puts the kids on the line to blackmail her into coming home, is sometimes the only adult in sight.

Tom's flophouse accommodation, his binges, his walks around Winnipeg to become reacquainted after the flesh-pots of Vancouver are treated in a direct-cinema style. The scenes in the hospital and at the house are presented with the more standard television conventions of domestic realism. Using yet another visual style, the camera explores Tom's pictures with slow rhythms, with real sensitivity, and a sense of unfolding discovery.

By the time we accompany Tom and his sister to the father's small bungalow, we are really interested in these people. The meticulously neat and professional carpenter's workshop is a surprise, the contents of the drawers heart-breakingly private. No sentimental music mars the moment of Tom's response to the bench and tools. Later, he sits by his blind father's bed and draws his hands, not his face. Then, as sarcasm and silences give way to wary affection, the son at last draws his father's face. The old man, touching the drawings, says, "Do you think you can sell them?"—a sign of real worth to him. The son takes them back silently. The father, sensing Tom's withdrawal, asks savagely, "Why do you hang around?" Tom, wordless and in tears, says quietly, "I don't know." The last scene is also wordless, with an obtrusive voice-over that does not quite spoil our sight of the son pushing his father in a wheelchair under the trees while both enjoy breaking the hospital rules as well as each others' company. The camera pulls back, leaving the two figures to disappear quietly behind a tree. *The Prodigal's* biblical allusion is apt. The father will not recover, but the difficult, painful reconciliation eases his way.

The drama output of CBWT, really of Don S. Williams, was varied indeed. The programmes available, the others being unfindable or else destroyed, showed a wide range of subjects: ranged from a thriller, *Every Night at 8:30* (1971), co-produced with Edmonton; the unlikely friendship of an Indian and an elderly widow, in *Melda and the Ducks* (1976); a controversial short series about a cowboy riding into town to get a job as assistant inspector of the city water works, a satire on television and local politics that appealed only to university students (1976); *The Larsens* (1976), combining social satire and suspense; *Puptent* (1978), about the male mid-life crisis; *Leona* (1978), a profile of forty years of marriage, and a really wonderful, imaginative and moving version of the Royal Winnipeg Ballet's adaptation of George Ryga's play, *The Ecstasy of Rita Joe* for *Musicamera* (1974); various half-hours in the children's series *The Magic Lie*, introduced by W. O. Mitchell and co-produced in Winnipeg and Vancouver, Edmonton, Halifax, Montreal, and Toronto. (Every producer found his own budget but cooperated on the anthology content. The executive producer was Philip Keatley).

Heretical to my other love (Canadian theatre) though this may be, I found the ballet of *The Ecstasy of Rita Joe*, which makes extensive use of Ryga's lyric passages and major characters, far less didactic, far more moving than the play and, at under an hour, just the right length. In other words, it is a real improvement on a famous but flawed piece of theatre.[8] The ballet worked for dramatic as well as choreographic reasons: Ann Mortifee's music and her pure voice both articulated Ryga's fine lyrics and served as another solo instrument in the orchestra. In this version, although the famous speech by David (Chief Dan George, who originated the role) about Rita Joe as a dragonfly breaking free of the chrysalis is intercut with her rape and death, the basically pessimistic tone of the play is altered. Instead of being unrelentingly a victim, Rita Joe gives full expression to her "ecstasy"—her capacity for joy and passionate love as well as her later self-hatred and helplessness are expressed fully in the dance. As a dramatic character, she is appropriately inarticulate; in dance, the subtext finds physical as well as verbal expression. Williams did an extraordinary job of imaginatively realizing the potential of the precise use of film, tape, specialized lenses, back projections, matt chromakey and graphics, mobile sets, symbolic objects looming large, poetic and atmospheric lighting, all balanced by sequences of grainy, documentary-style photography. All of these visual elements enhanced rather than upstaged the choreography. Ana Maria de Goriz is a superbly sensitive actress as well as dancer, and she was ably supported by Salvatore Aiello as Jamie Paul and the Royal Winnipeg's corps de ballet.

CBC Winnipeg is not palatially housed, though the 1985 cuts left empty desks and underused studios. Williams obviously preferred to work on location. Thus his films, like Kroeker's, give the viewer some sense of the broad, flat streets and narrow sidestreets that stretch on and on to a vanishing point, Late Victorian buildings, the river, the French fact, the Métis, Indian, Chinese, and Slavic faces, the old Jewish community, the youngish population, and the wide skies opening up to an easterner's eye between every building.

Some of William's dramas were less successful than others. *Puptent* was painful in its adolescent-middle-aged fantasizing, literally minded, with rather dull fantasy seductions; no better and no worse than many other plays on a similar theme featured on the network in the late 'sixties and through the 'seventies. *Every Night at 8:30* was an improvement, but some of its exposition and pop psychology is very dry as dialogue. Moreover, it telegraphs its punchline about halfway through. Still, the ancient and terrifying fear that a mask may one day merge with the face beneath it still has archetypal power.

The Larsens, by playwright Sharon Pollock, is really two plays in one hour. The first shows us in moving detail what it is like to depend on an old-age pension cheque that inevitably runs out before the end of the month. To solve this dilemma, a feisty old man, Teddy (played by George Waight), and his wife, Katie (Jane Mallett, who does the planning), shoplift and then sell goods to gullible young men and sceptical pawnbrokers. Her admonition, "Wrap up well to prevent bruising," makes sense later when we find out that Teddy "faints" and falls down escalators to create diversions while she makes off with the goods. The second half of this hour is about a parolee the couple take in under a government rehabilitation programme for the money, so they won't have to shoplift; he teaches them how to do a scam on a crooked real-estate developer. This part is played for laughs. The tone and structure of the piece swing uneasily between sharply observed social comment, comedy, and didactic dialogue set in an often anecdotal structure. One of the deft psychological touches is that Katie actually loves the excitement of ripping off the large companies, the crooks, and the government. *The Larsens* ends as the new parolee, who was in jail for forging cheques, starts to tell Katie about his technique while Teddy protests.

Melda and the Ducks (wr. Derek Goodwin and Wm. Marantz) is also a study of old age and loneliness. The pleasure here is in the clean storytelling, rounded characters, and fresh material. Melda is a widow with children who live far away. When we first meet her by the shoulder-high fence around the Assiniboine Park pond, she is feeding the ducks. An Indian takes her bag of crusts and starts to eat them while she is looking elsewhere. But he does not take her purse: she does not notice this, though we do. Indignantly, she snatches the crusts back. "Waste of good bread," he says, but she is oblivious to his hunger. To her horror, he talks about shooting the birds. "You a vegetarian," he asks. "Certainly not, but I like the ducks." "You don't eat them alive?" Now she is really angry and uncomprehending. She mentions her annuity. He points out that this is like his cheque. "Certainly not! My husband worked for it." "How do you think I got my UIC?" the Indian asks, and she has the grace to be embarrassed. He then asks her for a quarter, confirming her unconscious racism, but leaves, shouting across the pond, "It's only a loan."

On the next day, to Melda's surprise, he returns the quarter and brings wild rice, harvested by himself and his brother to feed her ducks. After a good chat, she takes Victor home, serves him tea, shows him the distant view of the park from her highrise and, finally, shares her big secret. Every winter, she "saves" one duck, which lives in her bathtub; Victor is appalled, even when she tells him that the females at least do survive. Every spring for ten years, "when the buds are sticky," she has returned them to the park. Her sentimental vision of a Heaven filled with ponds and ducks spurs Victor's reply: "He brings winter." Her rejoinder: "He also gives them wings. We're the ones who clip them." Victor observes that: "the buds are sticky" then, and, to ease her sorrow at parting from her little companion, he agrees to help return the duck.

The return is a disaster as Melda is apprehended by a park superintendent. Victor punches him in the eye, and both do-gooders wind up in court. Victor, in T-shirt and jeans with a headband and loose hair, looks like a typical, if not defiant, vagrant; he gets his "usual" thirty days. Melda is forbidden to go to the park for six months. Desolate, she sobs, "But if I can't visit my friends . . ." Cut to a mid-shot of her sitting in the antimacassarred living room and a call from Victor, who appears in a neat denim jacket and clean shirt with his hair tied back, bringing her a present. The last shot is of Melda, deeply happy, peering at her beloved ducks through a powerful telescope. This short half-hour is just what regional drama should be: a good story, skilfully told, with a strong local flavour.

Written by John Hearn, *Leona* gets its strength from fine, well-directed performances, skilful editing, and passages of painfully effective if somewhat nakedly emotional rhetoric as the marriage goes stale. The situation is familiar: two passionate young lovers become two uncommunicative adults as she puts him through endless years of medical school while having three children. Worn down and worn out, she doesn't hear his loving approaches until she shuts him up for good in a wounding monologue; he has an affair. The climactic quarrel is very good indeed—exacerbating and difficult to endure. The marriage does not end, however, because she then has an unfeigned and quite terrifying mental breakdown. With guilt and love, he slowly brings her back to normal life. Unexpectedly she comes to dominate the relationship verbally and emotionally, mothering him unmercifully when the children have left, chattering on about her friendship with Paul, her golf partner and then campaign manager. Eventually she becomes senile and regresses, confusing Paul and that "boy," her aging husband; at last, she slips peacefully into death. He opens the curtains to the daylight, and frantically takes his diplomas from his office wall. Then gasping for breath, flees into the open air. The camera stays inside, trapped behind the wrought iron doors.

The best aspect of the play is the tight editing. Each scene is intercut in various ways, seamlessly yet clearly. Questions posed at the end of one scene are answered five years later in the next. Without being gimmicky, the technique keeps the audience alert. I have never seen this kind of narrative handled as surely or effectively.

Before moving east to Montreal, it is perhaps worth remarking that there are different kinds of counterbalance against a too homogenized view of ourselves. Since our problem is "too much geography," it is logical, looking over three decades of drama, that certain places have imprinted the mind's eye: the Chilcotin, Gibson's Landing, the East End of Montreal, Cabbagetown before and after gentrification, huge combines in a prairie night, Elmira Raceway, Hatch's Mill, Iroquois Creek and Tamarack, the neat houses and slag heaps of Sudbury, sunrise on Lake Winnipeg, and a foggy island in Georgia Strait. Some actors convey the same sense of roots. I have mentioned the Beamsville-Parry Sound-Kicking Horse Pass-Maritimer comic constellation of Michael Magee, Don Harron, Dave Broadfoot, and Max Ferguson. Gordon Pinsent, Daniel Pilon, Frances Hyland, Monique Mercure, Ed McNamara, Robert Clothier, Chief Dan George, George Clutesi, Ruth Springford, William Needles, Hugh Webster, R. H. Thompson, Bruno Gerussi, Geneviève Bujold, Margot Kidder, August Schellenberg, Douglas Rain, and Merv Campone also let their national or regional backgrounds show in their speech patterns when they are not behind the masks of characters from other parts of the country.

Vancouver and Montreal are very different regional centres. For a country of this size, with a small population and marked regional diversity, in an ideal or even a sensible broadcasting environment, the CBC would have at least three major English-language production centres. Toronto would be one because it is within reach of the largest concentration of anglophone population, because of its large-scale theatrical, radio, and film activity, and because it has become an uncommonly interesting city. Vancouver would be another as an essential countervailing influence to the centrist tendencies of Toronto, as a completely different kind of landscape and mindset, as access to the West, and as a place which has enough highly trained, creative people and physical facilities—or, at least, did until 1986. Finally, we would have Montreal, with its access to, and what should be cooperation with, Radio-Canada, its superb facilities and its large pool of bilingual talent. Montreal is both a window on our other half and a beautiful and exciting city. This is not to denigrate the regional centres in Calgary, Edmonton, Regina, Québec, Halifax, St. John, and St. John's, most of which have produced interesting local drama. But to function as a country rather than as one City State, that is, as Toronto with satellites, we need checks and balances to our narrowing perspective.

It might be noted that in Britain, the BBC Scottish, Welsh, and Birmingham industrial Midlands regions produce their own drama for their own particular audiences without any necessary expectation that the drama will go out on the network as a whole. Because there is no centralized network in West Germany, drama can be made for and by the various regions. In Italy, networks are assigned according to political persuasion; both national and regional voices are heard. Belgium, like Canada, has two languages to contend with; Switzerland has four. Subtitles are taken for granted in Europe.

MONTREAL

We have already seen what the West can do, both locally and for the network. As an independent production centre, Montreal has effectively disappeared since the early 1970s. Yet, it was the source of two anthologies in almost continuous succession over a span of nearly a dozen years. *Shoestring Theatre* (1960-67) was the first, replaced by *Teleplay* (1967-71), which evolved from a Centennial script contest into an eclectic, often experimental half-hour very much like *Shoestring*.

In the late 'fifties, when late-night reruns and rock shows were unknown, it was still possible to do a show before or after the eleven-o'clock news that was thirty-five or forty-seven (or however many) minutes needed. Very occasionally, writers even created a two-parter presented over two weeks. There were many adaptations of short stories, one-act plays, readings with music, and so forth. Ken Davey was the executive producer and vigorous defender of *Shoestring Theatre* and *Teleplay*. Mac Shoub was both writer and story editor for the series until 1963, when it rather lost its way. Only three producers worked on *Shoestring Theatre* in the early days: Guy Beaulne, Jacques Gauthier, and Roger Racine. Altogether, it was a very homogenous operation, yet the plays spanned a wide variety of styles of presentation, tone, and subject matter. Many familiar Montreal actors appeared in them; in fact, nearly every major network performer appeared in one or two.

The two anthologies met several objectives. They developed or gave work to scriptwriters: old hands like Mac Shoub, who supplied many adaptations, newer playwrights like George Ryga, David French, William Bankier, Kay Hill, Dennis Donovan, Lyal Brown, Douglas Bowie, James B. Nicol, poets Al Purdy and Leonard Cohen—who suggested a television version of his one-act play *The New Step*, in 1964—among many others. According to *Closed Circuit* (14/5/69), a CBC in-house magazine, by the late-'sixties CBMT had by then done nearly two hundred half-hour plays, many of them Canadian. A few of the shows were rebroadcast on the national network: on

Playbill in 1959, in swaps with other stations, on *Pick of the Week* in 1968 and 1969.

Pat Pearce assessed both anthologies with her usual perspicacity in a review of *Teleplay* in the Montreal *Star* (19/9/67): "It's turned up some excellent potential in both writing and production . . . Bowie, Donovan . . . [Michael] Sinelnikoff, [Brian] O'Leary . . . and revealed a large core of known and unknown players. It has often been very tiresome indeed. But like 'Shoestring' before it, there's a glimpse here and there of people really trying to be expressive." Mind you, in another column she pointed out that it was not necessarily logical to match inexperience with inexperience, as happened when *Teleplay* was also given the job of training producers as well as finding new writers.

Because these two regional anthologies were broadcast fairly late at night, the producers were free to try some difficult and controversial subjects. Thousands of Montrealers were willing to look in at whatever they were offering: poetry, thrillers, comedies, experiments. Clearly, the anthologies were fertile ground for new ideas for the larger network. As Ken Davey spelled out in *Closed Circuit*, *Teleplay* defined itself as a showcase for "high quality material with a heavy emphasis on Canadian writers with story lines, preferably indigenous to this region." The shift towards Canadian content also reflected the country's growing late-'sixties nationalism. Approximately a hundred kinescopes of *Shoestring Theatre* and *Teleplay* are roughly catalogued and well preserved by the archives section of Radio-Canada— which seems to have been far better equipped for its task than English CBC where no true archives have ever been maintained. Among them were some interesting oddities from both anthologies. One was Wm. Bankier's *Pea Soup and Porridge* (24/1/65, prod. Jacques Gauthier), which opened with the wooing scene between Princess Catharine of France and Henry V of England in Shakepeare's *Henry V*, where she asks, "Is it possible that I should love the enemy of France?" The character of Chorus in Shakespeare's play notices the camera, laughs, and tells us that we are going to see, not that play but what goes on behind the scenes. All the dialogue is in blank verse, so that lines from Shakespeare and Bankier himself are blended in one set of flexible rhythms. The play is actually a mildly intriguing study of French-English tensions in Montreal in the 'sixties. Despite the cultural miscues which create problems behind the scenes for unhappy roommates Keith and Raymond, they are reconciled (as in *Henry V*) and the play goes on.

By contrast, David French's *The Willow Harp* (12/12/63) was a tight thriller with rather lyrical interludes. Al Purdy's *Point of Transfer* (n.d.) was a consciously ironic study of the joys of domesticity and the glamorous life of a prostitute. The encounter between wife and street woman, which results in their decision to exchange roles, is the basic focus of the plot, with the husband a glowering cipher. *Point of Transfer* had the problems of some *Shoestring* television drama; the production values were adequate, but broad acting and a rushed pace suggested too little time to rehearse. Kay Hill's *Cobbler Stick to thy Last* (n.d.) was set in eighteenth-century Cumberland County, Nova Scotia. It is a gentle tale, full of good interplay between the characters, as a widowed cobbler makes some sort of relationship with his new mail-order bride.

Teleplay was equally eclectic. It was a training ground for producer Gary Plaxton; Jean-Pierre Sénécal; Michael Sinelnikoff, a story editor with *Festival*; Jack Nixon-Browne, who later directed all kinds of taped series for the CBC. Some new writers appeared as well. Douglas Bowie wrote *Who Has Seen the Lone Ranger* (11/9/67), starring John Kastner, about a student prank that results in tragedy, a perennially topical theme. Dennis Donovan's *Ghosts of Montreal* was, according to the publicity, "a satirical look at historic sites in Montreal as seen by the ghosts of many well-known figures of the past." Suzanne Finlay, who had been a story editor for *Festival* and went on to be the story editor of *The Beachcombers*, was the script consultant in 1969, when the season included Malcolm Marmorstein's *Will the Real Jesus Christ Stand Up?*, broadcast near Easter, and Robertson Davies's *Overlaid*, a perennial television favourite. One of the last programmes, *When the Bough Breaks* (14/4/71, no credits on the kinescope or in the catalogue) outlined the effects of an accidental or deliberate pregnancy on the rocky relationship of two people living together.

In 1970, five of the teleplays were picked up for 10:30 p.m. summer viewing on the full network, scattered over three months. The best of the kinescopes I managed to see in Montreal was *Laurie* (9/1/70, rebroadcast on the network (9/9/70, wr. Dennis Donovan, prod. Jack Nixon-Browne). A high school teacher has been sleeping with a thirteen-year-old girl who clearly enjoys it. All their scenes together are treated playfully, lyrically, with the couple enjoying Montreal in winter. When her parents find out about the affair, however, they lay criminal charges, and she is pressured into helping the prosecution. The teacher goes to prison; prison life quickly takes over the foreground. No excuse is offered for the teacher's conduct, but no judgments are made by the playwright: we are left to make up our own minds. As the scenes unfold, we find that Laurie really thinks she loves him, and it is never made clear whether he knew, or asked, whether she was under fourteen. He in turn draws on happy memories of her when the terror and confusion of prison get to him. The abuse of trust is not downplayed, but the audience is persuaded to believe in the depth of their feelings for each other. It is an unusual and controversial approach to the subject.

The tension rises unbearably when the teacher, now in prison, discovers that it is a mark of pride to be beaten by the guards so cleverly that "there was no split skin . . . but I couldn't keep my food down for a week," as his cellmate, Big John, says. Voices off from other prisoners, whose remarks and quips define them as gay, add to the small cast of characters. The climax comes when the guard comes to beat the new prisoner almost as a routine initiation. Big John fights to defend his cellmate until the guard tells him that "he's a child molester . . . he takes little kids and does things to them." The camera pans to John's face filling slowly with disgust as the off-camera prisoners chant: "Big John Vogel fighting for a pervert . . . wait 'til the boys hear about this." John then repeats to the teacher an earlier line: "There's no rush. We're all going to the same place. We can fix it there." Intercut close-ups of teacher pleading, Vogel savouring the future, then the teacher on his bunk in shadows and terror. The last shot is a closeup of the 1962 Criminal Code as a voice-over reads that the penalty for seducing girls under fourteen is life imprisonment and whipping. The effect of the whole piece is unexpected. It questions a prison system that metes out cruel and unusual treatment to sexual offenders. It also raises the question that an affair between a teacher and an underage child, though unethical and unprofessional, may not necessarily harm either individual.

Montreal, CBLT and CITY, Toronto, Winnipeg, Vancouver, St. John's, Edmonton, Halifax could all enrich our television drama with more plays by and for local audiences. The best could be shared, indeed some are, on a regional exchange basis. We need both regional drama and network drama from the regions, not because it is good politics or because the mandate specifies that the CBC interconnect region to region, but because the talent, will, and commitment are there, because many of the programmes already made then have been good, and because the regions can and should both energize and counterbalance the centre.

NOTES

1. "Portrait of a President" (Al Johnson), in the Canadian *Forum*, February 1977, 24-32.
2. Canadian *Forum, Ibid.,* 32.
3. Ibid., 28.
4. *Radio Frequencies are Public Property*: public announcement and decision of the Commission on the applications for renewal of the Canadian Broadcasting

corporation's television and radio licences; Decision CRTC 74-70, 31 March 1974.

5. The move to centralize decision making in Ottawa, made in the mid-'sixties, presumably followed Queen Victoria's logic when she chose the little town as the capital of the new Dominion because it wasn't Toronto or Montreal (points east and west being presumed to be too far away).

6. Which appears confusingly in the CBC records under a non-existent title "Close Shave at Kamloops" in *Tales of the West*, an anthology that did not exist under that title.

7. The CBC production of Pirandello's *Six Characters in Search of an Author* also substituted images on the monitors for the "characters" who break up a rehearsal for a television play, asserting their "reality" over that of the actors and in the absence of a creator.

8. *The Ecstasy of Rita Joe and other Plays*, Talon Books, Vancouver: 1969.

V
CONCLUSIONS

11

WRAPUP
—*The View from Within*—
Interviews with John Kennedy

Before my final overview of the CBC harvest, wheat and tares alike, it seems fair to present the most important behind-the-scenes voice on this aspect of the network's activity over the last decade, that of John Kennedy, Head of CBC Drama. The following are excerpts from formal interviews taped nearly a year apart (19/6/84 and 11/4/85); they were shaped in turn by many informal conversations from the outset of this project. Without Kennedy's co-operation, this book could not have been drawn on anything like so rich a background: he gave me *carte blanche* to approach his producers, directors, and technical staff, suggested sources of information, and was unfailingly courteous throughout the process. As can be seen in what follows, he is direct and articulate about what he does.

Kennedy has no background as an artist (his word). A unit manager and administrator for many years, he is a marked contrast to his predecessor, John Hirsch. This has certainly not prevented him from setting his own stamp on the department. More original Canadian scripts are being used, most of them contemporary. Women have been given the chance to produce, and their concerns have changed the flavour of CBC drama. Most important, he has given his people a chance to get on with creating programmes, kept his input constructive, and fought like a tiger over priorities and allocation of resources with upper management. This is a fact of which many of his department are unaware. He listens carefully, works incredibly hard and has taken more and more risks as time goes on, although his chances of persuading the programme planners to go along with him diminish as funds shrink.

I have left this section as a lightly edited transcription of some of our conversations to give the reader the sound of the man's voice. John read the complete transcript, corrected one or two factual errors and authorized its use as it stood—which may tell you something about both his candor and his discretion when "on the record."

MJM: There has been less censorship in your tenure than in the previous two decades. As a drama head, have you ever been under severe pressure?

JK: No, I don't think I've ever been leaned on. I have had some powerful conversations with my colleagues in the network about this project or that project, about our treatment of various subjects. But it seems to me that in the course of the development of a programme the discussions that go on between the head of the department, the producer, sometimes the writer, often the director are such that we usually agree. I find that that process of discussion and debate over the extent to which a programme should take this direction or that direction is one that is reasonable enough. [The result is] that when it is finally done it is seen to be work that has had an honest and fairly reasoned approach. It may be partly as a consequence of that, that the piece has some kind of integrity—and integrity tends to win. Integrity is important, it's critical, and it may be that integrity that has caused us not to have to make [concessions].

In some ways, television has become more conservative, rather than more permissive. I mean it's permissive on the superficial level. You can get away with saying "God damn" on television, although there is a substantial part of the audience that finds that extremely offensive, where you couldn't have done so fifteen years ago. But, at a fundamental level, to me it seems to be more conservative—or perhaps we are just becoming a little more sophisticated about how to deal with some issues that would theoretically be sensitive. I can remember in *For the Record* doing a programme about euthanasia which people feel strongly about. The programme was quite well done: *A Question of the Sixth*. It inspired a lot of talk, but I don't think I would describe it as a lot of controversy, and certainly not as a "lot of heat." It didn't put pressure on me, certainly, either from the public at large nor from the CBC internally.

MJM: 1968 seems to have been the watershed "bad" year.

JK: Yes, I remember that year.

MJM: Want to talk about it?

JK: No.

[Regarding the change-over after John Hirsch submitted his resignation, Kennedy commented:] It's not quite as cataclysmic as reviewing all of the projects under way and starting all over again. There were several projects in the works when John [Hirsch] left, and that's normal, because we produce on a constant basis over twelve months. We produce well in advance of the moment at which we broadcast and therefore in that season [1977-78], *For The Record* was in continuing development, *Gift to Last* was in continuing production, *King of Kensington* was in production. I came in as associate head in the fall of 1977 and had that time from the fall until March [1978], when John left, to work closely with him.

[On the contrast between this orderly transition and the one Hirsch had inherited, Kennedy commented:] It was certainly much more orderly. John had an advantage going for him—he came with a brief, i.e. the drama study that had been done by Peter Herrndorf, Thom Benson, and Bob McGall at Gene Hallman's request and it was, in fact, a road map: a set of generalized objectives that the CBC wanted to achieve; a set of criticisms about what had happened in the relatively immediate past; some observations of what had happened in the more distant past. It represented an attempt first, to position drama as a high priority within the CBC, and second, to send out a signal that major changes were to be made, new directions were to be undertaken. And John Hirsch, being an extraordinarily bright and energetic man, was going to be the guy who effected all those changes.

A head of department can be very "hands on" in some projects and less so in others. Again, it would depend on the instincts of the person who was [head] at the moment as well as on other factors, like the availability of time to do the "hands-on" work. I do not behave in the same way, for example, as John Hirsch in terms of running the department, nor I suspect do I behave in the same way as any other area head in terms of my participation in any particular programme.

I watch rough cuts of shows and sit down with the producer, the director, the editor afterwards, and discuss the show fairly intimately at that stage. But my particular style is that they're the ones doing the work, they're the ones who most intimately know the piece, they know every frame. I know what I've seen through rushes and the rough cut and may have a sense of what is missing, but I wouldn't want to intervene directly. I don't know the show. The notes [verbal or written comments] I give tend to be the sort that require them to think a little bit about what it is that they have done. But it remains with them to do it. It's not my show.

MJM: Was there a perception that Hirsch had links in the theatre scene which had been ignored by the CBC?

JK: Yes, there was definitely subtext there, reflected by his fairly large scale importation into television of people with whom he'd worked in the world of theatre, among them Muriel Sherrin, Bob Sherrin, Henry Tarvainen. There were several who came in at that stage.

[On the director training programme:]

If you make a major undertaking to get into the business of training and development, you have to make that a fairly long-term undertaking, [with] work opportunity and, to some degree, an opportunity to fail. It was Hirsch's contention, and I agree with it, that there had to be an opportunity to fail. One had to expect that, statistically, there were going to be failures among those projects. But the CBC's financial position basically meant, and this is too crudely put, that we had to broadcast the failures. That dismayed Hirsch, dismayed several of the people involved. Through that and, I guess, a combination of other things, the programme stream that John was using to develop these people, which was called *Peepshow*, disappeared. It was not particularly popular with audiences. . . .

In looking back on it, it represented more of a failure than it did a success from the broadcaster's point of view. So much so, that attempts to encourage anthology drama using people newish to the medium were not met with a great deal of enthusiasm.

Judge began as a director development project. It was a project that was developed, first, to try and find a programme that would be reasonably easily accessible to audiences, non-threatening to the network and the CBC, and at the same time would create enough demands on those people who were going to put it together so that they would be stretched in television terms. To my mind, it has been largely successful on several fronts. The one that pleases me most is that those people who have been associated with it as directors have all become staples of the director market in this country. . . .

I next asked him about the extent of his own freedom in relation to his heavy responsibility.

Freedom is a function of knowing what is expected of you, of being more or less left to deliver it within fairly clear guidelines. There may be financial guidelines or content guidelines or style guidelines, any kinds of guidelines, but at least you have a fairly clear road that you're expected to travel and you're asked to "travel it, please." Whether you travel it in a Mercedes, a Porsche, or a Bug is pretty much up to you. Those circumstances, assuming that the chief driver, in this case the area head [of television drama], is reasonably self-confident enough to be able to make decisions, represents to me the healthiest of all possible worlds.

To the extent that somebody else is saying "You must drive it in a Bug, or a Chev, or an Acadian," it becomes more difficult. If and when planning becomes planning by "title" [genres or particular kinds of content], I think that we are doomed. Because then, as the objectives become unclear, as the persons setting them try to do so at much too intimate a level, failing to understand that energy begins from below, never from above, or rarely from above—we have a problem, everybody has a problem. That's when I think bureaucracy looks as though it is running around. Because then, as the objectives become unclear, and uncertainty and stress are put on the various bureaucrats, there is a human tendency on the part of the people who actually make the programmes to say to those higher up, "Well, okay, if you're going to make the decision, you go ahead and make the decision." [which is a good summary of what had happened during the late 'sixties and early 'seventies—MJM]. This is very dangerous from a creative point of view. It's also dangerous from the standpoint of anybody standing outside looking in, who says, "Well, this one clearly can't make the decision, and I wonder who else can." Then you start to get into, "Just how complex is this organization?" and, "Where are decisions made?" and "Where does the buck stop?" Up to now, more or less, it's been possible for the buck to stop in this office, with respect to drama progamming done for CBC. That is constantly under pressure.

MJM: It is still almost impossible to make a living as a television writer in Canada.

JK: It is more possible today than it was ten years ago, but it is very difficult; very few people are going to make a living writing for television.

MJM: Yet a fair number of *For the Record* scripts have been ideas submitted by non-professional playwrights.

JK: That's true, and I hope it will continue to be true. I can't remember, but there are around four thousand writers who are members of ACTRA, and in drama terms, the opportunities for those people have been in CBC TV or radio drama. That's it. And you're right, a lot of people starve in that process. I have currently in my office 250 proposals, in one form or another, from across the country, from people who have written before, all trying to take advantage of the broadcasting fund [Telefilm]. Well, the odds are that 190 of those are not going to be recompensed for their efforts to the extent that they can afford a mortgage, four kids and a dog.

If you are in a dwindling work opportunity situation, you continue to decrease the number of ideas, the number of individual voices that can be heard. Anthology is another manifestation of this. Clearly, the drop in anthology has reduced the number of individual voices. More often, the voices can only be heard in the context of the format. In a series like *Seeing Things*, you're writing for "Lou Ciccone" and "Marge" and "Heather Redfern." That is acceptable, good, exciting, but it isn't an opportunity.

MJM: But if a television drama does not start with a good script, it seldom works.

JK: Correct.

MJM: In a series with one writer, there is usually a high level of achievement. I'm thinking of Philip Hersch with *Wojeck*, Paul St. Pierre with *Cariboo Country*, Gordon Pinsent with *A Gift to Last*, and even *Seeing Things*, in the sense that all the scripts are reworked by Del Grande and Barlow.

JK: Well, I think that's a perfectly valid observation. However, it has some ramifications; one of them being that you can only do so many of these things, and television is a business that keeps crying out for more. Having found success, or something that is perceived to be a success, one is always under some pressure to deliver more and more and more of it. And the difficulty there is that a writer can only do so much before a certain automaticity sets in. You lose the integrity that you had, which caused the series to be a success in the first place.

I'm particularly thinking of the writers and the creative control over the piece. Now whether that is exercised by the writer or by some other member of the unit, is, to me, really neither here nor there. But rigorous attention to why we're doing this in the first place, and what it is about, has to be paid in order to get it done well. The moment that your attention span is diluted by too much volume, too many things to do in one day, exhaustion sets in, you begin to pay less and less attention to it. If you do that with a child, the child starts to drink out of the Javex bottle. You must pay attention.

MJM: Then your basic argument is that the characteristic CBC pattern of doing eight, ten, twelve programmes in a series per year is a major contributing factor to its quality.

JK: Absolutely, absolutely. I don't believe that we have the mechanism, and we certainly don't have the climate to do the kind of volume that our neighbours to the South do. I also contend that nobody else in the world except our neighbours to the South do the kind of volume that the North American audience wants to see on television.

MJM: But this policy does cost the corporation in potential sales to the U.S. market. I would assume that it also brings the audience numbers down in Canada, because our audiences have developed North American viewing habits. Most do not check with their TV guide to find out what the CBC is coming up with this month. Even the strategy of filling a given time slot with similar programmes has not broken the basic expectation. Except for its sitcoms, the CBC has to find and build an audience with new publicity every eight to ten weeks, not only for new programmes but also for returning series or anthologies.

JK: There's a lot that is absolutely accurate about that. It's a problem that we continually face, and I think we're going to face into the distant future. If you look at audience statistics on various kinds of programmes, using comedy as an example, *Hangin' In* is considered much more "demanding" than *Three's Company*. It is much more demanding, and that may have to do with understanding what the limits are to the numbers of people you could ever attract to *Hangin' In*. The other side of that coin is, if you try to do something that is distinctive, different than what else is available on the box in this country, you accept that the greater portion of your work is going to be closer to "demanding" than it is to "relaxing."

We are determined to have CBC drama contrast, be different, be distinctive, with either a small "d" or a capital "D," and it is a colossal challenge from any number of points of view, not least of which is that up until recently, [the volume of CBC drama output] was quite restricted because of finance and therefore the number of work opportunities was quite restricted. Let me put it another way. If we gave drivers as much opportunity to learn to drive as we give writers an opportunity to learn to write, there would be havoc on the streets. But there is an increased opportunity now, because of the government's initiative for more work to be done, [the Telefilm fund] and therefore for more people to have an opportunity to do it. To me that is *au fond*, fundamentally a very good thing.

MJM: At the 1984 conference of the Association of Canadian Theatre History, [playwright and scholar] James Reaney talked about the need for lots and lots of drama of all kinds, in both theatre and television as the necessary precondition for the maturity of the dramatic forms in this country. In connection with the idea that quantity is necessary for quality to appear, has the half-hour anthology, which used to provide so much practice for directors and good entertainment for audiences, disappeared entirely?

JK: It's more or less disappeared. Going back to *Boys and Girls* (1983), it made a brief appearance. But I suspect that it will come back.

[On programme planning:]

The context in which we produce, is that we use the same creative resources in the craft areas as several other programme types [like Arts and Sciences or Children's programming]. Therefore, some kind of phased planning has to happen. Just to give you an idea, two months ago—this is now mid-June [1984]—I had my first serious and fairly detailed conversation with the network directorate about the television season 1985-86, which is eighteen to twenty-four months from now. The detailed submission [containing] "offers" to the network for that year must be completed here by mid-August. I have to tell them that I am going to do x of this and y of that at these times, in order to meet the broadcast requirements of the network for drama.

MJM: You have to make a lot of guesses.

JK: I have to make a *lot* of guesses.

MJM: . . . about creative burnout in personnel, about the availability of freelance directors, actors . . .

JK: Absolutely!

MJM: . . . whether or not a script now in the outline stage will be ready on time . . .

JK: Yes. There are a lot of guesses, but my brain, and the brains of several other people are the computers here—and they come with at least a modicum of judgment. We have done this before in various capacities. One does know what can happen and, at least as critically, what can't happen. So it's possible. It's complex, it's frustrating, it's sometimes debilitating, but it's possible, and we do it.

[On CBC copshows and sitcoms imitating American formula:]

I think there is some of that. However, there's another way to look at it. Use *Delilah* as an example, because it was a failure. However, it was only one failure, and perhaps you could put your finger on two or three others. Most broadcasting enterprises, and certainly those in Great Britain and the United States, would simply forget it and go on. It's a function of the numbers of things they do: "What the heck, another failure. So we'll try again and we'll have another success." In our case, partly because of the small volume that we can do, we tend to remember these failures for a longer time—if we all knew how to do it brilliantly, we'd all be rich and famous and we'd be able to teach the world a few things. Although I think we may have taught the world a few things over the years, we have not become rich and famous.

There's another point that I think we have discussed before. It's less and less possible to fail. The stakes seem to be getting higher and higher. The volume has not changed in ten years—substantially, it's been around seventy hours. The mix is not—wait a minute now, I was going to say the mix has not changed substantially, but in some ways it has changed substantially in that time, from stage plays to original work being the most noticeable. I certainly feel, personally, the pressure to succeed every time out, and that is extraordinary pressure.

Anthology is a hard sell in this world. There is a generally accepted perception that series with continuing characters are what draw audiences. By definition therefore, without continuing characters anthologies don't draw audiences as well. I don't share that opinion. I mean I share it in one sense. There are endless examples of series with continuing characters that draw audiences: *Dallas, Flamingo Road, Falconcrest*, any number of series, including *Seeing Things*, draw good audiences, so I would not want to discount that. However, our only example of continuing anthology at present on the network has been *For the Record*, which has drawn reasonable audiences over a long period of time. I cannot believe that it's simply that its mandate is known by all and sundry. It seems to me that we could devise, quite easily, a set of objectives for an anthology series which would attract decent audiences.

MJM: The early 'seventies had *Canadian Short Stories* [later retitled *To See Ourselves*]; that disappeared too.

JK: That did disappear, except that it had a recent rejuvenation through independent production sources, specifically Atlantis, who managed to produce six half-hours on CBC [*Sons and Daughters*], one of which won an Oscar. And indeed, the future holds some promise of a mild resurgence in anthology. I put it that way because my expectations are modest. My ambitions aren't necessarily, but my expectations are. I believe that we do anthology very well. In fact, I believe that in many cases we do it better than most other things. Our outlet for that kind of work in the last five or six years has been mostly the so-called TV movies. Going back as far as *War Brides* and *You've Come a Long Way Katie*, through to *Chautauqua Girl*; and recently with *Gentle Sinners* and *Charlie Grant's War*, which is an original script, I think that in the next few years we will be able to increase the amount of sixty-minute anthology, "mini-movies," to make them more palatable internally.

MJM: Perhaps, finally, the spotlight has shifted from news and current affairs.

JK: Yes, much is expected. I mean government policy puts the finger on drama, corporate policy puts the finger on drama, television policy puts the finger on drama. And I have these three fingers in my ear, which are wiggling with various degrees of intensity, and that has something to do with what I am talking about.

MJM: The viewer demands it too.

JK: Absolutely.

MJM: In fact, over the years, the consumption of television fiction has risen in most societies (and certainly in Canada) even though, in the budget squeeze the number of hours offered by the CBC has dropped— effectively, the pressure for more is now coming from all directions.

JK: Yes. Now the task is to get on and do it.

When I interviewed John Kennedy again, the avalanche of Conservative budget rationalization, including trimming of CBC jobs and initiatives, had swept by. Morale throughout the corporation over the winter had sunk to record lows. Now it was the Spring of 1985: time to ask a few questions about the effect of the cutbacks on the future of CBC television drama, and fill in a few gaps from our previous conversation.

MJM: One of the things I would like to explore is regional drama and the relationship of regional drama to the centre. I know that the regions have some production facilities, but they are very limited.

JK: That depends on the region. Vancouver has better production facilities, in some ways, than Toronto. They are much less extensive today as compared to six months ago, because of budget cutbacks. The facility is still there, but the personnel are not. That is a large and capable production centre. It is also the most extensively equipped in terms of production personnel outside of Toronto.

MJM: It looks to me as if there is a very clear pattern that hasn't changed much since 1960, which runs something like: "There will always be some drama on the network originating from Vancouver." It can be *Cariboo Country* or *The Manipulators* or *The Beachcombers*, but after 1960 there is only a year when there isn't something consistent scheduled from Vancouver. Is that policy? Is that accident?

JK: It's probably a little of both. Vancouver is a major production centre, and it also happens to be the home of a good sized group of very talented people. In any group of talented people like that, you tend to get a flood of ideas. I would think that as long as those people are there and the facility is there the ideas will keep coming.

MJM: No one said, "politically, we must always have representation from the coast" plus the fact that there were good people who stayed because they were given the chance to make programmes.

JK: I never heard those words in those terms. On the other hand, ever since I've been involved in drama there has been that energy coming out of Vancouver. If it is policy, it's good policy.

MJM: Is the idea that there should be more than one centralized voice, perspective, or point of view?

JK: Yes. The general mandate [is] to reflect one part of the country to the other. Drama is a form through which that is done. [Thus] it is important to produce in Vancouver as well as elsewhere, because drama is affordable and there is the expertise which can be brought to bear. Drama tends to get done in various places around the country, not just in Vancouver.

[On CBC brass on television drama and on regional problems:]

[Last Fall] I sat down with [regional] directors of television and confirmed that they were interested in doing an anthology series in drama. [Programmes would originate in] various parts of the country which would be broadcast on each of the CBC's owned-and-operated stations. While there is normally a regional exchange where programmes are used in Vancouver, Regina and so on, this time they would be broadcast by the various stations in a block so as to increase their profile. . . . As it stands it is an anthology [i.e. *The Way We Are*]. The only qualifier that we agreed on was that it would be played regularly in a specific time slot [thus giving the anthology some chance of building as audience.]

MJM: Is the regional exchange of programmes a swap situation? Does one region say to the next, "I've got some super stuff if you want it," or do they say "You have to take mine if I take yours"?

JK: I'm not sure what the language is that is used, but there is a great deal of exchange between the various Regions [some of it surprising. Judy Squires, the assistant to the regional director, St. John's Nfld. told me that *Land and Sea* (which has been going for twenty-five years and is chiefly about fishing and farming) is very popular in Saskatoon. We don't see much of this material in Ontario].

MJM: Winnipeg has been in the past an active drama centre. I gather that it is not active at the moment. Was it you who said it depended very much on the interest of the people who are actually there?

JK: Yes, which is to say that if you have a certain level of both interest and expertise, you can count on some ideas coming out of that. While Don Williams was there, he was very interested in drama, and there was a fair amount of drama done. It's not simply Don Williams' move to Vancouver which slowed up the production of drama in Winnipeg. That is one factor, but not the only one. If you measure the CBC budget over the last ten years, it has been dropping. Look at what has happened across the system. The money tends to be collected as a magnet collects iron filings, around certain types of programmes which aren't drama, like current affairs.

MJM: I've been taking a look at *The Beachcombers* and noticing the way it has infused new blood into itself over the past thirteen seasons to retain the same kind of audience it always has had. *The Beachcombers* does have a longevity record for CBC drama. What makes you decide whether or not [a series] still has a lot of life in it?

JK: There are external indicators: the audience tells us very clearly if we are doing a good job or not. They continue to tell us that we are [on *The Beachcombers*]. I think there isn't a moment, per se, when you tend to sit back in your chair and say, "Ah! It's time to do something about *The Beachcombers*." It is a regular part of the process to continually look at the show from an objective standpoint to try to see whether there are things you should be doing to keep it as accessible to the audience as it has been over these years. [Review of series] is an ongoing process.

MJM: About the negotiations that go on concerning programming slots and scheduling—*Judge* has been moved from a 9:30 pm.-slot to a 7:00 pm.-slot, from a weeknight to a weekend. Effectively, that time slot represents quite a different kind of demographic makeup in the audience. I thought that 7:00 pm. was the worst conceivable time for it to appear because it often featured such "adult" material. How is that sort of scheduling decided?

JK: I'm probably not the best person to answer that question, but I'll hazard an answer by way of background. Because there was a time slot available.

MJM: We discussed in our previous conversation how *Judge* was developed and what purposes as training ground and series entertainment it served. Since then, it has run for another two seasons. I came to think it was a really strong and promising series. Then I began to notice that certain programmes either did not appear or had to be shifted to a later time slot because of the subject matter. I understood the shifting perfectly, but it is also the fastest way to lose an audience.

JK: Yes, it is not the easiest way to build an audience.

MJM: Obviously, *you* didn't make the decision to put it on at 7:00 p.m.

JK: I don't ever actually make those decisions. I am permitted to fight about them but I don't necessarily win the fight. I wouldn't have put it where it was in the second season. The first season of *Judge*, as you may remember, was in the summer when *The Journal* hadn't begun. There was a time slot available and *Judge* ran there, at 10:30 pm. The second season, it was preceded by *Sons and Daughters* at 8:30 pm. That was not a great conjuncture of programmes, particularly when *Sons and Daughters* had a very sensitive, light focus and *Judge* came on with a programme on murder or rape. We received a letter or two from people who were disturbed by the juxtaposition. The other thing about *Judge* was that it was never considered to be a large audience draw "prime time, Sunday night at 8:00 p.m." work. It was much more modest that that. Therefore, it did not get priority scheduling.

MJM: Yes, I understand that, and yet it seemed to me that in the three years it had grown considerably. It had a firm, sure touch and some of the scripts were very good—the performances and the productions as well.

JK: I think academics can somehow think more in those subtle terms than schedulers. I think the schedulers' point of view is concerned with audience attractiveness.

MJM: But it seems to me to be a self-fulfilling prophecy. Nobody is going to want to watch a programme like *Judge* at 7:00 p.m. on a Saturday. Here is an instance where there is no point at all in my finding out about the ratings because the ratings have been fudged by the scheduling.

JK: They have certainly been affected.

MJM:—in the sense of the wrong thing at the wrong time in the wrong place.

JK: Occasionally that happens. Sometimes that happens because it is the last thing to be put into the schedule. CBC-TV has a very complex schedule. It has some things running for short periods of time, like *Judge*, which ran for six to eleven episodes, against something which runs for a long period of time like *Hangin' In*, every Monday night at 8:00 p.m. You're not going to give *Judge*, or anything else, Monday night at 8:00 because you're going to destroy the continuity of *Hangin' In*. However, if you're going to do *Judge* in the first place, you still have to figure out how to find a place for it in the schedule.

MJM: Are the 1985-86, 1986-87 seasons a time when you are going to find yourselves, in effect, sticking with what you have because there are so little money and resources to develop new material?

JK: None of these questions have black-and-white answers. The answer is yes/no. No, we are not going to get any more money. Therefore, we will be working in the same-sized bathtub in which we have been working. Within that, since there is little money available, we can make various kinds of decisions, but they are the kinds of decisions that are effectively, "*Seeing Things* goes," or "*Hangin' In* goes." The world is open to you as to the kinds of programming you can use only if you replace [existing material].

There are two other objectives. One is series. The network would like series with continuing characters. Therefore, in the face of that particular objective, there is not a lot of point in putting money into an anthology series without a leading character. That is the situation today. I think we've debated this before when you asked me why the mix is so eclectic. I said it was because there were different kinds of people who want to do different kinds of things. I would prefer to do that. There are other physical reasons—the difficulty of producing series. The Americans' experience [in having many flops for every survivor] can illustrate much better than any words I can find. Theoretically, it is possible to take that same amount of money that we have now and convert the entire CBC drama output to series. We wouldn't want to do that but you could take a stand and say, "Professor Miller, the answer to your question is yes, in the long term, we are going to turn this place around, and instead of doing six of this and seven of that, we are going to do four series." That is not, however, the answer to your question.

MJM: Part of my question arose from something you had said [in the previous interview] which is that you hoped you would never find yourself having to programme "to title." The network has said, "We really would like series, copshows, sitcom, or whatever." Do I understand that they have not said, for example, "We always have to have a sitcom in the mix"?

JK: No, to some extent they have said that. In fact, we are marrying the forces of the independents with the forces of the CBC in an attempt to do exactly that.

MJM: The cutback doesn't completely pull the rug out from under drama, then?

JK: No, in fact, it's amusing in a bizarre sort of way at the moment that the tape people (that is, those who work with tape all the time, technicians and the management types) are quite concerned that we're turning to film. The film people are quite concerned that we're turning to tape. If you pick up a television technological magazine, it says, "Tape is the thing, folks. That's where it all is—it's lightweight . . . fast, and now the image quality is rapidly gaining on film." Our thrust from drama's standpoint—remember that drama is only one user of the technology—is to use both film and tape because they serve different purposes.

MJM: You must be feeling pretty good about this past season as well.

JK: Yes, at least I was for the seven minutes that I allowed myself to revel, for really, by the time it goes on it's ancient history. Today, Denis Harvey was complimenting everybody on a wonderful season and he made this elegant, eloquent, complimentary speech, at the end of which I said, "Do we have to do it again?" However, I sit here trying to do it again.

Telefilm significantly increased the number of hours from 70 to 120 telecast by the CBC in the 1985-86 season. Most of the co-productions made possible by the new money involved CBC resources and promise of air time, plus some input into the project and money and input from other backers—often European television production companies or networks. The initiator of the project, an independent Canadian producer, would put the whole thing together and try to retain intact the original vision. None of this is easy but when it works, as it did for producer Kevin Sullivan with *Anne of Green Gables* (1986), the viewers as well as the industry are the winners. When it does not, the result can be as bad as *Check it Out!* Most of those projects have been arranged with the CBC—not with the privately owned networks or stations.

A year later there were more cuts, more layoffs. However, given television drama's eighteen-month lead time, the effects would not be seen on our screens until 1988. Initially, when cutbacks occurred, it was not realistic to expect viewers to complain just when they are enjoying the feast that Telefilm and CBC ingenuity provided in 1984 and 1985 for the 1986 and 1987 season. Then by the time the starvation of the drama department did show up on the screen, people would have forgotten who to blame. Like everyone else, Kennedy wonders what will happen when Minister of Communications Flora MacDonald and the parliamentary standing committee on communications and culture for the CBC finally bring new legislation to the House. Governments of all hues have not had the political will to implement a coherent and effective broadcasting policy needed to preserve public broadcasting as a true alternative choice, since the first Fowler commission in the late 'fifties. It is not yet clear whether the Conservative government will or will not develop a coherent policy regarding public broadcasting. If it does, we may see the results—diverse, rich, and distinctive—on our screens to the end of the century and beyond.

12

THE HARVEST
The View from Outside

Walk with me from the fifth floor of 790 Bay street, where the Drama Department hangs its hat, out into the street. Where are we? Where have we been during this long journey? Was it worth the cost to all of us as citizens? To the people who put lifetimes of work into telling our stories on television? To you as viewer of those programmes? Are you convinced that we have created a huge number of competent programmes, many of them with something a little special for the eye, the mind, or the funny-bone? Will you believe or remember or find out for yourself that in thirty-five years there have been many superb television dramas? Or are you caught in the habit of mind so well described by Northrop Frye on television in a programme he called, "Journey without Arrival,"[1] one of the *Images of Canada* series:

"Those who first came here were looking for Cathay. What they found simply got in the way. Farther south, it was easier to adjust. The Spaniards found gold in Mexico and Peru; the Americans found the American Dream and settled for it instead. But here? . . . We have never lost the feeling that Cathay is always over the horizon, a journey away. We have never been a culture that believed in happy endings. The real ending is ironic because it's here."

In Canada, nothing has ever been self-evident. . . .

"There was another side to the nineteenth-century: the romantic sense of nature as a symbol or counterpart of man's inner life. The romantic expectation opened up chasms in the Canadian imagination. It's all very well to be thrilled by moonlit dark forests and nature's grand design. But the reality in Canada was all too often . . . terrifying. No-man's Land. Terra incognita."

Robert Fulford[2] (*Toronto Star*, 2/4/83) traced the well-known path of documentary through the NFB, CBC, collective theatre, poetry, novels, and paintings, all of which have in common "the belief that art functions best when it is verified by some historical or journalistic version of reality." He attributes this to the triumph of Egerton Ryerson and his Methodists (via, later, the United Church, the CCF and the NDP) over the nineteenth century. He continues: "Puritanism tends to view the imagination with suspicion. But puritanism is in love with education, and its belief in evangelical preaching translates easily into a respect for preaching through the media." "In the nineteenth century, the evangelicals dominating our schools had a central message, the Christian gospel. Today, their heirs have another message, which is a response to our political situation. It is less clear and it is not written down, but most of us carry it in our bones. It involves the resolution of conflicts, multiculturalism, and the spreading of human decency through institutions—in short, as the BNA Act says, 'peace, order and good government.'"

If our artists in film and television and theatre are to create what [Robertson] Davies calls 'great fable,' they may have to set out consciously to destroy the tradition that has in the past given us our most triumphant moments. Documentary, with its Methodist roots and its liberal sensibility, has in its way been good for Canada. But the task for the next generation of artists will be to move beyond it, toward something larger and more important." Fulford went on to say that instead of being a mirror of our limitations, the function of even a mass art form is to be an expression of a personal vision, to lead the imagination of an audience and open it up to possibilities. He quite fairly expects the popular arts to prophesy, extrapolate, and clarify existing feelings, yearnings, questions, and discontents. Regrettably, by 1987, he had lost hope. In an article called "Stage Fright," subtitled on the cover of *Saturday Night*: "Why Canadians Can't Create Drama for Television" (April 1987), he makes an excellent case for the observation that "Canadian television puts drama last on its list of priorities and very seldom gets to the bottom of it." He concludes, however, that "the network offers to viewers, not works of imagination, but reports . . . it never provides that heightened sense of reality that we find in narrative fiction," and even speculates that perhaps "those artists who choose to work in Canada lack the drive to communicate through television and films. It may be that they cannot imagine the Canadian characters and themes that would impress a national mass audience." This book takes another point of view.

Rick Salutin (*1837, Les Canadiens, Maria, Grierson and Gouzenko*) also thought of documentary style as "the curse of Canadian culture." His two provocative essays on it, written in 1977, several months apart,[3] led him to conclude that "we are already detached from our own experience. Far from being over-involved and over-identified, we hardly see enough of our own experience to recognize it. And what we do get is very often done in the detached documentary way. In this situation, not more but less detachment may be called for. A kind of anti-alienation effect may be on the agenda for Canadian culture at this point.''

Having heard from a media critic, journalists and a maker of television fiction, let's listen to a politician elected by, and responsible primarily to, the audience. The Honourable Flora MacDonald, minister of communications, is talking to the Parliamentary Committee on Communications and Culture (5 February 1987):

''I believe it matters most emphatically, Canadians are more than news-watchers. As a mass-audience cultural form, television, more than any other medium, provides a mirror for the society. And fiction and imaginative works can offer the truest reflection: more accurate often than the driest news program; more honest in portraying the emotions that underpin all our concerns than the most objective documentary.'' Amen, Flora.

To pick up Frye's argument again: ''And so we developed that curious streak of anxiety that distinguishes us from the other North Americans. Which we kept trying to sweep under the carpet. . . . It took Canadians a long time to get imaginative possession of their own space.

''The airplane gave us an outline—two dimensions instead of one; every part of the country imaginatively available at once instead of just centres connected at east-west ground links. Elsewhere, the plane may mean a loosening of bonds, a way of escape; in Canada it is a means of tightening the country into recognizable shape. . . . [We are] ringed by the world's great powers. . . . [Frye used a world map at this point.] Here is Canada in the middle.'' According to Frye, poet E. J. Pratt ''shows us that the Canadian attraction to the documentary form is still with us: apparently we are still in the process of taking inventories and rendering accounts. . . . We came into this century without any agreement on what kind of people we were or even whether a Canadian could be identified.'' Yet since World War II, Atwood, Birney, Purdy, and Reaney (among many, many others) have followed Pratt as ''map makers of the Canadian imagination,'' reaching well beyond the need to take inventories in their mature writing.

Early television served the country as plane and map maker; contemporary television is like the satellite. It can map the large patterns and focus on the small detail, using the spectra of infra-red and visible light, to reveal patterns of change and growth both visible and invisible to the unaided eye. Television technology in the 'eighties and 'nineties is such that there may be far more choice. (Governments and broadcasters here and abroad are anxious about the ''Open Skies'' that direct broadcast satellites are going to bring to most of the world before this century ends.)[4]

What might you see if you were trying to catch the hidden infrared emanations from the country as it is toward the end of the twentieth century? One of the catchwords of the "realistic and pragmatic" 'eighties in Canada was "modest," in scale and expectations. Canadians cringed at the kind of excesses we saw on American television during the "Lady Liberty" bash of the summer of 1986. Walking around in foam-rubber-rayed crowns vaguely reminiscent of the crown on the Statue of Liberty has never been our style. The Royal wedding of that summer did not express the national psyche either, much as we enjoyed watching it. Traditional pageantry, excessive, almost obsessive, glitter—that is not our way.

Murray McLaughlin's CBC music special, *Floating over Canada* (1/7/85, rebroadcast 1/7/86) provided us with an image which is more like ourselves. *Floating Over Canada* is a celebration of our landscape, our people, and our music, using a bush plane with floats as its narrative link, thematic motif, and overview. That is one of the things we are—many disparate voices scattered over several thousand miles, talking and singing in several languages, a cowboy and a chanteuse tied together by a float plane, a camera, a singer-narrator, and the CBC.

In writing this book, I have discovered some of the connecting paths, even roads between the spaces Frye rightly sees as still undefined. Themes appear and reappear: the difference as well as the dignity of being Indian, or Métis; intermittent explorations of multicultural roots and the cultural clashes between generations or between old and new immigrants; a genuine hatred of war; a sense of the separate identities that divide us into urban and rural, East and West, Maritimer and inlander, (too seldom) manager and worker; an emphasis on the individual caught in social structures that, without particular malevolence, injure private rights; a remarkable number of portraits of extraordinarily strong women—rarely famous, and often survivors rather than winners or power figures, but in the drama specials of the 'seventies and 'eighties so many of them. On the other hand, by comparison with German, Swedish, or British television drama, our television is remarkably apolitical in the formal or self-conscious sense. Much of the best of the drama written specifically for television is populist, nationalist, with a thrust for social change, frequently emphasizing individuals who rally a group into collective action. The vividly personal dramas are more often focused on ethical dilemmas than on relationships, although there are many exceptions—*Teach Me How to Cry, How to Break a Quarterhorse, Loyalties, Gentle Sinners.*

Our view of ourselves seems to emphasize that, as a culture, we make it by persistence or luck rather than vision, a perception reinforced by a tendency to demythologize our historical heroes by treating a good many of our fictional ones ironically or comically. The values of hard work, tolerance for differences among us, and the efficacy of collective good will are pervasive in our series and drama specials. Our television also shows us, far more unsparingly and consistently than American television does, our sins of commission and even of omission. A lot of us don't like that, and tune in instead to the latest miniseries melodrama. Yet over and over, at least 15 to 20 per cent of us will choose programmes that do not guarantee to relax or reassure us, whether the programme was found on *Folio, Q for Quest, Eyeopener, Wojeck, Cariboo Country, Programme X, For the Record,* or an episode from other series and miniseries.

We are ambivalent about authority and authority figures, yet we acquiesce to them. We are almost proud of the absence in our national and provincial life of grand gestures, yet our cameras are in love with our infinitely varied landscapes, from the fence corners of small rocky pastures to the endless miles of rolling tundra. English Canada, Newfoundland excepted, does not think of itself as eloquent. Yet this book is filled with examples (a tiny fraction of the whole) of eloquent words, gestures and settings. Our television is permeated through and through with a documentary flavour, yet the personal, quirky voice has always been heard, even in the years of confusion and low morale: Keatley, St. Pierre, Till, Prizek, Newman, Weyman, Allen, Israel, Shoub, Petersen, Salverson, Almond, Williams, Cahill, Orenstein, Hersch, Gardner, Hart, Gough, Blandford, Woods, Evdemon, Hebb, Bonnière, Lowe, Sandor, Pinsent, Barlow, Del Grande, Long, Sarin, McEwen, Locke, Frank, Siegel, Dorn, Soloviov, Gauntlett (Hugh and Doris), Sinclair (Lister and Alice), Ferry, Grimaldi, and the literally dozens of other cinematographers, designers, writers, directors and producers whose individual vision can be traced, even in this most collective of arts.

There is no surprise in that. If I have persuaded the reader that we have a lot of television drama worth investigating and that some is first rank, then it follows that creative people have managed to do good work all over the country, despite formidable odds within and without the corporation.

At the same time, the satellites of other countries can now easily displace our own pictures, interpretations, visions about the shape of our landscape with broadcasts made elsewhere. If we do not take care, we will continue to make the world's most advanced satellites so that others may more easily shape our imaginations. If this does happen, however, it will not be because some other culture decides to do so; it will be because, as a nation, we permit it. CBC television drama has, for most of its years on the air, provided a recognizably different alternative among the ever increasing number of choices available to us. Yet as the minister reminded that Parliamentary Committee:

"Sixty percent of what the English-speaking audience watches in peak viewing hours is fiction: Comedy, suspense and drama, whether in individual programs or in series. Within this category, however, in 1984, 95% of the programs available were foreign; only 5% were Canadian. . . .

A nation's fictional repertoire is the lifeblood of its culture. We should never underestimate the impact of dramatic television programming for, contained in it, we find the surest expression of our cultural values as well as our collective memory.''

From 1952 to?

In the 1950s, television drama to a large degree took over from radio the roles of educator; patron of actors, playwrights, designers, and musicians; training ground; innovator in the new medium itself; challenger of accepted values and ''community standards'' of taste; nightly story teller, and window on a rapidly changing society.

In the days of live television drama there was a lot of interplay between Britain, the United States, and Canada in scripts, directors, in new forms, and in technology. However, unlike their colleagues in England or the United States, the men flying by the seat of their pants in the control-rooms of live television in early CBC days were working without any theatrical or filmic tradition. There was one important and influential exception in the National Film Board. The NFB. The NFB also provided some of the best directors and producers: Sydney Newman, Ron Weyman, Ron Kelly, Claude Jutra, Robin Spry, Donald Brittain. Successes like *The Open Grave*, with its techniques borrowed from direct cinema and its improvised dialogue; *Wojeck*, with its commitment to social action, flavour of the personal documentary, gritty texture, and displacements of focus away from the star; and the long-lived docudrama anthology *For the Record* reflect in different ways the influence of the NFB tradition. Whether CBC television also influenced the NFB I leave to the specialists in film to determine.

Linear or complex narrative structures, expressionist, constructivist or realist design, presentational or representational conventions were all to be found on CBC TV in its first fifteen years. Then, with film, more easily edited tape and (perhaps) changing taste, ''realism'' took over—what senior designer Rudi Dorn calls, ''design by Eaton's Catalogue.'' Too many scripts in the Toronto-based series and sitcoms imitated American formulae. Fortunately, in the last few years, one sees a fresh freedom and range in style—and, occasionally, a brand new playfulness, often sadly lacking in this culture—as in *For the Record*'s *Hide and Seek*, a look at the information age through the sidelong glance of science fiction, in the last season of *Seeing Things* (1987), in several of the dramas in *Some Honourable Gentlemen*, some of the topical dramas of *For the Record*, and drama specials like *Canada's Sweetheart: The Saga of Hal Banks, They're Drying Up the Streets* or *Chautauqua Girl*.

The CBC has had a tradition of being self-reflexive, of looking at itself and what it does in fictional terms: for example *The Open Grave* with its look at how television actually covers a fast-breaking story; *The Paper People*'s ironic look at documentary portraiture; *Blind Faith* on televangelism; *Freedom of the City*, which layers the conventions of television news, features and fiction. The CBC record in censorship is not perfect by any means. Nevertheless, many of the *For The Record* dramas and CBC drama specials could not be broadcast, even yet, on the commercial networks in the United States.

Until cable, satellite and DB satellite, video cassette recorders and video cassettes came along, the rule of the communications game was that privately owned television was supposed to deliver the desired ethnographic mix of age, sex, and income to advertisers. Publicly owned television was supposed to inform, educate, and entertain everybody, though not necessarily all at once. More in theory than in practice as the 'sixties wore on, CBC television was supposed to address pluralistic audiences, leaving the mass audience to the private broadcasters.[5] That is the message of the 1968 mandate and the 1974 CRTC decision. Budget cuts, starting in 1976, government displeasure, the inability of the CRTC to enforce its own rules, the rise of cable and pay TV send quite a different message. A third message is the establishment of Telefilm which encourages independent television production but demands that CBC, CTV or networks operated by private stations also provide money and airtime. The CBC provides most of that broadcast support—the crucial investment for independent producers like Kevin Sullivan, Anne Wheeler or the Atlantis group to give us the timeless delight of *Anne of Green Gables*, the explorations of friendship and culture clash in *Loyalties*, and the evocative, moody, quirky short stories in the *Sons and Daughters* anthology. Perhaps we will see a fourth, more positive message if the Caplan-Sauvageau, ministerial and committee support for a second national television service becomes reality. The minister outlined it this way:

> "If the first need is for more and better Canadian mass-audience prime-time TV, the second need is for TV programming of an alternative kind. What does this mean? Let's call it popular programming of a specialized nature, programming that will appeal to all of the people some of the time. What does this include?
> —quality programming for children;
> —programming for other age groups, such as senior citizens;
> —a national showcase for regional productions;

—making our visible minorities visible on the TV screen;

—lower budget original and innovative fiction programming;

—"issue"-oriented programming;

—programming directed to concerns primarily affecting women;

—the arts, not just in performance, but for discussion—for example, a good movie-review show;

—the best of non-North American television, not otherwise available, from Australia, the U.K. and other Commonwealth countries, of particular interest to Canadian minorities;

—documentaries, for which we have a special aptitude, and an abundant supply of excellent material;

—repeats of the best Canadian programming from other networks.

The new service would address the needs of an audience which, while not mass-market at any specific time, is substantial and can include almost all Canadians. Currently, the major Canadian networks usually shoot for an audience of at least a million during prime-time hours. But an audience of 3- or 400,000 is still significant. Accordingly, I am increasingly convinced of the merits of a separate TV Canada service such as the Task Force recommended."

In her presentation, the minister identified in a general way some of the most obvious gaps in our range of choices. Let me define, amplify, and add to the list.

And Now? What's Missing?

When it addresses or stirs up controversy, the corporation is expected to present a balanced view of an issue in the course of its whole schedule, not necessarily within a single programme. This means that dramas do not have to be bland. Sponsors are still necessary and eagerly sought, yet in its first three decades, if the subject was too problematic, the corporation was usually willing to go ahead and present it. Those days are gone: producer Bill Gough (*Charlie Grant's War*) told me in 1985 that he could not put a script which would be a natural for television, like David French's lyrical and bitingly funny play *Salt Water Moon*, on the air because it would have to be presented in one uninterrupted eighty-five-to-ninety-minute sweep, and such imaginative opportunities are no more. From being rich in the kinds of drama we can enjoy, we have become poorer than all European countries, Britain, or even the United States, where ever-struggling PBS can try that sort of thing on rare occasions. Our "contrast" is fading.

What else is missing from what is lovingly (ironically?) known as the "programme mix?" Let us first note that with *The National/The Journal* on 10-11:00 p.m. every week night and other efforts in current affairs and drama there is significantly more Canadian content on CBC prime time since the mid-'seventies—even though drama output shrank dramatically from 1977-82. Still, the focus has narrowed. In the past, but no longer functioning in these roles, CBC drama has been many things to many people. For twenty years, it was a window on contemporary and classical drama from around the world. Since 1978, we have not seen current Pinter, Beckett, Rabe, Shepard, Ayckbourn, or rebroadcasts or remakes of many excellent British or any European television plays—not, at least, on English CBC. Intermittently, CBC drama used to be a window on francophone Canada. Yet during the most serious political crisis of our history, during the years that led up to the Québec referendum, we heard no Québec voices in translation helping English Canada understand some of the anger and joy shaking that part of the country. Tremblay's *Les Belles Soeurs* was the only exception in the 'seventies; since then, there has been a handful of hybrid theatrical/television films like *The Crimes of Ovide Plouffe* and *Joshua Then and Now* and adaptations of Antonine Maillet's *La Sagouine* and David Fennario's *Balconville*. Yet there is no consistent committment to present one founding culture to the other. Playwright Michel Tremblay in particular is notably absent, even though his small-cast plays would seem just right for an imaginative director. What about the more obviously naturalistic Gratien Gélinas, or the prolific Marcel Dubé who did get on CBC English TV often twenty years ago? How about showing us the cause-célèbre of *Les Fées en Soif*,[6] a biting feminist satire on the Church and the Quebec woman, which Premier Levesque banned from publication? Rémouillard, Laurendeau and Languirand found their way to *Festival*: where are the contemporary equivalents now?

One of the most serious casualties of the 'seventies cutbacks was experiment in the forms of television. I am speaking here, not about a training-ground or even "Shoestring" operation, but about more experiment in the non-naturalistic possibilities of the medium and in demanding, non-topical subjects. Our vision in this area is in danger of suffering middle-aged myopia.

Where is full-scale regional drama? I do not have in mind overly didactic "historical" drama like most of the dramas in *The Undaunted*, or nice friendly half hours like *The Way We Are* or *Red Serge*, or an interesting anomaly like David Barnett's *Catalyst Television*. I am not thinking of a crisp, episodic, fast-moving set of anecdotes like *Heritage Theatre* (summer 1986) in which Pierre Berton demonstrated once again that he is a good on-camera story-teller and that entertaining television can be made for next to nothing with good scripts, unit sets, imaginative lighting, intelligent direction and strong performances; and certainly not plays that star the scenery rather than the situations or the characters. What I miss is fully flavoured, fully funded, mature regional drama of the kind theatres all over the country are developing. They should be either originals for television, (not records of plays like *Joey*, however valuable to the theatre historian) although adaptations of serious plays with the energy and inventiveness of *Les Canadiens*, *Quiet in the Land*, or *Talking Dirty* could be as exciting as *Ma!* demonstrated. Filming a play or two on location in the regions with crews and concepts flown in from Toronto is not what I mean.

Commissioning more scripts from experienced television, film, or theatre writers like Bolt, Pollock, Ryga, Findley, Thompson, Cameron, Whitehead makes sense. It would make better or at least different television if occasionally they, not the producer, came up with the basic idea.

How much is the technology dictating the forms of television? Is this the equation: film ˋ expensive, so better be safe? The very uneven, often surrealistic (sometimes Canadian) videos accompanying the "new" music late at night on the regular networks and in odd corners of *Muchmusic* are rather like what three- to five-minute impressionistic morsels could be. True, most music videos are pumped full of derivative images which bore both the eye and the mind, but some are succinct short stories, a few are very disturbing, and the occasional one is very funny indeed.

Once upon a time, the CBC had a *Shoestring* slot where, at a comfortably late hour of the night, people both experienced and new to the business were allowed to learn things and try things, with miniscule budgets and ingenuity. It was a seedbed, not a hot-house, where those making the programmes were allowed to fail more or less without penalty and ratings were irrelevant. Why could we not have such a slot for the kind of surprises that such drama can bring, where a warehouse, a theatre, or an unused corner could be used, sometimes live, sometimes taped or filmed, to present drama of all kinds, where experiments in video might turn up from the more accessible artists in that medium? Why not a "second stage," to use the regional theatre analogy? If people would stay up until after 11:00 p.m. for a half-hour of surprises in Montreal in the 1950s and 'sixties, why not now? Johnny Carson ruled the airwaves then, too; reruns of old movies or copshows are not particularly strong competition for viewers tuning in.

I find myself wondering why Parliamentary hackles rarely rise any more, whether the subject is racism, incest, drugs, televangelism, or old people's homes. Subjects like those have been treated adequately, often very well, by the CBC. Yet it is not as easy now to get the M.P. from Prince Albert to fulminate, as when John Diefenbaker represented the vocal derision and hostility of some of the population—although in 1986 there was a hot debate in the letters pages of the *Globe and Mail* and *Macleans'* about whether the CBC was biased toward the political left, as a University of

Calgary study suggested. The study's methodology, motives and conclusions have all been both vigorously attacked and defended. I will simply note that since CBC news and current-affairs programming is likely to be critical of whoever is in power, and since the NDP has yet to form the government in Ottawa, it would seem inevitable that certain methods of content analysi would arrive at that conclusion. I have serious doubts that one could arrive at any conclusion about the political bias of the CBC Drama Department at present, using the forms of content analysis on which most sociologists and semiologists rely. Nevertheless, despite the country's so-called move to the right and despite deep suspicions harboured in the past about the "pinkos" of the CBC, controversy about drama is in short supply. Has the audience grown up, or is the CBC, as Kennedy says, more conservative?

Is this in part because we are not presenting "high culture" in the same sense *Festival* did? As always through the centuries, it is a study in the uses of power like *Macbeth*, or *Antigone*, or *Endgame* that gets under people's skin. Shaw and Shakespeare and Molière are enjoyed by all kinds of people. Has the CBC management started to undervalue the intelligence of its viewers in recent years? Do they think everyone fits into the neat paradigms of demographic profiles? Are they unwilling to give a challenging programme or anthology a chance to find an audience? It took at least three years for *Festival* to build a following, according to Robert Allen and Mario Prizek.

Why don't we have a mix of modern international and home-grown drama in our television diet? They could be shot on tape or film, in colour, or black-and-white, in a studio or on location, at fifteen minutes or two hours in length. Britain's Channel 4 shows the way here. Television should show perspectives on personal, social and political situations, from within and without the country, through various forms of drama. Why should our scriptwriters look only at the contemporary equivalents of hockey and maple syrup? With the many changes in our lives resulting from events during the latter part of this century, is there no Canadian perspective to be aired on decades of tension in the Middle East, the corruption of international trade unionism, the third wave of feminism, the greying of western society?

The CBC has perceived intermittently that its primary task is not simply to reflect the world within our borders but to form myths and create mythical figures like (yet unlike, because they are Canadian) Britain's George Smiley, or the United States' Hawkeye Pearce. We appreciate characters who present: "what we feel, not what we ought to say"—to quote Edgar, one of the survivors in *King Lear*, as he looks first at a stage filled with the dead and the myths they have made, and then at us. How about Therese Casgrain; Nellie McClung, and Emily Murphy (both with a little depth this time); Wilfred Grenfell; Bible Bill Aberhart; "Red" Hill; Sieur de La Salle; Generals Frontenac and Brock; Agnes MacPhail; Courier-de-bois Radisson (who deserves better than the inept CBC series from the mid-'fifties); rebel and Métis frontiersman Gabriel Dumont; Wally Floodie, the real tunnel king in "The Great Escape;" Buzz Beurling, the loner hero of Malta; political beacons like Tommy Douglas and brave rascals like explorer Etienne Brulé. The list is endles and, of course, growing year by year.

Up to 1987 myths have not been our style. We (unintentionally) trashed *Riel* with an incoherent plot structure and dialogue full of strident polemic, we cut Van Horne and John A. down to size in *The National Dream*, mercilessly lampooned the "Flush of Tories" who manhandled the country after Macdonald died, and presented Sam Hughes with every unattractive flaw intact. Never in our series television has the CBC tried to build up a heroic "hunk of manhood," either fictional or historical. Women? Where is television's Hagar or Joanna Donnelly when we need her?

Yet we do not always flatten our fictional heroes. Canadian television drama has created some great popular figures, most of them in series: John Vernon's Wojeck, Chief Dan George's Ol' Antoine, Gordon Pinsent's Sergeant Sturgess, Al Waxman's King, Bruno Gerussi's Nick, and Louis Del Grande's Louie Ciccone. We are still waiting for a woman to star in a series that is not a sitcom. The anthology and docudrama chapters, however, cite many characters, both men and women, who made indelible impressions.

If there were reruns on a second channel, as suggested by the Caplan-Sauvageau Report, based on proper archival holdings, we would be able to enjoy again the earlier performances of some of our world-class actors in very demanding classical roles. The almost universal ignorance about our television drama which I have tried to redress in this book may in turn be connected with another phenomenon I have observed. It is not at all clear to me why directors, writers, actors, designers, cinematographers and, particularly, producers cringe when I apply the perfectly legitimate descriptive (or, heaven forbid, evaluative term) "art" to some CBC television drama. Structuralist critics, semioticians, and even experts in popular culture would challenge my use of the word on theoretical grounds. CBC "artists" (the word, again, is John Kennedy's) are not reacting on these grounds, yet in a completely typical Canadian way, they all reply, "Who, me? What, this?" As we have seen and as the people themselvs admit, CBC English television as a collectivity has never had a sense of its own achievements, tradition, and history.

Canadian series that depend on borrowed stereotypes—mostly sitcoms and copshows, since we avoid daytime or nighttime soaps—fail miserably to entertain because they are imitative, and the "real thing" is just the flick of a switch away. The best CBC series take the elements of familiar formula, invert them, subvert them with parody, or invent different conventions altogether. *Wojeck* broke the rules just about every time. *The Manipulators* flaunted the conventions of form and subject matter strikingly in its first season. Episodes in *The Beachcombers, King of Kensington, Hangin' In, Sidestreet,* and occasionally *The Great Detective* did that in the 'seventies. Even in a tight thirty-minute slot *Judge* sometimes found a complex human dilemma in the cracks and crannies of the law. *Seeing Things* was, arguably, the most successful distinctive hybrid of the lot.

Money is scarcer than ever. Yet it could be that the trend to narrow-casting—making programmes or even devoting channels to specialized audiences—will free the CBC from the "c" rations of the ratings game. As a culture, we might find the renewed courage to make the distinctive choices tailored to the pluralistic audience Pierre Juneau called for in the 1974 CRTC decision which have almost disappeared by the mid-'eighties. While in the United States viewers complain bitterly about reruns, Canadians are starved for Canadian reruns of most of our good programming. It would be nice to see the contemptuous connotations of the title *Rearview Mirror* (now defunct) exchanged for a prime time showcase of vintage CBC television, with drama as its centrepiece.

Finally, at the risk of reiteration, the CBC has both the responsibility and the joy of being the principal shaper of this country's imagination. For most of its history, it has performed this function—a dry word for so vital a task—adequately, well, or very well. In so doing, the CBC has also created a body of work in which it should take great pride, guard with care, and rebroadcast as part of the natural order of its programming; items from all thirty-five years, particularly in drama. The corporation should also use a selection of its best work to systematically educate its newer creative people in the accomplishments and techniques of their predecessors.

Excellence in that staple of prime time series television at the CBC means very good, often allusive writing with an ear for different modes of speech, rounded characters, focus on characters other than the protagonists, and carefully detailed development of familiar characters. It means camera work that actually reflects the specific requirements of script and direction, and original music that is used with discretion—often as counterpoint, never as emotional hype. It means a director, producer and editor willing to let a scene develop and take time instead of constantly moving on to the next joke or plot twist. It means recognition that silence is not the sole province of radio drama. When these elements are present, a viewer's sharp eye and ear will often be rewarded with parodic, lyric, or superbly ironic overtones.

Our better series television, like the rest of our better television drama, is influenced by anthology. CBC series drama at its most distinctive takes ambivalent moral stances when required; it challenges its audience with open or downbeat endings instead of our tired old friend "poetic justice," has tackled topical subject matter quite regularly, and usually sketches its protagonists in several shades of irony and deprecation instead of the brilliant colours of crusaders, omniscient wise men, and flamboyant villains.

On the other hand, CBC television drama can, should, and sometimes does, lift the heart and reaffirm the best in us. It provides something for the eye, the ear, the mind, and the heart. In a country the size of ours, with proven talent and accomplishments, we can have any form of television system, any kind of technology and any kind of programming we need. At a seminar on the 1967 television drama *Paper People*, Timothy Findley said: "I don't believe we've ever been searching to see who we are. I think we've been avoiding seeing who we are."[7] Despite all my strictures about omissions, censorship, self-censorship, internal and external constraints, and the regional-centralist dispute, I do not agree. We do, intermittently, see who we are. At present our own stories are still told by the singer of tales in the corner of the living room. But if we do not continue to hear and see our own faces, or our own back yard, or reflect our own imaginative perspective on the world, then soon when we look into the window of our television screen, we will see instead a one-way mirror. All the faces behind that looking glass will belong to someone else. When you look in a mirror and see someone else's face, something crucial to your survival is breaking down. If this does happen, all of us will begin to forget who we are.

In the end it is you who has the power as consumer, viewer, and voter to make the decisions about what you will see and what you will not. You will write the ending to this book, reader, not I, not the critics, not the CBC or the independent producers or the politicians. You.

NOTES TO CHAPTER 12

1. The *Globe and Mail*, (6/4/76) quoted the script at some length.
2. Editor of *Saturday Night* and co-instigator, with economist Abraham Rotstein, of the Committee on Television, a high-profile group that included Patrick Watson, Allan King, Morris Wolfe, and Mary Jane Miller, who were so disturbed by the state of the CBC and its programming in 1973 that they intervened against the renewal of its licence.
3. Dated June 1977, in *Marginal Notes: Challenges to the Mainstream*. Lester and Orpen Dennys, Toronto: 1984, 97.
4. A look at Peter Lyman's *Canada's Video Revolution: Pay-TV, Home Video and Beyond* (James Lorimer and Co. in association with the Canadian Institute for Economic Policy, Toronto: 1983) will acquaint the reader with the technology, some suggestions for working out policy to deal with it before the "Wave" hits,

and the timely reminder that "New technology won't go away"—one of Lyman's subheadings.

5. "The commission agrees with the CBC that the national broadcasting service should remain a popular service and that it should guard itself against becoming the preserve of esoteric *minorities*. However, the commission is of the opinion that a preoccupation with mass audience concepts, stimulated by the contemporary North American marketing environment, is inappropriate for a publicly supported broadcasting service." Conclusion 5, p.18 of "Radio Frequencies are Public Property," 31 March 1974, CRTC Decision 74-70.

6. (By Denise Boucher), now translated into English by Alan Brown as *The Fairies are Thirsty*, Talonbooks, Vancouver: 1982.

7. On *"The Paper People,"* in *Canadian Drama*, 63.

APPENDIX A

Note that a series anthology could appear in different viewing seasons, e.g. January-March 1978, and October-December of 1978, yet have a one year calendar date. By television reckoning, that's a two-year run.

1952-1953 Sunshine Sketches
1952-1955 CBC Theatre
1952-present Wayne and Shuster
1953-1957 CBC Television Theatre
1953-1954 CBC Playbill (also known as Ford Theatre Playbill)
1954-1955 Performance
1954-1955 Scope
1954-1958 On Camera
1954-1961 General Motors Presents
1955-1960 Folio
1958-1960 The Unforeseen
1958-1959 Playbill (from regions to network), scheduled very irregularly
1959-1960 Studio Pacific, Spectrum, Pacific 8 and Vancouver Television Theatre
1959-1967 Shoestring Theatre (Montreal)
1959-1969 Festival
1960-1961 First Person
1960-1961 R.C.M.P.
1960-1967 Cariboo Country
1961-Jake and the Kid (repeated in 1963—both were summer seasons)
1961-Tidewater Tramp
1961-1964 Playdate
1961-1964 Q for Quest
1962-Scarlett Hill
1963-Room to Let
1963-1965 The Serial
1963-1966? The Forest Rangers
1964-1966 The Show of the Week
1965-Seaway

1965-The Eyeopener
1966-1968 Wojeck
1967-1969 Quentin Durgens, M.P.
1967-Hatch's Mill
1967-1971 Teleplay (Montreal)
1968-Toby
1968-1969 Pick of the Week
1969-1970 MacQueen: the Actioneer
1970-1971 Corwin
1970-Canadian Short Stories
1970-Adventures in Rainbow Country
1971-Paul Bernard Psychiatrist
1971-Program X
1971-1973 The Manipulators
1971-1972 Sunday at 9
1971-1974 To See Ourselves
1972-Purple Playhouse
1972-1973 Delilah
1972-1973 Police Surgeon
1972-The Whiteoaks of Jalna
1972-present The Beachcombers
1973-CBC Drama '73
1973-Nightcap
1973-1975 The Collaborators
1974-The Play's the Thing
1974-The National Dream
1974-1975 Opening Night
1974-1975 Performance
1975-1976 Festival (of a sort—the title reappears briefly for a few specials)
1975-1979 Sidestreet
1976-1979 King of Kensington
1976-Camera '76
1976-Front Row Centre
1976-Here to Stay
1976-Royal Suite
1976-1978 Performance
1977-1985 For the Record
1977-Catalyst Theatre
1977-Custard Pie
1978-1979 A Gift to Last
1978-1981 Flappers
1978-1979 The Newcomers

1979-Nellie, Emma, Ben, & Daniel
1979-1981 Ritter's Cove
1979-1981 The Great Detective
1980-The Phoenix Team
1980-Duplessis
1980-Marquée
1980-1984 Homefires
1981-The Winners
1981-1987 Seeing Things
1981-1987 Hangin' In
1982-Empire Inc.
1982-1984 Judge
1983-1984 Backstretch
1983-Vanderberg
1983-1985 Snow Job
1983-1987 Fraggle Rock
1984-1986 Some Honorable Gentlemen
1984-present Danger Bay
1985-present Night Heat
1985-present Check it Out
1986-1987 Airwaves

The CBC records are contradictory and fragmentary. These dates have been reconstructed from scattered sources and may contain some inaccuracies. They are, however, more accurate than any other print source available to the reader as of 1987.

APPENDIX B

WHERE TO FIND MATERIAL, AND A CAUTIONARY NOTE

Scholars doing research into the remaining Kinescopes and tapes of television drama or radio drama should first approach the Public Archives of Canada—National Film, Television and Sound Archives Branch. If a request is made ahead of time, the P.A.C. will make reference copies of existing material for listening or viewing at the archives in Ottawa. Not only has their acquisition of video and audio protection and reference copies accelerated significantly in recent years, with sufficient warning, they are willing to make copies of material not yet protected if they can get at them. Virtually the entire collection is accessible if the number of programmes identified is reasonable. An incomplete but useful catalogue recording the series available and number of hours surviving, together with a brief general description of the kind of material covered by the series, can be obtained from the same source. Write to Ernest J. Dick, Head of Collections Development, 395 Wellington St., Ottawa K1A 0N3, Ontario, Canada.

Provincial archives are another good source of background materials, photographs, papers etc. and some recordings.

Most of the surviving scripts from the CBC collections are now to be found in the archives of York University, Downsview, Ontario. Regrettably, only an experienced researcher will be able to place the script in the production process—i.e. as a first draft or last clean copy before production alterations. The collection has few production-as-broadcast, i.e. revised scripts. Some scripts and other materials have found their way into the library at McMaster University. Those which I have checked are duplicated at York.

APPENDIX C

The following list, in alphabetical order by series, and alphabetical by programme within a series, contains a selected list of those Canadian programmes which I watched critically and in detail, in many cases several times. However, several no longer exist and I was only able to report on them from existing paper records. In some cases, I had to rely on my viewing notes taken many years ago. Together they form the basis of my critical analysis. Many, though by no means all, are analyzed in detail in the book. Several appear under more than one entry because they were produced more than once or repeated under a different title.

A Gift to Last

Militia
Regimental Party/The Proposal
The Catholics
The Great Tamarack Fire
The Closing Episode

(The) Beachcombers

Boomsticks
Easy Day
Evil Eye
Fraser Red
Happy Birthday Molly
Here Comes the Groom
Independence Day
Jo-Jo
Molly's Reach
Our Champion
Runt O' the Litter
Steelhead

The Hexman
The Highliners
The Sea is our Friend
The Search

Cariboo Country

All Indian
Frenchie's Wife
How to Break a Quarter Horse
    (Shown as a *Festival* Special)
Infant Bonaparte
Morton and the Slicks
Ol'Antoine's Wooden Overcoat
One Man Crowd
Sarah's Copper
The Education of Phyllistine (reedited
    and shown a one-hour *Festival*
    special)
The Sale of One Small Ranch
The Strong Ones

Camera '76

Kathy Kuruks is a Grizzly Bear
The Insurance Man from Ingersoll

CBC Drama 1973

Bird in the House
Lighten my Darkness
Our Ms. Hammond
Welcome Stranger

*CBC Theatre / CBC Television
   Theatre*

Coventry Miracle Play
Fortune My Foe
Hilda Morgan
One John Smith
Othello
Teach me how to Cry
The Acrobats
The Blood is Strong
Whiteoaks of Jalna

Corwin

Any Body Here Know Denny?
What Do You See When the Lights
   Go Out?
Who Killed the Fat Cat?

Eyeopener

A Borderline Case (censored)
Hear Me Talkin' To Ya'
Sara and the Sax
The Black Madonna
The Blind Eye and the Deaf Ear
The Dutchman (telecast or shelved?)
The Golden Bull of Boredom
The Guard

The Lonely Machine
The Trial of Joseph Brodsky
The Tulip Garden
The Wanting Seed
Uh-huh?

Festival

A Doll's House
A Handful of Grass
A Remnant of Harry
A Resounding Tinkle
A Very Close Family
Affaire
An Ideal Husband
Antigone
Arms and the Man
Ashes to Ashes
Bousille and the Just
Cathy Come Home
Cats Paw
Columbe
Cradle of Willow
Death of a Salesman
Diary of a Scoundrel
Education of Phyllistine (The)
Enemy of the People
Fifteen Miles of Broken Glass
Galileo
Grand Exits
Hedda Gabler
Heloise and Abelard
I Spy
Ivan
Ivanov
Julius Caesar
Juno and the Paycock
Ladies in Retirement
Let Me Count the Ways
Man Alive
Mother Courage
Mrs. Dally has a Lover

Neighbors
No Two People
Noises of Paradise
Othello
Queen After Death
Pale Horse, Pale Rider
Penny for a Song
Private Memories of a Justified Sinner
Reddick
Roots
Sergeant Musgrave's Dance
Sister Balonika
Six Characters in Search of an Author
The American Dream
The Basement
The Birthday Party
The Blue Hotel
The Brass Pounder from Illinois
The Close Prisoner
The Drama of Peter Mann
The Duchess of Malfi
The Dybbuk
The Endless Echo
The Family Reunion
The Gambler
The Innocents
The Killdeer
The Labyrinth
The Lark
The Luck of Ginger Coffey
The Master Builder
The Offbeats
The Offshore Island
The Paper People
The Police
The Queen and the Rebels
The Slave of Truth
The Supplicant
The Murder
The Three Musketeers
The Three Sisters

The True Bleeding Heart of Martin B
The Wild Duck
The Wolf
The Write Off
Today is Independence Day
Twelfth Night
Two Terrible Women
Uncle Vanya
Volpone
Waiting for Godot
Yerma
Yesterday the Children Were Dancing

First Performance

'55 In Revue
At My Heart's Core
Black of the Moon
Cousin Elva
Ice on Fire
Janie Canuck
Monserrat
O'Brien
Panic at Parth Bay
The Colonel and the Lady
The Discoverers
The Doll's House
The Inheritance
The Man Who Caught Bullets
The Seeds of Power
Time Lock

Folio

Hedda Gabler
Honey and Hoppers
John A. and the Double Wedding
Soviet Portrait
The Black Bonspiel of Wullie McCrimmon
The Case of Posterity vs. Joseph Howe

The Concert
The Devil's Instrument
The Hand and the Mirror
The Small Rain
The Trial of James Whelan
The Unburied Dead
Under Milk Wood
Ward Number Six

For the Record

A Far Cry from Home
A Matter of Choice
A Question of the Sixth
A Thousand Moons
Ada
An Honourable Member
Becoming Laura
Blind Faith
By Reason of Insanity
Cementhead
Certain Practices
Change of Heart
Dreamspeaker
Dying Hard
Every Person is Guilty
Final Edition
Hank
Harvest
Hide and Seek
High Card
Home Coming
I Love a Man in Uniform
Je Me Souviens
Kathy Kuruks is a Grizzly Bear
Lyon's Den
Maintain the Right
Maria
Moving Targets
One of Our Own
Out of Sight, Out of Mind
Ready for Slaughter

Reasonable Force
Rough Justice
Running Man
Slim Obsession
Snowbird
Someday Soon
The Boy Next Door
The Front Line
The Insurance Man From Ingersoll
The Winnings of Frankie Walls
The Tar Sands
Tools of the Devil
Where the Heart Is

Ford Startime

Clearing in the Woods
The Crucible
The Importance of Being Earnest
Tiger at the Gates

Front Row Centre

Hedda Gabler
His Mother
Of the Fields Lately
One Night Stand
Other People's Children
Raku Fire
Sarah
Saturday, Sunday, Monday
The Eye of the Beholder

GM Presents

Ashes in the Wind
A Business of His Own
A Feast of Stephen
Flight into Danger
Riel Parts 1 and 2
Teach Me How to Cry
The Apprenticeship of Duddy Kravitz

The Blood is Strong
The Eyeopener Man (3 of them, one
   in General Motors Presents, two
   on Playdate)
The Grown Ones
The Haven
The Man Who Ran Away
Zone

Hangin' In

various episodes from every season

Hatch's Mill

Temperance
The Prophet

Journalistic Dramas

A Thousand Moons
Kathy Kuruks is a Grizzly Bear
Man from Ingersoll
Nest of Shadows

Judge
various

King of Kensington

Cathy's Hobby
Cathy's Last Stand
Christmas Show
Diabolical Plots
Down but not Out
Gestalt of Kensington
Moving On
School Days
The Check-up
The Real Mrs. King
The Teacher

MacQueen

Home is Where the Heart Ain't

On Camera

A Women's Point of View
Big League Goalie
Explorations
Stagecoach Bride
The Bottle Imp
The Canadian Mirror
The Golden Age

Opening Night

Captain of Kopenick
Freedom of the City
Head Guts and Sound Bone Dance
You're Alright Jamie Boy

Peepshow

A Country Fable
Death
Festering Forefathers and Running
   Sons
''Microdramas''

Performance

Baptizing
Get Volopchi
Going Down Slow
Mandelstam's Witness
Raisins and Almonds
Red Emma
Ten Lost Years
The Betrayal
The Good and Faithful Servant
The Last of the Four Letter Words
The Man in the Tin Canoe

The Middle Game
Trial of Sinayevsky and Daniel
Village Wooing

Playbill

Ad Lib
Laurie
Metronome
They are all Afraid

Playdate

Black Bonspiel of Wullie
   McCrimmon
In Good Time
Men Don't Make Passes
Not For Every Eye
Stop the World and Let Me Off
The Broken Key
The Cell 5 Experience
The Cowboy and Mr. Anthony
The Critic
The Eyeopener and the Wages of
   Zinn
The Prisoner
The Thirteenth Laird
The Stoneboat
Willow Circle
With my Head Tucked Under my
   Arm
Valerie

Programme X

A Day That Didn't Happen
An Evening with Gordon Pinsent
An Evening without James Reaney
   [One Man Masque]
An Evening With Kate Reid
Ashes for Easter
Banana Peel

Black Ship
Boss
Charlie Who?
Concerto for Television
Evening with Barbara Hamilton
Generation Game
Gerber's Girls
Lemonade
Musical Chairs
Nothing to Declare
One's a Heifer
Parallel 68
Sniper
Ten Women, Two Men and a Moose
The Couch
The Double-Jointed Turned-On
   Picnic
The System

Q for Quest

An Early Grave
Bedlam Galore for Two or More
Burlap Bags
Crawling Arnold
District Storyville
Dreams
Eulogy
Gallows Humour
Indian
Kim
Mission of the Vega
Night of Admission
Oppenheimer's Transcripts
Picnic on the Battlefield
The Bathroom
The Brig
The Evolution of the Blues
The Man on His Back
The Neutron and the Olive
The Suitcase
The Trial of Lady Chatterly

The Wounded Soldier
Two Soldiers

Quentin Durgens M.P.

A Case for the Defence
It's a Wise Father
The Road to Chaldea
You Take the High Road

Scope

Don Giovanni
Hamlet
O Canada
The Odds and the Gods

The Serial

Cariboo Country (all)
McGonigle Strikes Again
Strangers in Ste Angèle (one episode)

Seeing Things

all episodes

Shoestring Theatre

Cobbler Stick to thy Last
How Beautiful with Shoes
Pea Soup and Porridge
The New Step
The Willow Harp
Point of Transfer

Sidestreet

Between Friends
Big Brother
Holiday with Homicide
Just Another Day

Once a Hero
Stakeout
The Holdout
The Rebellion of Bertha MacKenzie
With this Ring

Some Honourable Gentlemen

A Passion of the Patriots
A Flush of Tories
Grierson and Gouzenko
Sam Hughes' War

Specials and Miniseries

A Place of One's Own
A Population of One
Anne's Story
Artichoke
Bethune
Billy Bishop Goes to War
Charlie Grant's War
Chautauqua Girl
Christmas Lace (CTV)
Coming out Alive
Crossbar
Death of a Nobody
Empire Inc.
Every Night at 8:30
Gentle Sinners
Love on the Nose
Horse Latitudes
I am a Hotel
I Married the Klondike
Joey
Kaleshnikoff
King of Friday Night
Last Call
Leona
Love and Larceny
Love and Maple Syrup
Ma!

Melda and the Ducks
Noh Drama
Nellie McClung
Peer Gynt
Point of Departure
Pup Tent
Reddick II
Rexy
Riel
Rimshots
Sam Adams I and II
Samuel Lount
Sarah
Seer
Separation (CTV)
Shellgame
Skid Row
Stacey
Striker's Mountain
The Albertans
The Baron of Brewery Bay
The Canary
The Day my Grandfather Died
The First Night of Pygmalion
The Good and Faithful Servant
The Hill
The Jesus Trial (OECA)
The Larsens
The Last of the Four-Letter Words
The Life of Edward Alonzo Boyd
  (CTV)
The Making of the President
The Man Inside
The Man Who Wanted to be Happy
The Masseys
The Megantic Outlaw
The Prodigal (CKND)
The Wordsmith
They're Drying Up the Streets
To Serve and Protect
Today I am a Fountain Pen
Tramp at the Door (CKND)

Turning to Stone
Tyler
War Brides
You've Come a Long Way Katie

Studio Pacific

How Beautiful With Shoes
Some are so Lucky
The Clubman
The Treasure

Sunday at 9

A Small Remedy
Twelve and a half Cents
Vicky

Sunshine Sketches

The Mariposa Light Power and
  Census Station

Teleplay / Montreal

Ghosts of Montreal
Laurie
When the Bough Breaks

The Collaborators

Dreams of Things
Episode #1 (no title)
Whatever Happened to Candy

The Manipulators

Spike in the Wall
The Double Bind
The Flock
Turn to the Wind
X-Kalay

The Play's the Thing

A Bird in the House
And Then Mr. Jones
Back to Beulah
Brothers in the Black Art
Friends and Relations
How I Met my Husband
The Bells of Hell
The Man from Inner Space
The Roncarelli Affair
The Servant Girl

The Unforeseen

Pastorale
The Ikon of Elijah
The Witness
When Greek meets Greek

To See Ourselves

Hair Apparent
McIvor's Salvation
Pity the Poor Piper
The Coronet at Midnight
The Ninth Summer
The Painted Door

Where the Action Is

Cataline and the Judge Boy
Spud Murphy's Hill
The Kamloops Incident
Tumult with Indians

Wojeck

After All, Who's Art Morrison?
All Aboard for Candyland
Another Dawn, Another Sunrise,
    Another Day
Listen, An Old Man is Talking
The Last Man in the World
They're Dancing in the Streets

BIBLIOGRAPHY

Allan, Andrew, *A Self-Portrait*, Macmillan of Canada, Toronto: 1974.

Arlen, Michael, *The Camera Age: Essays on Television*, Penguin, London: 1982.

Atwood, Margaret, *Survival: A Thematic Guide to Canadian Literature,* Anansi, Toronto: 1972

Audley, Paul, *Canada's Cultural Industries: Broadcasting, Publishing, Records and Film*, The Canadian Institute for Economic Policy Series, James Lorimer and Co., Toronto: 1983.

Bakewell, Joan and Garnham, Nicholas, *Structures of Television,* 2nd. ed., BFI monograph #1, London: 1978.

Barnouw, Erik, *Tube of Plenty: The Evolution of American Television,* Oxford University Press, London: 1975.

Barnouw, Erik, *The History of Broadcasting in the United States*, Vol. III (from 1953), Oxford University Press, New York: 1970.

Barris, Alex, *The Pierce-Arrow Showroom Is Leaking: An Insider's View of the CBC*, Ryerson Press, Toronto: 1969.

Barsam, Richard M., *Nonfiction Film: A Critical History*, Dutton Paperback Original, New York: 1973.

Bennett, Tony, Boyd-Bowman, Susan, Mercer, Colin, and Woollacott, Janet, eds., *Popular Film and Television*, Open University, British Film Institute in association with Open University Press, London: 1981.

Berger, John, *Ways of Seeing,* BBC and Penguin, London: 1972.

Bigsby, C.W. ed., *Approaches to Popular Culture,* Edward Arnold, London: 1976.

Brandt, George ed., and intro., *British Television Drama,* Cambridge University Press, Cambridge: 1981, p. 21.

Briggs, Asa, *The History of Broadcasting in the United Kingdom* Vol. IV, *Sound and Vision*, Oxford University Press, London: 1979.

Carpenter, E. and Marshall McLuhan, eds., *Explorations in Communications,* Beacon Press, Boston: 1960.

Caughie, John ed., *Television: Ideology and Exchange,* BFI monograph #9, London: 1978.

Cawelti, John, *Adventure, Mystery, Romance: Formula Stories as Art and Popular Culture,* University of Chicago Press, Chicago: 1976.

Collins, Richard, and Porter, Vincent. *WDR and the Arbeiterfilm: Fassbinder, Ziewer and Others.* BFI Television Monograph. London: BFI, 1981.

Conrad, Peter, *Television: The Medium and Its Manners,* Routledge and Kegan Paul, Boston: 1982.

Davis, Douglas and Simmons, Alison, *The New Television: A Public/Private Art,* MIT Press, Cambridge, Mass.: 1977.

Douglas, Peter, *Television Today,* Osprey, London: 1975.

Dyer, Richard, et al. *Coronation Street*. BFI Television Monograph. London: BFI, 1981.

Eliot, Marc, *American Television: The Official Art of the Artificial: Style and Tactics in Network Prime-time*, Anchor Press/Doubleday, Garden City, New York: 1981.

Ellis, David, *Evolution of the Canadian Broadcasting System: Objectives and Realities 1928-1968,* Government of Canada, Department of Communications, Ottawa: 1979.

Ellison, Harlan, *The Glass Teat*, Ace edition, New York: 1973.

Ellison, Harlan, *The Other Glass Teat,* Ace edition, New York: 1983.

Esslin, Martin, *The Age of Television*, W. H. Freedman and Co., San Francisco: 1982.

Frick, N. Alice, *Image in the Mind: CBC Radio Drama 1944-1954,* Canadian Stage and Arts Publication Ltd., Toronto: 1987.

Frye, Northrop, *The Bush Garden: Essays in the Canadian Imagination,* Anansi, Toronto: 1971.

Frye, Northrop, "Across the River and into the Trees," *The Arts in Canada: The Last Fifty Years*, ed. by W. J. Keith and B. Z. Shek, University of Toronto Press, Toronto: 1980.

Gabriel, Juri, *Thinking About Television,* Oxford University Press, London: 1973.

Gianakos, Larry James, *Television Drama Series Programme: A Comprehensive Chronicle, 1950-75*, Scarecrow Press, Metuchan N.J.: 1978.

Gielgud, Val, *Years in a Mirror,* Bodley Head, London: 1965.

Gitlin, Todd, *Inside Primetime*, Pantheon Books, New York: 1985.

Greenberg, Bradley S. et al. *Life on Television: Content Analyses of U.S. TV Drama,* Ablex Publishing Corporation, Norwood N.J.: 1980.

Greenfield, Jeff, *Television: The First Fifty Years,* Harry Adams Inc., New York: 1977.

Hall, Stuart, "Encoding and Decoding Television Discourse," Paper for the Council of Europe Colloquy, University of Leicester, on "Training in the Critical Reading of Televisual Language," Centre for Cultural Studies, University of Birmingham, Birmingham: 1973.

Halliwell, Leslie, *Halliwell's Television Companion* with Philip Purser, Granada: 1982.

Hardin, Herschel, *A Nation Unaware: The Canadian Economic Culture*, J. J. Douglas Ltd., Vancouver: 1974.

Hardin, Herschel, *Closed Circuits: The Sellout of Canadian Television*, Douglas and McIntyre, Vancouver: 1985.

Hartley, John and Fiske, John, *Reading Television*, Methuen and Co., London: 1978.

Head, Sydney W., *Broadcasting in America: A Survey of Television and Radio,* second ed., Houghton Mifflin Co., New York: 1972.

Hindley, Patricia M., Martin, Gail and McNulty, Jean, *The Tangled Net: Basic Issues in Canadian Communications,* Douglas and McIntyre, Vancouver: 1977.

Hoggart, Richard, *Speaking to Each Other: Essays by Richard Hoggart about society,* Chatto and Windus, London: 1970.

Hood, Stuart, *On Television,* Pluto Press, London: 1980.

Hunt, Alfred, *The Language of Television: Uses and Abuses*, Eyre Methuen, London: 1981.

Jackson, R. L., "A Historical and Analytical Study of the Origin, Development and

Impact of the Dramatic Programmes Produced for the English Language Networks of the Canadian Broadcasting Corporation," Wayne State, M.A. Thesis, 1966, unpublished.

James, Clive, *Visions Before Midnight: Television Criticism from the Observer 1972-76*, Picador, London: 1982.

James, Clive, *The Crystal Bucket: Television Criticism from the Observer 1976-79*, Picador, London: 1981.

Johnson, A.W. (CBC President) *Touchstone for the CBC*, CBC: 1977.

Johnson, A.W (CBC President) *Canadian Programming on Television: Do Canadians Want It?*, CBC: 1981.

Juneau, Pierre (President of the CBC), *Introductory remarks* to the House of Commons Committee on Communications and Culture, 20 May 1986, CBC.

Koch, Eric, *Inside Seven Days: The Show that Shook the Nation*, Prentice Hall/Newcastle, Scarborough, Ont.: 1986.

Lane. Allen, *The New Priesthood: British Television Today*, Penguin, London: 1970.

Legris, Renée, *Dictionnaire des auteurs du radio-feuilleton québecois*, Fides, Montreal: 1981.

Legris, Renée et Page, Pierre, "Le Théâtre à la radio et à la télévision au Québec", extrait des "Archives des lettres canadiennes, publication du Centre de recherche en civilisation canadienne-française de l'Université d'Ottawa, tome V, *Le théâtre canadien-français*, Fides, Montreal: 1973.

Lord, Alfred, *The Singer of Tales*, Atheneum, New York: 1965.

Lyman, Peter, *Canada's Video Revolution: Pay-TV, Home Video and Beyond,* James Lorimer and Co. in association with the Canadian Institute for Economic Policy, Toronto: 1983.

McLarty, Lianne M. "Seeing Through Things" *Canadian Forum*, August-September 1985.

McLuhan, Marshall, *Understanding Media: The Extensions of Man,* McGraw-Hill Paperback, New York: 1965.

McNeil, Alex, *Total Television: A Comprehensive Guide to Programmes from 1948-1980*, Penguin Books Canada: 1980.

Miller, Mary Jane, Editor, *Canadian Drama*, Spring, 1983 (on Canadian television and radio drama) an editorial and a detailed analysis of Timothy Findley's *"The Paper People"*, pp 49-60.

Miller, Mary Jane, "Canadian Television Drama 1952-70: Canada's National Theatre", *Theatre History in Canada*, Vol. 5, No.1, Spring 1984, pp 51-71.

Miller, Mary Jane, *"Cariboo Country:* the CBC Response to the American Television Western", *The American Journal of Canadian Studies*, 14 #3, Fall 1984.

Miller, Mary Jane, "Television Drama in English Canada", a chapter in *Canadian Theatre: New World Visions*, edited by Anton Wagner, Simon and Pierre, Toronto: 1985.

Miller, Mary Jane, "Blind Faith and Pray TV—Canadian and American TV Looks at the same Issue", *Journal of Popular Culture,* vol 20 #1 Summer 1986.

Nachbar, Jack and John Wright eds., *The Popular Culture Reader,* Bowling Green University Popular Press, Bowling Green: 1977.

Nelson, Joyce, "Dumping Ground," *Saturday Night,* May, 1981.

Nelson, Joyce, "TV Formulas: Prime Time Glue," *In Search*, Department of Transport, Ottawa: Fall, 1979.

Newcomb, Horace, *TV: The Most Popular Art*, Anchor Books, Garden City, New York: 1974.

Newcomb, Horace ed., *Television: The Critical View,* Oxford University Press, New York: 1976.

Newcomb, Horace, and Alley, Robert S., *The Producer's Medium: conversations with creators of American TV,* Oxford University Press, New York: 1983.

O'Connor, John E. ed., *American History/American Television: Interpreting the Video Past*, Frederick Ungar Publishing Co., New York: 1983.

Paz, Octavio, *Claude Levi-Strauss: An introduction,* tr. J.S. Bernstein and M. Bernstein, Cornell University Press: 1970.

Peers, Frank, *The Politics of Public Broadcasting*, University of Toronto Press, Toronto: 1969.

Peers, Frank, *The Public Eye: The Politics of Canadian Broadcasting,* University of Toronto Press, Toronto: 1979.

Problems of Television Research: A Progress Report of the Television Research Committee, Leicester University Press, Leicester: 1966.

Quicke, Andrew, *Tomorrow's Television: an examination of British Broadcasting past, present and future*, Lion Publishing, Berkhamsted: 1976.

Reed, Alison and Evanchulk P.M., *Richard Leiterman* in the Canadian Film Series, ed. Piers Handling, Canadian Film Institute, Ottawa: 1978.

Rubin, Don and Cranmer-Byng, Alison, *Canada's Playwrights: A biographical guide,* CTR Publications, Downsview: 1980.

Rutherford, Paul, *The Making of the Canadian Media,* McGraw Hill Ryerson, Toronto: 1978.

Salutin, Rick, *Marginal Notes: Challenges to the Mainstream*, Lester and Orpen Dennys, Toronto: 1984.

Singer, Benjamin D. ed., *Communications in Canadian Society,* Copp Clark, Toronto: 1972.

Schwartz, Tony, *Media: The Second God,* Anchor Press/Doubleday, Garden City, New York: 1983.

Self, David, *Television Drama: An Introduction,* Macmillan, London: 1984.

Sklar, Robert, *Prime Time America: Life On and Behind the Television Screen,* Oxford University Press (paperback), Toronto: 1982.

Slade, Mark, *Language of Change: Moving Images of Man*, Holt Rinehart Winston of Canada, Toronto: 1970.

Smith, Anthony, *The Shadow in the Cave: A Study of the Relationship Between the Broadcaster, His Audience and the State,* George Allen and Unwin, London: 1973.

Stuart, Sandy, *Here's Looking at Us: A Personal History of Television in Canada,* CBC Enterprises, Toronto: 1986.

Sulman, Arthur and Youmans, Roger, *How Sweet It Was: Television: A Pictorial Commentary*, Bonanza Books, New York: 1966.

Sutton, Shaun, "The largest theatre in the world," The Fleming Memorial Lecture, to the Royal Television Society, 2 April, BBC publication: 1981.

Television as a Social Force: New Approaches to Television Criticism, Praeger, New York: 1974.

Thorbun, David, "Television as Melodrama" in *Television as a Cultural Force,* eds. R. Adler, D. Cater, Praeger, London: 1976.

TV Guide; The First Twenty-Five Years, compiled and edited by Jay S. Harris, New American Library, New York: 1978.

Twomey, John, *Canadian Broadcasting History Resources in English: Critical Mass or Mess?*, Canadian Broadcasting History Research Project, Toronto: 1978.

Weir, E.A., T*he Struggle for National Broadcasting in Canada,* McLelland and Stewart: 1965.

Wenham, Brian ed., *The Third Age of Broadcasting,* Faber and Faber, London: 1982.

Wilk, Max, *The Golden Age of Television: Notes from the Survivors,* a Delta Special, Dell Publishing Co., New York: 1977.

Williams, Raymond, *Drama in a Dramatized Society,* Cambridge University Press, Cambridge: 1975.

Williams, Raymond, *Television: Technology and Cultural Form,* Fontana Technosphere series, Fontana/Collins, London: 1974.

Wolfe, Morris, *Jolts: The TV Wasteland and the Canadian Oasis,* James Lorimer and Co., Toronto: 1985.

Woolfolk Cross, Donna, *Media-speak: How Television Makes Up Your Mind,* Mentor, New York: 1983.

Worsley, T.C. *Television: The Ephemeral Art,* Alan Ross, London: 1970.

DOCUMENTS

CRTC Bibliography: Some Canadian Writings on the Mass Media, Information Canada, Ottawa: 1974.

British Television Drama: 1959-1973, complete programme notes, ed. Paul Madden, Television Acquisitions Officer, National Film Archive, BFI: 1976.

Control of Subject Matter in BBC Programmes, a reprint of two appendices from the BBC to the *Report of the Joint Committee on Censorship of the Theatre,* H.M. Stationary Office: 1967.

Pay TV, published by *Cinema Canada* for the Council of Canadian Film makers, *Cinema Canada*, August, 1976.

Report of the Federal Cultural Policy Review Committee, chaired by Louis Applebaum and Jacques Hebert, Minister of Supply and Services, Ottawa: 1982 and the Summary of the Briefs and Hearings for that Committee.

Report from the Task Force on Broadcasting Policy, chaired by Gerald Caplan and Florian Sauvageau, Minister of Supply and Services, Ottawa: 1986.

Radio Frequencies Are Public Property: public announcement and decision of the Commission on the applications for renewal of the Canadian Broadcasting Corporation's television and radio licenses, Decision CRTC 74-70, 31 March 1974.

Strategy for Culture, Canadian Conference of the Arts, Ottawa: 1980.

Survey of Canadian Attitudes towards Public Broadcasting , Environics Research Group study commissioned by the Friends of Public Broadcasting, October 1985.

Telefilm Canada : developing a Canadian film and television industry, Xerox hand-out at the Banff International Television Drama Festival, Banff: 1986.

PERIODICALS

The Globe and Mail
The Hamilton Spectator

The Montreal Gazette
The Montreal Star
The Ottawa Citizen
The Toronto Daily Star
The Toronto Telegram
The Vancouver Sun

BBC Lunch Time Lectures, specifically Frank Muir's "Comedy on Television", fifth
    series -3-BBC, London: 1966 and Huw Wheldon's "Television and the Arts"
    BBC Lunch-time Lectures, third series -3, 1964.
Canadian Drama, Vol. 9, No. 1, 1983 is devoted to Canadian radio and television.
Channels of Communications—U.S. trade publication intended for the industry.
    Editor Les Brown.
Cinema Canada—Canadian, the only continuous source of periodical Television
    reviews, news, biography and retrospective after Morris Wolfe's few years of
    television criticism in *Saturday Night.*
Contrast—U.K. 1961-64
Journal of Popular Culture, 1970 following, Bowling Green State University,
    Bowling Green, Ohio.
The Listener—a BBC publication of reviews.
Screen, particularly vol. 25 No. 2 1984—the issue is devoted to "Watching
    Television."
Sight and Sound —U.K.
Television Quarterly —U.S.A.
University Film Association Journal—U.S.A.
Note that the British Film Institute puts out a whole series of television monographs.
    I found particularly useful #12, *WDR and the Arbeiterfilm: Fassbinder, Ziewer
    and Others* by Richard Collins and Vincent Porter, and #13, a collection of papers
    on *Coronation Street,* both BFI Publishing, London: 1981.

PLAYS

Note—most of these plays have been adapted or rewritten for the stage. Some were
written first for the stage but appeared on television. It is very common in Canada to
find a script appearing on radio, television, in the theatre and on film in any
combination, as well as being a novel or short story: for example, *The
Apprenticeship of Duddy Kravitz* by Mordecai Richler has appeared in all five
media.

Bowering, George, *The Home for Heroes* in *Ten Canadian Short Plays,* ed. by John
    Stevens, Dell, New York: 1975.
Dube, Marcel, *Zone* in *Ecrits du Canada-Français,* Vol. II, Editions de la Cascade,
    Montreal: 1955.
Findley, Timothy, *The Paper People,* in *Canadian Drama,* vol. 9, No. 1, 1983.
Hailey, Arthur, *Flight Into Danger* in *Ten Canadian Short Plays*, ed. by John
    Stevens, Dell, New York: 1975.
Hill, Kay, *Cobbler Stick to thy Last* in *Ten Canadian Short Plays.*
Hood, Hugh, *Friends and Relations* in *The Play's the Thing: Four Original
    Television Dramas,* ed. Tony Gifford, Macmillan of Canada Toronto: 1975.

Israel, Charles E. *The Labyrinth*, Impact books, MacMillan of Canada, Toronto: 1969.

Joudry, Patricia, *Teach Me How to Cry* in *Canada's Lost Plays: Women Pioneers,* CTR, Toronto: 1979.

Languirand, Jacques, *Les Grands Départs*, Editions de renouveau pédagogique, Montreal: 1970.

Laurendeau, André, *Théâtre,* Collection L'Arbre, volume g-5, editions HMH, Montreal: 1970.

Malcolm, Ian, *God Save McQueen* in *Performing Arts in Canada*, Fall, 1969.

Moore, Mavor, *The Store* in *A Collection of Canadian Plays*, Vol. II, Simon and Pierre, Toronto: 1973.

Moore, Mavor, *Come Away, Come Away* in *Encounter: Canadian Drama in Four Media,* ed. by Eugene Benson, Methuen, Toronto: 1973.

Moore, Mavor, *The Roncarelli Affair* in *The Play's the Thing*, op. cit.

Munro, Alice, *How I Met My Husband* in *The Play's the Thing,* op. cit.

Nicol, Eric *The Man From Inner Space* in *The Play's the Thing,* op. cit.

Ryga, George, *The Ecstasy of Rita Joe and other Plays,* ed. Brian Parker, New Press, Toronto: 1971.

Ryga, George, *Indian* in *Ten Short Canadian Plays*, op. cit.

Salverson, George, *Hero at Hatch's Mill''* in *Invitation to Drama* rev. ed. edited by Andrew Orr, Macmillan of Canada, Toronto: 1967.

St. Pierre, Paul, *The Education of Phyllistine* in *Invitation to Drama.*

St. Pierre, Paul, *Sister Balonika*, The Book Society of Canada, Agincourt: 1973.

Woods, Grahame, *Twelve and a Half Cents* in *Camera 3*, eds. Ron Side and Ralf Greenfield, Holt, Rinehart, and Winston, Toronto: 1972.

Woods, Grahame, *Vicky*, Simon and Pierre, Toronto: 1974.

Note—for British plays to 1977 see Malcolm Page's bibliography, *Theatre Quarterly* vol. VII, No. 27, 1977 and supplement in *Theatre Quarterly*, No. 30, Summer, 1978.

Significantly rewritten and lengthened are W.O. Mitchell's two scripts adapted for theatre—*Back to Beulah* and *The Black Bonspiel of Wullie McCrimmon* in *Dramatic W.O. Mitchell.* Many of Paul St. Pierre's scripts for *Cariboo Country* appear as short stories in anthologies of his work.

INDEX